Ecology and Management of Coppice Woodlands

Ecology and Management of Coppice Woodlands

Edited by

G.P. BUCKLEY

Wye College
University of London

CHAPMAN & HALL
London · New York · Tokyo · Melbourne · Madras

Published by Chapman & Hall, 2–6 Boundary Row, London SE1 8HN

Chapman & Hall, 2–6 Boundary Row, London SE1 8HN, UK

Chapman & Hall, 29 West 35th Street, New York NY10001, USA

Chapman & Hall Japan, Thomson Publishing Japan, Hirakawacho Nemoto Building, 7F, 1–7–11 Hirakawa-cho, Chiyoda-ku, Tokyo 102, Japan

Chapman & Hall Australia, Thomas Nelson Australia, 102 Dodds Street, South Melbourne, Victoria 3205, Australia

Chapman & Hall India, R. Seshadri, 32 Second Main Road, CIT East, Madras 600 035, India

First edition 1992

Typeset in 10/12pt Sabon by Intype, London
Printed in Great Britain at the University Press, Cambridge

ISBN 0 412 431106

A catalogue record for this book is available from the British Library

Library of Congress Cataloging-in-Publication data available

Contents

Contributors

J.P. Barkham
School of Environmental Sciences, University of East Anglia, Norwich NR4 7TJ

C.E. Bealey
2 White Row, Wick Lane, Downton, Wiltshire SP5 3NL

John A. Booker
Oakhollow, West Walberton Lane, Walberton, Arundel, West Sussex BN18 0QS

A.H.F. Brown
Institute of Terrestrial Ecology, Merlewood Research Station, Grange-Over-Sands, Cumbria LA11 6JU

G.P. Buckley
Wye College, University of London, Wye, Ashford, Kent TN25 5AH

H. F. Cook
Wye College, University of London, Wye, Ashford, Kent TN25 5AH

I.P.F. Cummings
School of Environmental Sciences, University of East Anglia, Norwich NR4 7TJ

Julian Evans
Forestry Commission Research Station, Alice Holt Lodge, Wrecclesham, Farnham, Surrey GU10 4LH

Martin N. Evans
5F, South Bank Terrace, Surbiton, Surrey, KT6 6DE

R.J. Fuller
British Trust for Ornithology, The Nunnery, Nunnery Place, Thetford, Norfolk IP24 2PU

F.B. Goldsmith
Ecology and Conservation Unit, University College London, Gower Street, London WC1E 6BT

John Gurnell
School of Biological Sciences, Queen Mary and Westfield College, University of London, Mile End Road, London E1 4NS

Martin Hicks
School of Biological Sciences, Queen Mary and Westfield College, University of London, Mile End Road, London E1 4NS

J.N. Greatorex-Davies
Institute of Terrestrial Ecology, Monks Wood Experimental Station, Abbots Ripton, Huntingdon PE17 2LS

K.J. Kirby
English Nature, Northminster House, Peterborough PE1 1UA

R.H. Marrs
Institute of Terrestrial Ecology, Monks Wood Experimental Station, Abbots Ripton, Huntingdon PE17 2LS

P.L. Mitchell
Botany School, Downing Street, Cambridge CB2 3EA

G.F. Peterken
Joint Nature Conservation Committee, Monkstone House, Peterborough, PE1 1JY

P.R. Ratcliffe
Forestry Commission, Northern Research Station, Roslin, Midlothian, Scotland EH25 9SY

P.A. Robertson
Game Conservancy Trust, Fordingbridge, Hampshire SP6 1EF

Ruth M. Tittensor
Walberton Green House, The Street, Walberton, Arundel, Sussex BN18 0QB

J.A. Thomas
Institute of Terrestrial Ecology, Furzebrook Research Station, Wareham, Dorset BH20 5AS

Susan J. Warr
Departments of Biological & Geographical Sciences, Polytechnic South West, Plymouth, Devon, PL4 8AA

M.S. Warren
2 Fairside, Higher Ansty, Dorchester, Dorset DT2 7PT

Steve Whitbread
Department of Biology, The University, Southampton SO9 3TU

Acknowledgements

The contributions in this volume are based on a meeting of the British Ecological Society's Forest Ecology Group, held at Wye College, University of London, in March 1990. I am particularly grateful to Dr P A Robertson for help in organizing the meeting, and to the British Ecological Society for financial support. A number of the contributors to these pages shared the task of refereeing and editing, as did the following: Dr S W Burrage, Dr P W Bright, Dr M J Copland, Dr J Mansfield, Dr J Mitchley, Mrs E A Simmons, Dr T A Watt and Dr A Whitbread.

Field visits to Kings Wood, Ham Street Woods and Blean Wood in East Kent enlivened the proceedings of the original meeting, and were organized with the cooperation of the Nature Conservancy Council and the Forestry Commission. Special thanks go to Norman Day, Ian Embury, David Maylam and Robert Petley-Jones for their graphic illustrations of points discussed then, and which recur repeatedly within these pages.

Preface

All coppice woodlands have one thing in common – they are repeatedly cut down, in the very reasonable expectation that the trees will regrow by themselves. Whether this regrowth is the result of new shoots sprouting from cut stumps or tree roots is unimportant: the point is that the new canopy forms rapidly and, for the main part, *vegetatively* from the old, without the need for any great management effort. The process is almost infinitely repeatable.

Some woodlands have been coppiced for long periods of their history, often for hundreds and sometimes thousands of years. These are therefore ancient woods in their own right, being descended from an earlier, primeval woodland cover; some are heavily modified by grazing or silvicultural selection; while others are much more recent plantations of commercially economic species. Within the rural economy, domestic markets traditionally demanded quantities of small-sized poles and brushwood for fuel, fodder, charcoal, thatch, fencing materials and hand tools, cut on short rotations of one to two decades. The large trees (standards) which were grown over this underwood for construction timber, or confined to other parts of the wood or hedgerow, were the only tall trees representative of a true forest environment.

Coppicing, therefore, has created woodlands with a highly specialized ecology, different from either the original, primeval woodland condition or the uniform forestry plantations which so often have replaced them. In coppices there is periodically more light; they contain fewer mature trees, less dead wood, and perhaps a greater diversity of plant and animal species which have their niche at the edge of the forest environment. In an extreme form, coppice woodlands are an intimate mosaic of small management units of disturbed, and relatively open, young woodland at close intervals of growth. This is in sharp contrast to the inherently more natural structures of high forest systems which, however, tend to depend for their ecological diversity on the scale and arrangement of felling operations.

The problem today is that most coppice is no longer commercial, and, increasingly, coppicing has come to be justified in terms of its aesthetic and scientific benefits. Sometimes these are self-evident, as in the displays of spring flowers which follow cutting, the birdsong of summer migrants, or the persistence of some rare butterfly species in patches of young coppice regrowth. In some areas, where traditional markets have virtually disappeared, the only woods undergoing active coppicing are already designated nature reserves or important sites of scientific interest: the remainder have been grubbed out, planted with commercial tree species, or, through neglect, have become tall forests.

The re-activation of coppicing in the remaining, less promising areas is difficult to justify, particularly where no cutting has been done for forty or fifty years and there is evidence of a decline in biological interest. Organizational problems such as a lack of skilled labour, the absence of marketing outlets, and the constant need to maintain the enthusiasm of conservation volunteers for a well-meaning management plan, tend to reinforce the status quo. These difficulties are unlikely to be overcome unless greater certainty can be had about the effects of coppice treatments, so that benefits and drawbacks of renewed management can be weighed together.

Having said this, there is often a good consensus of opinion of how to manage coppice for optimum biological diversity, with several contributors here arguing for the use of small felling coupes, the presence of standard trees to increase overall structural complexity and the juxtaposition of young growth stages. However, despite a long history of scientific study, there are still many ecological aspects of coppicing which are poorly understood. Particularly relevant here are investigations of how coppicing compares with alternative, and potentially less intensive forms of woodland management, such as conversion to high forest, small-group felling, wood pasture and non-intervention methods.

The task of this book is to show how coppicing determines the ecology of woodland, and in particular how the environment it creates influences the wider community of plants, birds, mammals and insects. Only when its effects are known in detail will we have a basis for assessing the special significance of coppices for nature conservation, and can begin to understand the implications for their future management.

G P Buckley

Part One

INTRODUCTION

– 1

Coppices in the lowland landscape

G.F. PETERKEN

The country is well wooded and shady, for the fields are all enclosed with hedges, oak trees, and several other sorts of trees, to such an extent that in travelling you think you are in a continuous wood.

Quoted in Thirsk (1967), from *Description des Royaumes d'Angleterre et d'Ecosse*, composée par Estienne Perlin, Paris. 1558.

1.1 INTRODUCTION

Coppices are generally perceived as isolated patches within a contrasting landscape. To a substantial degree this is a true reflection of their character: the modal coppice in lowland England is a compact wood of 5–50 ha set in open, cultivated farmland from which it is separated by sharp woodland margins defined by a ditch. This perception is reinforced by maps, which depict woods as blobs of green on a white background; by ownership patterns, which often separate ownership and management of woods from that of surrounding farmland; and by the structure of rural bureaucracy, which isolates responsibility for forestry from that for agriculture.

My contention in this introductory essay is that lowland coppices were once closely linked to the pattern of the landscape; that they have recently become more isolated; and that this increasing isolation must influence both their ecology and their role in nature conservation. Upland coppices (which mostly occupied the steeper slopes between the high moors and acid grassland of the hills, and the mires, neutral pastures and cultivated ground in the valleys) had different relationships with the landscape which are not considered here. Although most of my points relate specifically to coppices, some apply more precisely to ancient woods (most of which were coppices), and others to woodland in general.

1.2 SOME COPPICE CHARACTERISTICS

British woods were treated as coppice in the Neolithic. Some of the trackways across the Somerset peatlands were constructed of rough hazel wattle hurdles (Rackham, 1980). Less than a mile from the headquarters of the former Nature Conservancy Council at Flag Fen, archaeologists are uncovering foundations of Bronze age buildings which show evidence of both coppicing and pollarding. In the 11th century, Domesday Book shows, coppicing was a widespread and common form of management, and so it remained until late in the 19th century. By the 1960s, however, it had retreated to south-east England and had been reduced to an area of 30 000 ha or just 1.7% of all British woodland. As a result the country was left with a scatter of coppices that were no longer cut, the great majority of which were ancient woods.

Many coppices are remnants of the original-natural woodlands (Peterken, 1981; Marren, 1990). They are populated by species which have presumably descended directly from the primitive woodlands. Certainly they contain a disproportionately high share of the rare and local native plant and invertebrate species, many of which are known to have very limited powers of dispersal and colonization. Most coppices were not greatly modified by planting until the 18th century, so the distribution of species (including the trees) within them bore some relationship to natural factors such as soil texture and drainage. Around 1800 many coppices in south-east England were planted with sweet chestnut (*Castanea sativa*) and other species, but elsewhere in the lowlands most remained as mixtures of local trees and shrubs, directly descended from the forest communities of the primeval woodland.

Despite this direct link, coppices were not completely natural, only semi-natural. Their characteristics differed from natural woodland in several important respects (Peterken, 1991a,b).

1. In managed coppices gaps were created at the rate of 5–10% per year, compared with 0.5–2% per year for natural woodlands. Managed coppices thus maintained an unnaturally high amount of temporary open space and opportunities for early successional species.
2. Coppice structure consisted of even-aged compartments of 0.5–10 ha. The natural pattern is based on small groups occupying ground vacated by the fall of 1–10 canopy trees (up to 0.4 ha). Occasionally, much larger openings are formed by extensive blowdowns.
3. Very large, old trees were usually absent from coppices or confined to margins. They usually comprised one genus, oak (*Quercus robur* and *Q. petraea*). Natural stands mostly take the form of old-growth stands, with many large canopy trees of ages up to 300–500 years, and often consisting of many different species.
4. Coppices were usually dominated by light-demanding species, such as oak and ash (*Fraxinus excelsior*), or by shrub-trees such as hazel (*Corylus avellana*). They included a good stock of pioneers, such as

birch (*Betula* spp.) and sallow (*Salix* spp.), and various shrubs. Natural woodlands are usually dominated by shade-tolerant species, such as lime (*Tilia cordata*), beech (*Fagus sylvatica*) and elm (*Ulmus* spp.), and contain few pioneers and shrubs.

5. Coppices contained no snags (standing dead trees) or large down logs, whereas natural woods normally contain substantial amounts of both.
6. Coppices contained habitats which are unknown in natural woods, notably the permanent open space habitats of rides and glades.
7. Streams within coppices have usually been straightened and kept clean. Most soils have been drained by ditches. In natural woodlands the soils and streams are unmodified. Streams there contain a much wider range of microhabitats, including dead wood, debris dams, shoals and pools.
8. Coppices have generally been small and isolated habitats, which have been influenced by their surroundings. This isolation, which was the result of millennia of clearance and fragmentation of the original forest, probably had the effect of reducing the diversity of original woodland species within each wood over the centuries. Natural woodland originally formed continuous massifs, broken only by small areas of other habitats.

Coppices thus had a dual identity. They were cultural artefacts formed by a silvicultural system which had been part of the British scene since prehistory. They were recorded in ancient documents and on early maps, and in their earthworks they preserved evidence in intimate detail of our landscape history. However, many coppices on ancient woodland sites were also lineal descendants of primeval woodlands. They co-evolved with the communities which used them.

1.3 LOWLAND COPPICES IN THE PRE-1940 LANDSCAPE

In their heyday coppices consisted of several components: the coppiced underwood, the standard trees standing within the underwood, the grassy rides which separated the underwood into several blocks, and the ditches and streams which drained the wood (Figure 1.1). Each of these components had a distinct relationship with the landscape outside the wood and with each other (Figure 1.2). These relationships undoubtedly changed over the millennia and varied from place to place (Rackham, 1986; Taylor, 1983).

The coppiced underwood was generally bounded by a hedge surmounting a bank. Larger coppices were sometimes further subdivided by hedged banks. These coppice hedges were linked to the network of hedges around adjacent fields and to strips of scrub woodland which commonly formed the margins of streams and rivers. In 'old' countryside (Rackham, 1986) this network extended throughout the landscape, linking one wood with

Figure 1.1 Ryarsh Wood, Kent, in July 1954. This large ancient wood is still traditionally coppiced. The several age classes of coppice regrowth form a coarse patchwork divided by open rides. No standards have been retained in the most recently cut patches but elsewhere their prominent rounded crowns are scattered amongst the thickets of underwood. Cambridge University Collection No. 0080, Crown Copyright reserved.

another over many kilometres, and linking woods with other habitats that shared some of the conditions of the woodland interior, such as sunken lanes, abandoned quarries, pits, village closes, waste ground and gardens. In the countryside of open fields this network was originally punctuated by large gaps, but coppices were nevertheless generally part of a group of old inclosures on the parish margin, which were contiguous with streamside habitats and thence the village cores. Only in the last 200 years of inclosure did they become part of a complete landscape-wide network of hedges. In both the old countryside and the later forms of the 'new' countryside, the hedgerow network not only linked with and extended the coppiced underwood, but the two habitats shared a common structure and demo-

Figure 1.2 Panorama looking east down the Rother valley from a point west of Scotchfords Bridge near Mayfield, Sussex, in June 1956. On maps the woodlands appear to be isolated habitats but on the ground they are still linked by hedges, roadsides, valley grasslands and naturally sinuous streams and rivers. Cambridge University Collection No. SR92, Crown Copyright reserved.

graphy. Both formed populations of long-lived, locally native trees and shrubs, treated by repeated cutting, which allowed individuals to develop substantial stools and rootstocks, supporting vigorous young growth.

Within the coppices, standard trees were usually scattered throughout and supplemented by lines of pollarded trees on the external and internal banks. This mature timber habitat extended with the hedges into the countryside as a whole. Indeed, in many areas the density of hedgerow trees was greater in the landscape of small hedged closes than in the woods themselves. This habitat was consolidated by scattered trees standing within pastures, pollard willows growing beside ponds and streams, and most importantly by parklands, which were also closely linked into the network of field boundaries. Like the underwood and hedges, the treatment of trees within coppices and the rest of the landscape was roughly the same: many were pollarded and retained to great age, but others were cut at relatively small sizes and replaced by means of planting or natural regeneration. The standard and boundary trees in coppices were mostly oak, ash and elm, just like the hedgerow trees.

Rides within coppices were linear strips of grassland, tall-herb communities and shallow marshes which commonly incorporated several small ponds. Larger coppices often included small meadows or pastures, which formed an expanded portion of the ride system. Rides were formed mainly within the last 300 years, but they replaced earlier networks of small tracks. These rides were usually contiguous with grassland outside the coppices, for not only was a proportion of the general farmland treated as pasture for most of the time, but coppices were also often part of a group of old closes treated as pasture or meadow and were linked to streamside meadows. The close integration of woodland-grassland with farmland-grassland was assisted by the common practice of mowing rides as meadow, and the position of rides as integral parts of the network of farm tracks.

Coppices were located beyond the major floodplains on 'upland' sites. Even here they were mainly located on slopes or poorly-drained plateaux, leaving streamsides to pasture and meadow. The links between coppices and the drainage network were therefore almost entirely through minor streams and headwaters, and the strips of meadow which commonly bordered them. At both a broad scale and a parish scale, the landscape could be divided roughly into 'upland' districts of coppices, small closes, hedges, pastures, meadows and headwater streams, and low-lying districts of cultivated fields, villages and large floodplains without coppices.

The landscape of which lowland coppices were part was both diverse and reasonably stable. The diversity was expressed as an intimate intermixing at different scales of trees, scrub/young-growth woodland, pasture, meadow and cultivation, as exemplified by maps that show coppices as inconspicuous patches in a well-treed landscape (e.g. the 1611 map of Lawshall, Suffolk (Rackham, 1985)). The stability took the form of patterns of elements that (a) changed generally by evolution rather than revolution, and (b) included a proportion of elements that did not change in location or character for hundreds of years at least. At a small scale, the landscape was also dynamic, e.g. patches of coppice were cut, thereby generating a constantly changing pattern of open spaces and younger growth; hedges grew and were cut back; wood margins and hedges expanded when grazing was relaxed, and were cut back when it again increased. Likewise pastures reverted to scrubland and were restored.

Considered as a whole, habitats that comprised the lowland landscape had two important characteristics which, whilst not present everywhere, were far better developed formerly than in the modern lowland landscape:

1. High connectivity within each habitat element. Mapped separately, woodland, grassland, edge/young growth, and mature trees would each have appeared as a landscape-wide network with few blind ends, few detached fragments and no great distance between any detached fragment and the rest of the network. At intervals within the network expanded knots or nodes (e.g. the coppice within the woodland network, or the marsh within the network of small streams and ditches) formed concentrations of native, non-ruderal species which presumably functioned as source populations from which populations in the network as a whole were maintained.
2. Large numbers of contacts between habitat elements, particularly the interdigitation of grassland into woodland and strips of woodland and semi-woodland habitats into farmland. Such was the intermixing that in many districts it might have been better to regard the whole complex of habitats as a functional unit on a landscape scale, rather than a patchwork of separate elements.

The foregoing description applies typically to the great majority of the English lowlands and adjacent parts of the uplands, where woodland formed a small proportion of the land area in flat or gently undulating countryside. In traditionally well-wooded districts, such as the southern Welsh borderland, the Chilterns and parts of south-east England, the integration of coppices and the landscape was even closer, but the main features already described were still apparent. Such districts generally possessed a high proportion of old hedges, a low proportion of cultivated land, and a network of commons, which were often a form of wood pasture. Under these conditions, the landscape was not so much open ground through which distant woods were connected by a lacework of hedges, but an all-pervasive woodland which had been punctuated by clearance. Many of the hedges were, in fact, retained strips of the original woodland, which were themselves treated as coppice.

1.4 RECENT CHANGES

Changes since 1945 in forestry, agriculture and river management have substantially altered the external relationships of coppices. These changes were the culmination of trends which started long ago. For example, the total area of coppice in Britain declined from 233 000 ha in 1905 to 142 000 ha in 1947, and 30 000 ha in 1965, before enjoying a modest revival to 40 000 ha in 1980 at the prompting of conservation organizations and a revitalized fuelwood market. Several changing elements can be identified (Peterken and Allison, 1989):

1. Coppices have been cleared, mainly for agriculture, and thus distances between coppices have increased. The proportion that has been destroyed in this way has varied widely between districts, but (assuming that coppices followed the pattern of all broadleaved and ancient woodland) the range may be as large as 1–42%, with a general average of about 10%. To some extent this increasing isolation has been mitigated by the development of secondary woodlands on farmland, but this will not have helped species which lack the capacity to colonize these new woodlands.
2. The network of hedges has been greatly reduced by field enlargement and hedge removal. The area of this semi-woodland habitat has thus been reduced and the connectivity between woodland habitats across the landscape has largely been lost (e.g. the post–1611 fate of hedges and ancient coppice in Rawshall (Rackham, 1986). Here, too, there has been immense variation between districts, and the data for total losses rarely cover the entire post-war period, but substantial parts of the farmed landscape have lost over half their hedges. Moreover, many hedges that remain have become gappy lines of shrubs with little resemblance to a woodland habitat.
3. Mature trees in farmland have been greatly reduced. Hedge trees have been lost with hedgerow removal and felled without replacement even from hedges that have survived. Elm disease had a particularly severe impact on farmland trees. Total losses are again extremely variable between districts, but as much as 90% of all trees have been lost from particular parishes. To some extent hedgerow trees have been replaced by field corner plantings, roadside clumps, motorway landscaping and village trees, but these often take the form of small woods, which do not maintain the contiguity and connectivity of mature trees in the landscape. These changes have been reinforced by the felling of timber trees in surviving coppices. Substantial fluctuations in the number of farmland trees occurred in the past (Rackham, 1986) but the current densities are very low.
4. Streams have been ditched and rivers have been channelized and realigned. Marginal scrub and trees have been removed in a programme of river management which has been particularly active since the passing of the Land Drainage Act of 1930. These changes have accentuated the changes in items 1–3.
5. Most meadows have gone and pastures have been greatly reduced. Those which survive have been drained, fertilized and treated with herbicides. The extent of unimproved grassland in the lowlands has

been reduced by 97% (Fuller, 1987). Green lanes have been lost through disuse, hedge removal and ditching. Road verges have ceased to be mown or grazed and instead have become dumping grounds for waste and ditch spoil. Grassland habitats within coppices have thus been isolated from other surviving semi-natural grasslands.

6. Coppices in most districts have ceased to be regularly cut. This neglect has reduced the amount of young-growth scrub in the landscape (which has already been reduced by hedge removal), and allowed the uncut woodland to shade and often extinguish the woodland grasslands. Surviving open spaces within woods are small, not connected with open spaces in the wider landscape, and are no longer bordered by a dynamic mixture of young growth and temporary open space.
7. Massive amounts of fertilizers, pesticides and herbicides have been applied to the farmland matrix. This appears to have stimulated the growth of a limited number of herbs in hedgerows and reduced the numbers of woodland species sheltering there. Direct and indirect transfer of material and nutrients to woodlands also appears to be taking place. Ancient coppices had hitherto remained islands of oligotrophic soils but now they are evidently being eutrophicated, both on their margins and in patches within their interior.

These changes in the coppices interestingly reflect changes in society. The coppices were once an integral part of the farm and estate economy, providing significant and constantly available local employment, and contributing materials for use on the farm. Now, they have become irrelevant to most of the people living locally. Many have been let or sold for other uses, including forestry, nature conservation, public recreation and even dispersed housing. Those which survive as game coverts within large estates probably come closest to maintaining the traditional role of the coppice as an integral part of the physical and social landscape.

In the densely encoppiced districts, such as the Wyre Forest, Wye Valley and parts of the Weald, changes have taken a different form. Few parts of these large woods have been cleared, but many have been converted to timber plantations. In this form they have retained their woodland-grassland, albeit in the modified form of open rides. The small fields within the woodland matrix have often survived as pasture because they cannot conveniently be incorporated in arable farmland.

1.5 THE NATIVE WILDLIFE OF TRADITIONAL COPPICES

The development of the traditional landscape (by which I mean the landscape which evolved before 1900, based on local sources of energy and fertility) permitted fragments of the original woodland to survive as coppices. The fauna and flora of original-natural woodland was substantially modified: e.g. the larger vertebrates and the biota dependent on dead wood were greatly reduced, and the species associated with gaps, young-growth woodland and permanent open spaces must have substantially increased. Other original components, such as the shade-bearing flora, probably survived with little change. Indeed, the coppice fauna and flora still includes many species associated with mature, stable woodland, which appear in the modern landscape as 'ancient woodland indicators'. Their populations have presumably survived on or close to their present locations for hundreds or thousands of years.

Outside woodlands the traditional landscape provided many refuges for woodland species. Hedges were not only related habitats but many were relict margins of vanished coppices which retained a mixture of tree and shrub species, as well as some of the slow-colonizing species amongst the fauna and flora. It is no surprise that so many of the older natural history books speak of 'the flowers of woods and hedgerow' under one heading. Streamsides and road embankments provided sheltered, moist conditions in which woodland species thrived. Meadows and marshes also contained many woodland species, thriving there partly because (like coppices) these habitats were not heavily grazed in the growing season. Bracken-covered land on heaths provided seasonal shade comparable with that in coppice woodlands, and woodland plants such as bluebell (*Hyacinthoides non-scripta*) and wood sorrel (*Oxalis acetosella*) were often abundant there.

Permanent open spaces within the coppices developed a mixture of grassland, heathland, marsh and tall-herb communities. I know of no direct comparison of within-woodland and extra-woodland forms of these communities, but superficially the woodland forms seem to be poorer in light-demanding species, but enriched by an admixture of shade-bearing woodland herbs. They presumably developed, like the assemblages of later canals and railway embankments, by invasion from the surrounding farmland, a process which was facilitated by the traffic between wood and farmland (Woodruffe-Peacock, 1918). The outcome has been a substantial enlargement of the coppice fauna and flora. Indeed, it is probable that more species thrived within the traditionally managed coppice than in an equivalent area of original-natural woodland.

The coppice-and-farmland landscape provided a great deal of woodland-edge habitat and young-growth woodland or scrub. The edges occurred within farmland along hedges and streamsides; within woodland along ride margins and the boundaries between coppice compartments; and on the boundary between wood and farmland. Young-growth woodland was a permanent feature of larger coppices, where it provided ideal conditions for the development of a rich, vernal flora, and enabled the Lepidoptera,

Diptera and other invertebrate groups in the edge habitats to expand temporarily.

The general effect of the higher connectivity and lower isolation within the traditional landscape was probably to maintain high species diversity within each individual habitat patch. The 'dynamic stability' maintained by management (e.g. the continuity of the woodland habitat combined with perpetual structural change associated with coppicing) would have further diversified the fauna and flora. Furthermore, new habitats generated by changes in farming would have been rapidly enriched. Thus secondary woodlands, enclosure hedges, grasslands formed on former arable and new rides within coppices were all able to develop appropriate faunas and floras. This was possible not only because the landscape-wide network of habitats minimized isolation, but also because traditional agriculture did not modify site characteristics as much as modern agriculture. Undoubtedly, some specialized groups were unable to benefit, such as the fauna dependent on large, sheltered, rotting logs and snags, but for much of the woodland fauna and flora the traditional lowland landscape maintained an extended woodland habitat, which probably went a long way to maintain the richness of the woods themselves.

1.6 THE EFFECTS OF RECENT CHANGES ON WOODLAND WILDLIFE

The effects of recent changes on the wildlife of the lowland landscape is well documented for habitats outside woodlands. Even at the scale of 10 km^2, it is clear that grassland species, such as *Fritillaria meleagris* and *Orchis morio* have been greatly reduced, and that marsh faunas and floras have been devastated. The effects of fragmentation on the fauna and flora of lowland heaths have been known since the classic study by Moore (1962). It is hardly surprising that such changes should be obvious, for they are the outcome of habitat destruction. The changes in woodland wildlife, however, are less obvious, for they are the consequence more of habitat change than decisive destruction. Nevertheless, some points can be established and others can reasonably be conjectured.

Woods with a tradition of coppicing have probably become less diverse. For example, 30 species have been lost from the long list of vascular plants in Bedford Purlieus (Peterken and Welch, 1975), most due to the loss of species characteristic of grassland and wet open spaces, such as lesser pond sedge (*Carex acutiformis*), meadow-sweet (*Filipendula vulgaris*), pennywort (*Hydrocotyle vulgaris*), marsh cinquefoil (*Potentilla palustris*) and saw-wort (*Serratula tinctoria*). More species of open spaces would have been lost without the sustained efforts of conservation organizations to maintain the open rides. Only a few shade-bearing plants disappeared, e.g. giant bellflower (*Campanula latifolia*), violet helleborine (*Epipactis purpurata*) and polypody (*Polypodium vulgare*), and these were probably only present as small populations. Rackham (1980) analysed the losses from five coppices for which records are available from the 18th century or earlier. Here, too, there were few losses of shade-bearing herbs, except from one wood which was mostly cleared. Otherwise, the main losses were of grassland, scrub and wood-margin species, which suffered from both overgrowth of open spaces within the coppices and the destruction of habitats on their margins and adjacent land. In recent decades there has been a major decline of woodland butterflies, such as the fritillaries, which depended on open spaces within the woods and habitats in farmland. This decline in butterflies has been mitigated (or perhaps postponed) in some large coppices which were converted to plantations (see Chapter 13), but even here the long-term effect on the macrolepidoptera as a whole has been damaging, as the century of records from Abbot's Wood, East Sussex (Hadley, 1982), indicates.

Woodland species which do not depend on open spaces and young growth, such as the shade-bearing flora, have probably been least affected, but they must have been reduced by woodland clearance. Those which colonize slowly or not at all have presumably lost 10% of their woodland population via clearance since 1945, and a much higher proportion of their extra-woodland lowland population. The losses could be greater if small populations have also been eliminated by the conversion of coppices to conifer plantations although the evidence is that such planting has no immediate effect at the scale of whole sites (Peterken and Game, 1984).

One would also expect that recent changes have led to increasingly hostile influences from the farmland matrix in the form of drainage, nutrient transfer from fertilizer drift and roosting flocks of farmland birds, and nutrient increases from rubbish dumped in woods and the detritus from pheasant rearing. Eutrophication effects, in the form of patches of nettle (*Urtica dioica*), cow parsley (*Anthriscus sylvestris*) and other species, are apparent on wood margins and interior patches near paths and rides. There is a possibility that coppices and other lowland woods are being submitted to a sustained nitrogen rain, which is changing the balance in the ground flora (Falkengren-Grerup, 1986).

Two further effects of modern agricultural practices can be anticipated. (a) The increased isolation of all woodland species suggests that the diversity of individual former coppices will decline. Local extinctions will continue as populations fluctuate with changing management and natural factors but they will not be so readily counteracted by corresponding colonization. This is particularly likely to affect the true woodland species, which include many slow colonizers and many species that occur as small populations, all of which tend to become genetically less capable of dispersal (den Boer, 1990). (b) Increasing isolation is also likely to weaken the capacity of woodlands to act as sources for stocking new woodlands, hedges and ditch-sides.

1.7 DISCUSSION

This analysis of changes in the contents and circumstances of woods with a tradition of coppicing raises many interesting questions about the origins of the natural fauna and flora of Britain, changes in the long-term interaction between people and their environment, and the implication for populations of long-term isolation, but for me the principal issue is our response in the future. From the landscape-scale perspective several important points emerge:

1. Although unmanaged woodland provides habitats which are poorly represented in coppice, such as dead wood and a multi-layered structure, coppices that are no longer cut pass through a long period in which open spaces are very limited, and unmanaged woods contain no permanent open spaces at all. If coppicing cannot be sustained or restored, there is therefore a case for converting former coppices into managed broadleaved high forest stands, which at least retain their open spaces and, if a normal age-class distribution is maintained (see item 3), maintains a supply of young-growth. Neglected coppice takes a long time to acquire a rich invertebrate fauna associated with dead wood: despite decades of coppice neglect, there are still very few places which harbour a rich dead-wood fauna (see Chapter 6).
2. Conservation policies should be based on the assumption that species must be conserved in the sites they now occupy. Formerly, we might have been justified in expecting any local extinctions to be made good by re-colonization from nearby populations, and this is still possible in well-wooded districts, but in most places woods are now so isolated from each other that they must be managed as self-contained units.
3. Following on from item 2, it is important to maintain a cycle of conditions within each site for all species. In traditional coppices this is best achieved by maintaining the coppice cycle, but in coppices which are converted to plantations it is important to maintain a normal age-class distribution. Many woods are too small to sustain a continuous supply of all habitat conditions, i.e. they are smaller than the 'minimum dynamic area' (Pickett and Thompson, 1978), and these should therefore also be enlarged by planting locally native species on adjacent ground.
4. Woods should be regarded as grassland and wetland reserves as well as woodland reserves. Management should take as much trouble to main-

tain these habitats within woodlands as it does to manage the stands themselves. As in upland conifer forests, the key to optimal nature conservation policies within lowland woods lies as much in ride management and minimal drainage as in the treatment of the stands.

5. The remaining well-wooded districts in which the intricate habitat mosaic has been maintained within and outside the woods should be managed as landscape-scale reserves, in order to provide substantial refuges for all lowland species.
6. Outside woods we need to restore the range of habitats within agriculture. Extensification of agriculture could have several benefits for woodland species if nutrient inputs to woods were thereby reduced and existing quasi-woodland habitats, such as hedges, wet meadows, survived or were restored. Intensification combined with habitat re-creation would provide opportunities to design new connections, which should incorporate not only plantations, but also grasslands and a diverse and natural range of streamside habitats. If the connections and contact points with existing woods were well designed, connectivity could be restored in the landscape, and individual woods could become less isolated. The moderately mobile species would again be able to move around the landscape and the sedentary species may well have a better chance of expanding their isolated populations.

ACKNOWLEDGEMENTS

I am grateful to Keith Kirby and Ian McLean for helpful discussion and comments. Bill Wright, editor of the *Quarterly Journal of Forestry*, allowed me to reuse, with minor changes, sections from a recent paper.

REFERENCES

den Boer, P.J. (1990) Density limits and survival of local populations in 64 carabid species with different powers of dispersal. *Journal of Evolutionary Biology*, **3**, 19–48.

Falkengren-Grerup, U. (1986) Soil acidification and vegetation changes in deciduous forests in southern Sweden. *Oecologia* (Berlin), **70**, 339–47.

Fuller, R.M. (1987) The changing extent and conservation interest of lowland grasslands in England and Wales: a review of grassland surveys 1930–84. *Biological Conservation*, **40**, 281–300.

Hadley, M. (1982) The decline of the indigenous macrolepidoptera of Abbot's Wood, East Sussex. *Entomologist's Record*, **94**, 92–6.

Marren, P. (1990) *Woodland Heritage*, David and Charles, Newton Abbot.

Moore, N.W. (1962) The heaths of Dorset and their conservation. *Journal of Ecology*, **50**, 369–91.

Peterken, G.F. (1981) *Woodland Conservation and Management*, Chapman and Hall, London.

Peterken, G.F. (1991a) Ecological issues in the management of woodland nature reserves, in *Scientific Management of Temperate Communities for Conservation* (eds I.F. Spellerberg, F.B. Goldsmith and M.G. Morris), British Ecological Society Symposium no. 31, Blackwell, Oxford, pp. 245–71.

Peterken, G.F. (1991b) Managing semi-natural woods: a suitable case for coppice. *Quarterly Journal of Forestry* 85, 21–9.

Peterken, G.F. and Allison, H. (1989) Woods, trees and hedges: a review of changes in the British countryside. *Focus on Nature Conservation*, no. 22, Nature Conservancy Council, Peterborough.

Peterken, G.F. and Game, M. (1984) Historical factors affecting the number and distribution of vascular plant species in the woodlands of central Lincolnshire. *Journal of Ecology*, **72**, 155–82.

Peterken, G.F. and Welch, R.C. (eds) *Bedford Purlieus: its history, ecology and management*, Symposium No. 7, Monks Wood Experimental Station, Huntingdon.

Pickett, S.T.A. and Thompson, J.N. (1978) Patch dynamics and the design of nature reserves. *Biological Conservation*, **13**, 27–37.

Rackham, O. (1980) *Ancient Woodland*, Edward Arnold, London.

Rackham, O. (1985) Ancient woodland and hedges in England, in *The English Landscape: past, present, and future* (Ed. S.R.J. Woodell), Oxford University Press, Oxford, pp. 68–105.

Rackham, O. (1986) *The History of the Countryside*, Dent, London.

Taylor, C. (1983) *Village and Farmstead*, George Philip, London.

Thirsk, J. (1967) *The Agrarian History of England and Wales, IV. 1500–1640*, Cambridge University Press, Cambridge.

Woodruffe-Peacock, E.A. (1918) A fox-covert study. *Journal of Ecology*, **6**, 110–25.

2

Coppice forestry – an overview

JULIAN EVANS

2.1 INTRODUCTION

Coppice is a forest crop raised from shoots produced from the cut stumps (called stools) of the previous crop, and coppicing is the operation of regenerating crops in this way. It can usually be repeated many times and is a useful means of regenerating broadleaved trees within short time intervals (less than 30 years) to produce small roundwood. It is the main way of managing underwood and has been widely used in small woods for centuries.

The main forms of coppice and associated terms are listed in Table 2.1. Most broadleaved species will coppice, most conifers will not, with the main exceptions of monkey puzzle (*Araucaria araucana*) and coast redwood (*Sequoia sempervirens*). A few broadleaved species do not coppice vigorously or only have the capacity to do so when the stump is fairly small: Italian alder (*Alnus cordata*), beech (*Fagus sylvatica*), birch (*Betula* spp.), cherry (*Prunus avium*) and some poplars (*Populus* spp.).

The ability of broadleaved trees to regenerate themselves from the cut stump, i.e. to coppice, has had a great influence on woodlands in Britain. Evidence of the use of coppice products for numerous rural needs can be found throughout the Bronze Age, Roman and Saxon periods and by 1250 coppicing was almost universal, even in such large woodland areas as the Forest of Dean (Rackham, 1980). Working native woodland in this way ensured a supply of small-sized, easily handled material for firewood, trackway construction, building materials and fencing.

Since the 12th century the history of coppice working has been well recorded, and it is clear that until 150 years ago it was the most widespread silvicultural practice. Coppice with standards dates from medieval times, with the retention of standards for timber being required by law in the 16th century. During the 17th and 18th centuries coppice not only continued to supply building and fencing materials and domestic firewood but was also increasingly in demand for charcoal for the iron and glass industries and, in the case of oak (*Quercus* spp.), to supply bark for tanning.

In the middle of the 19th century the importance of coppicing began to

Table 2.1 Coppice types and terminology (modified from Evans, 1984)

Type	*Description*	*Comments*	*Examples*
Simple coppice	Crop consists entirely of coppice, all of which is worked on the same cycle (even-aged)	May consist of only one species (pure) or several (mixed)	Nearly all sweet chestnut (*Castanea sativa*) is worked as pure simple coppice
Coppice with standards	Two-storey forest; coppice (underwood) with scattering of trees (standards) being grown to timber size	Standards may be of seedling origin (maidens) or develop from a stump shoot left for the purpose (stored coppice); standards retained for a period of 3–8 coppice cycles; oak is much the commonest standard species	Oak (*Quercus*)–hazel (*Corylus*)
Stored coppice	Tree or stand of coppice origin as a result of growing coppice on beyond its normal rotation	Many woodlands resembling high forest are stored coppice, owing to decline in coppice working this century	Much of the oak around Dartmoor and Exmoor
Short-rotation coppice	Arbitrarily designated as coppice worked on a rotation of less than 10 years to produce stick-size material	Provides material for many rural crafts. Recent interest in production of biomass for energy using poplars (*Populus* spp.) or willows (*Salix* spp.)	Osiers (*Salix viminalis*, etc.) grown for wicker work, hazel and underwood worked for spars, pea and bean sticks, etc.
Pollards	Trees cut off at 2–3 m above ground so that the shoots which sprout are not in danger from browsing	Regenerative mechanism identical to coppice. Formerly component of 'wood-pastures' now little practised in traditional form	Riverside willows; similar effect seen in 'topped' poplars and alders (*Alnus* spp.) used for screening and shelter
Underwood	General name for all coppice or woody shrubs occurring under another tree crop		Hazel component of oak–hazel coppice with standards

Figure 2.1 Harvesting 16-year-old sweet chestnut coppice at Kings Wood, Challock. (Photograph: J. Evans.)

decline as many traditional products were superseded. This decline accelerated after the First World War as rural electrification programmes and other more convenient energy sources finally supplanted firewood, so that by the mid-1950s regular coppicing, apart from that of sweet chestnut (*Castanea sativa*), had become rare. The singular exception of sweet chestnut coppice was due to the demands of the hop-growing industry, for long, straight, durable poles which could be produced conveniently on an 8–10-year rotation. Substantial planting of chestnut coppice, often on poor agricultural land but occasionally to upgrade mixed woodland, began in the 1820s and became widespread 30–40 years later when prices for chestnut were high. For example, in 1868 best quality cants realized £90/ha (£3000 at 1990 prices) on the Cowdray Estate, West Sussex (Roberts, 1929).

Prices for chestnut coppice fell rapidly at the end of the last century, owing to the large increase in supply from the many newly established coppices and to the declining demand of the hop industry as wire supports replaced poles. However, a new market emerged using cleft chestnut from 12–16-year-old crops for fence palings that has continued to this day (Figure 2.1). Recently, interest in other types of coppice has been re-awakened owing to a resurgence in rural crafts, e.g. thatching, an increased demand for firewood, and as a conservation measure to maintain the woodland flora and fauna that has long been associated with coppice working in Britain.

2.2 PRESENT STATUS AND EXTENT OF COPPICE

Although active coppice management, apart from chestnut, has declined it is still a significant forest type (Table 2.2). This table also shows that the total area of woodland in Great Britain that has been influenced by past coppicing is substantial and that 95% of actively worked coppice is in England. The analysis for England by species, Table 2.3, shows the dominance of sweet chestnut. Table 2.4 provides details of currently worked coppice by species from Evans (1984).

Table 2.2 Area (ha) of coppice by country and type (modified from Evans, 1984)

	England	*Wales*	*Scotland*	*Great Britain*
Simple coppice	25 711	1849	4	27 564
Coppice with standards	11 473	80	15	11 568
Stored coppice* (estimated)	48 100	17 900	5900	71 900

*Now classified as high forest but clearly identified at present as of coppice origin. Data from survey of 'other woodland' types and a subjective estimate of extent in both Forestry Commission and Dedicated and Approved woodland. *Source*: Locke (1987).

Table 2.3 Area (ha) of worked coppice in England by type and species (from Locke, 1987)

	Principal species of coppice						Total
	Sycamore (*Acer pseudo-platanus*)	Ash (*Fraxinus excelsior*)	Sweet chestnut (*Castanea sativa*)	Hornbeam (*Carpinus betulus*)	Hazel (*Corylus avellana*)	Other species	
With standards	115	193	5 275	1 697	1 465	2 728	11 473
Coppice only	2 297	1 184	13 816	1 716	1 573	5 125	25 711
Total	2 412	1 377	19 091	3 413	3 038	7 853	37 184
% of coppice total	7	4	51	9	8	21	100

Table 2.4 Details of currently worked coppice

Species	Area (ha)* Simple coppice	Area (ha)* Coppice with standards	Soils and sites	Occurrence	Rotation (years)	Markets and uses†	Comments
Alder (*Alnus glutinosa*)	<200		Rarely on clays; valleys, fens and plateau sites with high water-tables	Southern England especially Surrey, east Sussex and Wiltshire	10–20	Turnery (brush-heads, chair-legs)	
Ash (*Fraxinus excelsior*)	1554	193	Moist, fertile loams, not heavy or strongly acidic	Wales, east Midlands especially Leicestershire, and southern England especially Kent, Sussex, Avon and Gloucestershire	10–25	Turnery, thatching sways, scythe and tool handles, split rails	Rarely with standards; stools may not coppice until second season after cutting
Birch (*Betula* spp.)	<500		Mainly acid, sandy heaths	All regions of England but especially Kent, Sussex, Surrey, Hampshire, Berkshire, north-western and eastern districts	15–23	Turnery (brush-heads, cotton-reels), horse-jumps	Coppicing generally not important; most birch arises from natural seeding; coppicing poor if trees over 25 years old
Hazel (*Corylus avellana*)	1630	1465	Wide range of soils	Southern and eastern England especially Dorset, Hampshire, Wiltshire and west Sussex	6–10	Thatching spars, bean and pea sticks, hurdles	Large area of over-mature hazel coppice no longer worked
Hornbeam (*Carpinus betulus*)	1716	1697	Acid soils of moderate clay content	South-eastern and eastern England especially Kent, Surrey, Sussex, Hertfordshire, Bedfordshire, Essex,	15–35	Firewood	

Lime (*Tilia cordata*)		<300	Acid soils, often loess-rich over boulder clay	Mainly Gloucestershire, Herefordshire and Worcestershire, Essex, Suffolk	20–25	Turnery	
Mixed species	5700	2200 (approximate)	Variable	Predominantly south-eastern England but significant areas in eastern, south-western and north-western England	Variable	Firewood, pulpwood, other uses as for individual species constituting the mixture	
Oak (*Quercus* spp.)		<300	Moderately acid soils, loams to clay-loams	Wales, all regions of England except north-east but especially Hertfordshire, Hampshire, Sussex, Cornwall	18–35	Fencing (round, cleft or sawn), tanbark, charcoal	Formerly widespread, much converted to high forest by storing coppice
Sweet chestnut (*Castanea sativa*)	13 816	5275	Acid soils, loamy sands to clay with flints; warm sunny localities	Kent, east and west Sussex, small areas in Essex, Hampshire, Surrey, Gloucestershire and elsewhere		Cleft palings for fencing. Hop poles, fence posts	Much the most important commercial coppice crop still worked
Sycamore (*Acer pseudoplatanus*)	2380	119	Wide range of soils	Southern England but especially Kent, east Sussex, Dorset and eastern England	10–20	Turnery	Rarely with standards

* Actual figures from Locke (1987).
† All species marketable for firewood and pulpwood, though some may be preferred, e.g. hornbeam for firewood, white woods for pulp.

2.3 SOME ASPECTS OF COMMERCIAL MANAGEMENT

2.3.1 Fertilizing coppice

In forestry, coppicing has been considered a likely candidate to cause net nutrient export from a site and therefore possibly to initiate a decline in soil fertility. Because the crop is cut on short cycles and, because regeneration makes use of the stump shoots (coppice), coppicing can be carried on almost indefinitely. Removals from a site by coppicing typically average 2–4 tonnes (dry)/ha/year. The continuous removal of produce over many centuries may have led to a net loss of plant nutrients, especially phosphate, and this has been used by some to explain the term 'tired' coppice, i.e. weak coppice regrowth. It is suggested by Rackham (1967) to be a reason for the decline of coppice yields in the Middle Ages, and is claimed to be a disadvantage of the coppice system in France (Auclair, 1982).

Because of this persuasive prima-facie case for expecting nutrient problems in long established coppice, foliar analysis surveys and, later, fertilizer experiments were carried out by the Forestry Commission in the early 1980s in both sweet chestnut and stored oak coppice. The results four years after treatment were reported by Evans (1986). Application of phosphate significantly increased foliar phosphate concentration for both species but did not lead to a significant improvement in stem-diameter growth. Liming of sweet chestnut, at a rate of 2 t/ha, significantly depressed growth. Remeasurement of the experiments in 1990, eight years after fertilizer application, has not changed the earlier conclusions.

2.3.2 Thinning and dressing

Traditionally coppice shoots are not thinned or pruned (dressed) during the lifetime of the crop, though this practice is common in the tropics (Evans, 1982). It is possible that a single thinning at a suitable point in the cycle could redistribute increment to desirable stems, leading to better quality produce and a shorter rotation length. Not only can the more butt-swept outer stool shoots be removed, but deliberate thinning accelerates the natural process of shoot number reduction from over 150 per stool in the first year to 5–15 viable stems at 15 years of age. The value of thinning and dressing is being investigated for sweet chestnut by the Forestry Commission.

2.3.3 Storing coppice

Owing to neglect of coppicing, of large areas of former coppice have become stored and almost identical to high forest (see Table 2.2). As a silvicultural technique for converting coppice to high forest storing has certain drawbacks:

1. It is inappropriate with hazel (*Corylus avellana*).
2. Stored stems may be more prone to swept (curved) boles.
3. There is anecdotal evidence that in ash and sweet chestnut the timber defects of 'blackheart' and 'shake', respectively, are more prevalent.
4. Growth may be inferior to maiden trees of the same species.

If shoots are thinned after it has been decided to store coppice an acceptable 'high forest' crop can often be formed. Both oak and sycamore (*Acer pseudoplatanus*) respond well to this silviculture (Penistan, 1986; Stern, 1989).

Large areas of coppice of all species have suffered neglect. With the exception of hazel over about 40 years, observation shows that all the principal coppice species, especially sweet chestnut, oak, ash (*Fraxinus excelsior*) and lime (*Tilia cordata*) do not lose the capacity to coppice until biological old age.

2.4 GROWTH AND YIELD

Although there have been few investigations of the productivity of long-rotation coppice, the global figures of 2–4 t/ha/year apply for most species (Begley and Coates, 1961). Most work has concerned sweet chestnut (Begley, 1955; Ford and Newbould, 1970). Recent research into pure sweet chestnut coppice (Rollinson and Evans, 1987) revealed little effect of site on yield in southern Britain and high correlations between age and number of shoots, volume and fresh weight per hectare. Within the range of 5–30 years and where sweet chestnut accounted for more than 90% of the crop, age alone proved an accurate predictor of volume or weight per hectare ($R^2=0.95$).

Application of these relationships to pure coppices of other species would be of interest. Based on the evidence from sweet chestnut, it seems possible for the forester to use the age:biomass:shoot density relationship to calculate the optimum stool density for a given rotation, or for a particular pole size or market.

Yield from poplar coppice grown in short rotation biomass crops is typically 10–12 t/ha/year on 2–4 year rotations.

2.5 MARKETS

Markets for coppice products are generally firmer now than they have ever been since the Second World War. Straight stems can be sold for pulpwood, while an irregular but continuing outlet for firewood allows disposal of poorer quality material. Prices for sweet chestnut have never reached those of the mid-Victorian time in real terms, but are nevertheless high enough to yield a return comparable with productive conifer plantations. In November 1989 cants of chestnut marketed by the Forestry Commission attracted prices ranging from £200/ha to £2150/ha, depending on quality and location.

Demand for hazel spars for use in thatching is once again substantial, but is still unlikely to sustain more than 3000 ha of worked coppice compared with the 40 000 ha or more once managed when hurdles, wattles and other rural products were in demand. Numerous thatching teams now work in Britain and use, very roughly, 6000 spars per thatching. Whether plastic substitutes or imports of hazel make significant inroads into local supplies will depend very much on the availability of skilled workers willing to cut both active and stored coppice areas.

2.6 CONCLUSIONS

The 38 000 ha of worked coppice in Britain today is the lowest area since early medieval times, reflecting the declining markets for coppice products during the present century. Markets for coppice products generally remain uncertain. Short-rotation energy crops of poplars and willows have the potential to yield 12 t/ha/year, but new markets need to develop for biomass.

Large areas of coppice have, through neglect, become 'stored' although (with the exception of hazel which begins to collapse if neglected beyond 40 years) these stands can be successfully re-coppiced, even after 100 years of neglect. They can also be converted to high forest by thinning, though the quality of timber produced is generally poorer than from maiden crops. Finally, the maintenance of coppicing, or its resumption to continue countryside traditions and crafts as well as for conservation, will depend critically on improved market outlets.

ACKNOWLEDGEMENTS

I am particularly grateful to Simon Humphreys of Silvanus, Pat Dineen, Weald Forest District, Forestry Commission and Alastair West of the Thatching Advisory Service for provision of up-to-date marketing information and general comments on prospects for coppice.

REFERENCES

Auclair, D. (1982) Present and future management of coppice in France, in *Broadleaves in Britain* (eds D.C. Malcolm, J. Evans and P.N. Edwards), Forestry Commission/Institute of Chartered Foresters, Edinburgh, 40–6.

Begley, C.D. (1955) *Growth and Yield of sweet Chestnut Coppice*. Forestry Commission Forest Record, **30**, HMSO, London.

Begley, C.D. and Coates, A.E. (1961) Estimating yield of hardwood coppice for pulpwood growing, in *Forestry Commission Report on Forest Research 1956/60*, HMSO, London, 189–96.

Evans, J. (1982) *Plantation Forestry in the Tropics*, Clarendon Press, Oxford.

Evans, J. (1984) *Silviculture of Broadleaved Woodland*, Forestry Commission Bulletin, 62, HMSO, London.

Evans, J. (1986) Nutrition experiments in broadleaved stands II. Sweet chestnut and stored oak coppice. *Quarterly Journal of Forestry*, **80**, 95–104.

Ford, E.D. and Newbould, P.J. (1970) Stand structure and dry weight production through the sweet chestnut (*Castanea sativa* Mill) coppice cycle. *Journal of Ecology*, **58**, 275–96.

Locke, G.M.L. (1987) *Census of woodlands and trees 1979–1982*. Forestry Commission Bulletin, **63**, HMSO, London.

Penistan, M. (1986) Oak in Wessex: an account of field studies 1982–84. *Forestry*, **59**, 243–57.

Rackham, O. (1967) The history and effects of coppicing as a woodland practice. *Proceedings of Monks Wood Experiment Station Symposium*, **3**, 82–93.

Rackham, O. (1980) *Ancient Woodlands*, Edward Arnold, London.

Roberts, T. (1929) The growing and conversion of chestnut in West Sussex. *Quarterly Journal of Forestry*, **23**, 220–9.

Rollinson, T.J.D. and Evans, J. (1987) *The Yield of sweet Chestnut Coppice*, Forestry Commission Bulletin, **64**, HMSO, London.

Stern, R.C. (1989) Sycamore in Wessex. *Forestry*, **62**, 365–82.

Part Two

THE PHYSICAL ENVIRONMENT

– 3

Growth stages and microclimate in coppice and high forest

P.L. MITCHELL

3.1 INTRODUCTION

An important distinction among silvicultural systems is between coppice, where repeated crops of small-sized wood are regenerated at short intervals from cut stumps, and high forest in which single-stemmed trees grow from seed or transplants to their full height. For centuries coppicing was the usual method of managing woodland in the lowlands and coppiced woods are now generally regarded as valuable for nature conservation (Rackham, 1980; Peterken, 1981). There is therefore much interest in comparing worked coppice, neglected coppice and high forest systems as options for woodland management, particularly where several objectives (all or any of nature conservation, wood production, recreation, landscape or the production of game) must be satisfied, within a financial constraint.

When comparing stands of trees under different silvicultural treatments or of different species it is important to take into account the changes that occur in all stands during the rotation (Mitchell and Kirby, 1989). This can be done by comparing stands at the same stage of growth.

3.2 GROWTH STAGES

An even-aged plantation passes through several stages during the rotation, i.e. from planting to felling (Figure 3.1). At first the young trees are small and separate and have little influence on the site. As they grow the individual crowns meet and then coalesce to form a closed canopy – this starts the thicket stage. The green canopy moves upwards as the lower branches die. The first thinning initiates the pole stage during which transient gaps in the canopy caused by thinning are closed by rapid growth of neighbouring trees. If the plantation is not felled at the normal time (i.e. at the age of maximum mean annual increment, in volume or value) then it progresses to a further stage which I shall call 'veteran' to avoid the over-worked word 'mature'. Here the growth of the trees slows down and they are unable to close gaps in the canopy caused by thinning or individual deaths. Eventually all the trees start to die.

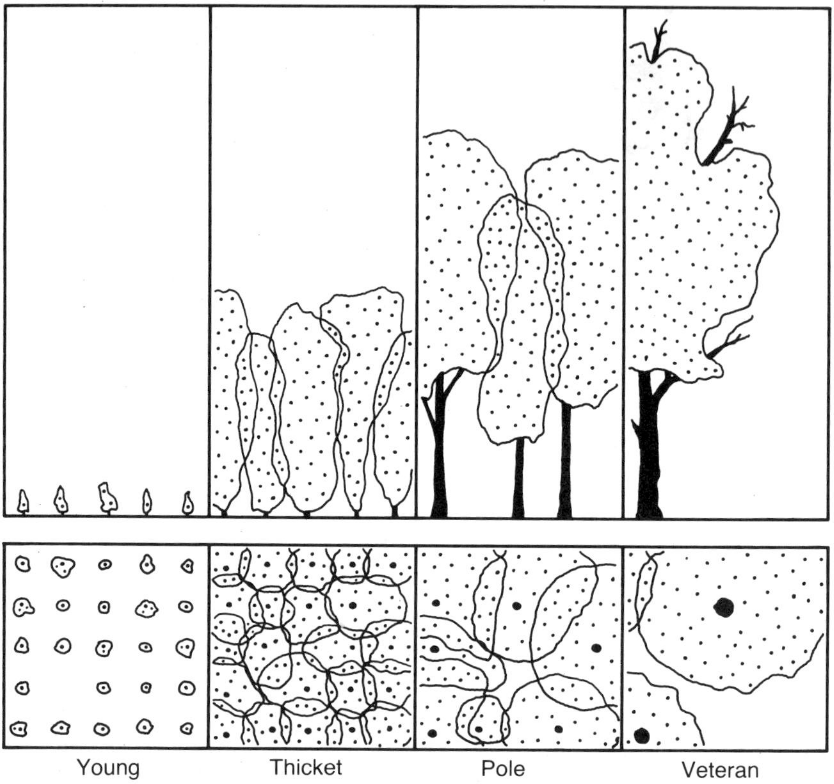

Figure 3.1 Representations of the young, thicket, pole and veteran growth stages in a plantation of broadleaved trees; in profile above (10 m × 25 m) and in plan below (10 m × 10 m).

This sequence of changes is well known and has been used in work on ground vegetation (Ovington, 1955; Hill, 1979), birds (Rose, 1979; Moss, 1979) and deer (Ratcliffe, Hall and Allen, 1986). It is most clearly seen in even-aged plantations where canopy closure is more uniform, of evenly spaced trees, and where the first thinning initiates the pole stage. As well as these general stages others are often defined for particular sites, tree species or applications. Such is the importance of canopy closure that the period around it may be defined as a separate stage (Ratcliffe *et al.*, 1986; Chapter 9). It is possible to extend the concepts of 'stand', 'rotation' and 'growth stages' to apply them to silvicultural systems other than clear-felling and to natural forest (Mitchell and Kirby, 1989; Whitmore, 1975 (forest growth cycle); Ellenberg, 1988).

Worked coppice passes through the young stage and the thicket stage only. On very short rotations, e.g. 2–3 years for the production of chestnut (*Castanea sativa*) walking sticks (Crowther and Evans, 1984), the canopy may only just close at the end of the rotation or for the final year. Neglected coppice progresses to a sort of pole stage but as self-thinning of shoots on the stool takes place all the time (Crowther and Evans, 1984) it is difficult to define a transition point, and self-thinning probably never leaves gaps in the canopy. Silvicultural treatment to 'store' a coppice and convert it to high forest, by thinning the shoots to leave one or two on each stool, clearly introduces a pole stage.

For fair comparisons between stands the stocking needs to be taken into account as well as the growth stage. Stocking may be measured as density (trees per hectare) or basal area (the sum of the cross-sectional area of the trunks at breast height (1.3 m) in square metres per hectare) according to the size of trees and the purpose of the assessment. Stands progress through the early stages more quickly if more densely stocked. For example, Low (1974) reported a variation of 10–13 years for the time of canopy closure within a species at spacings of 0.9 m × 0.9 m to 2.4 m × 2.4 m. Similarly, vigorous chestnut coppice at 2000 stools/ha closed canopy after three years (Mitchell and Woodward, 1987) but at 600 stools/ha canopy closure was after seven years (Ford and Newbould, 1970). At the pole stage, stands can differ widely in stocking according to their rate of growth and the thinning regime.

Changes, in ground flora, animal populations, etc., during the coppice rotation are usually reported in relation to the age of regrowth. Unless stands are similar in species, stocking and vigour of growth it is difficult to make a fair comparison age for age. The age of canopy closure should be recorded: it is a reference point in the rotation of worked coppice and age could be expressed relative to it.

3.3 MICROCLIMATE

The environmental factors comprising the woodland microclimate are solar radiation, thermal radiation, temperature, relative humidity and wind. All these factors are relevant in the short term to the heat and water balances of all organisms and to the rate of photosynthesis in plants; they have long-term effects on growth, development, reproduction and completion of the life cycle. They vary in a complex and interrelated manner in open sites and further complexity is added by the trees in woodland sites.

Stands of trees moderate the fluxes of radiation, air temperature, humidity and wind speeds within them (Geiger, 1965; Ellenberg, 1988). The more plant material there is in the canopy (e.g. leaves on deciduous trees) and the thicker the canopy layer, the more the space below is insulated from the effects of radiation fluxes and changes in temperature and humidity above, and the more the wind speed is reduced.

These general effects result from the ability of the canopy to absorb or reflect solar and thermal radiation and to absorb the energy in air movement. This in turn depends on the physical properties of the canopy: the density (in number per unit volume) of leaves, twigs and branches, their size and distribution and especially the orientation of leaves. Differences between stands arise mechanistically from differences in canopy structure and not because it is species A rather than B. Differences among growth stages are to be expected in so far as they reflect differences in the completeness of the canopy (closed in thicket and pole stages), its uniformity (thicket is more uniform than the pole stage, which is more uniform than either young or veteran stages), and its height above ground and thickness (for which all stages differ). There seems to be no reason to expect differences between coppice and high forest, stage for stage, unless the details of canopy structure (e.g. single-stemmed versus multiple-stemmed trees), become important. For example, the measurements of air temperature (as an average of integrated totals for 2–4 week periods) in four chestnut coppices of different ages showed that where the canopy was incomplete the 24 hour average temperature was 2–3°C higher in late spring and early summer and 1–2°C higher for the rest of the year than under a closed canopy. The amplitude between day and night temperatures was greater for sites with lower or incomplete canopies (Mitchell and Woodward, 1987). This is much what would be expected in high forest stands at the same stages (Geiger, 1965).

Since microclimatic factors tend to vary in parallel, related to the presence and properties of the tree canopy, it is reasonable to concentrate on one factor of over-riding importance that changes greatly through the year and is markedly influenced by the canopy – solar radiation. Two aspects of solar radiation are photosynthetically active radiation (PAR, 400–700 nm wavelengths) and the ratio of red to far-red radiation (R/FR ratio). (In everyday terms PAR can be thought of as the quantity of light and the R/FR ratio as the quality of light from the point of view of plant growth.) The emphasis on PAR is justified because this is the only energy source for green plants, and ultimately for all organisms, and it is coupled quantitatively to plant growth through the amount intercepted (Cannell *et al.*, 1987). Leaves preferentially absorb PAR (including red) and scatter infra-red radiation (including far-red) so that the R/FR ratio is lowered; the ratio is a sensitive indicator of the degree of shade to which many plants respond (Holmes and Smith, 1975; Smith, 1982). This account considers only the microclimate of the ground vegetation but animals are affected by similar environmental factors and depend on plants for food and cover.

The soil and biotic environmental factors should not be forgotten when trying to understand the occurrence and performance of plants and animals in the woodland ecosystem. For example, shade tolerance and soil fertility interact (Ellenberg, 1988; Peace and Grubb, 1982) and there is the perennial question of whether it is shade or competition with trees for water and nutrients that determines the abundance of the ground vegetation (Ellenberg, 1988; Chapter 4).

3.4 PHOTOSYNTHETICALLY ACTIVE RADIATION

3.4.1 Coppice

Over a period of 18 months I made measurements of PAR in 12 coppice plots at three locations in southern England (Mitchell, 1984). The plots were chosen to include a series of ages in chestnut and a series of stool spacings in willow *Salix burjatica* 'Korso' see Table 3.1 for details. Using an electronic instrument PAR was integrated over time (Mitchell and Woodward, 1987). In each plot a sensor was placed between stools in a position that would be covered last by the extending canopy as the coppice regrew after cutting. It was set horizontally about 10 cm above the ground and kept clear of ground vegetation. The instruments were read at intervals of 2–4 weeks. The results are given as average daily totals of PAR for the period between readings, in $E/m^2/day$ where an einstein (E) is Avogadro's number (one mole) of quanta. (To put the numbers in context: the usable scale is 0–100 $E/m^2/day$; averages for southern England are 42 $E/m^2/day$ for June, 22 for the year and 4.8 for December (see Figure 3.5); values for individual days can range from 0.7 to 70 $E/m^2/day$.) Records of solar radiation (300–3000 nm) at the nearest meteorological stations were used to calculate the PAR incident on the canopy and hence the percentage transmission. A typical set of results is shown in Figure 3.2 for a chestnut coppice during the second and third season of growth.

The influence of the leaf canopy is clearly seen in the very low PAR recorded in the summer so that the highest values occurred in the spring. In the autumn canopy transmission rose sharply as the leaves fell, but because incident PAR was now low there was only a shallow peak in PAR under the canopy. Even in winter the canopy transmitted only about 50% of incident PAR. The differences between the summers of 1983 and 1984 were due to the canopy closing over the PAR sensor early in 1984.

Apart from a few months in the summer of 1984 canopy transmission was not constant. Since there was no change in the canopy during the winter this arose from the varying contributions of direct (sunshine) and diffuse (skylight) radiation, which depends on solar geometry and the weather (Anderson, 1964b). From the winter solstice as the sun gets higher in the sky, and for longer each day, more direct radiation penetrates through the canopy in which most holes tend to be near the zenith. Long-term, integrated measurements such as these give the most reliable estimates for canopy transmission. Spot measurements can be made under standard-

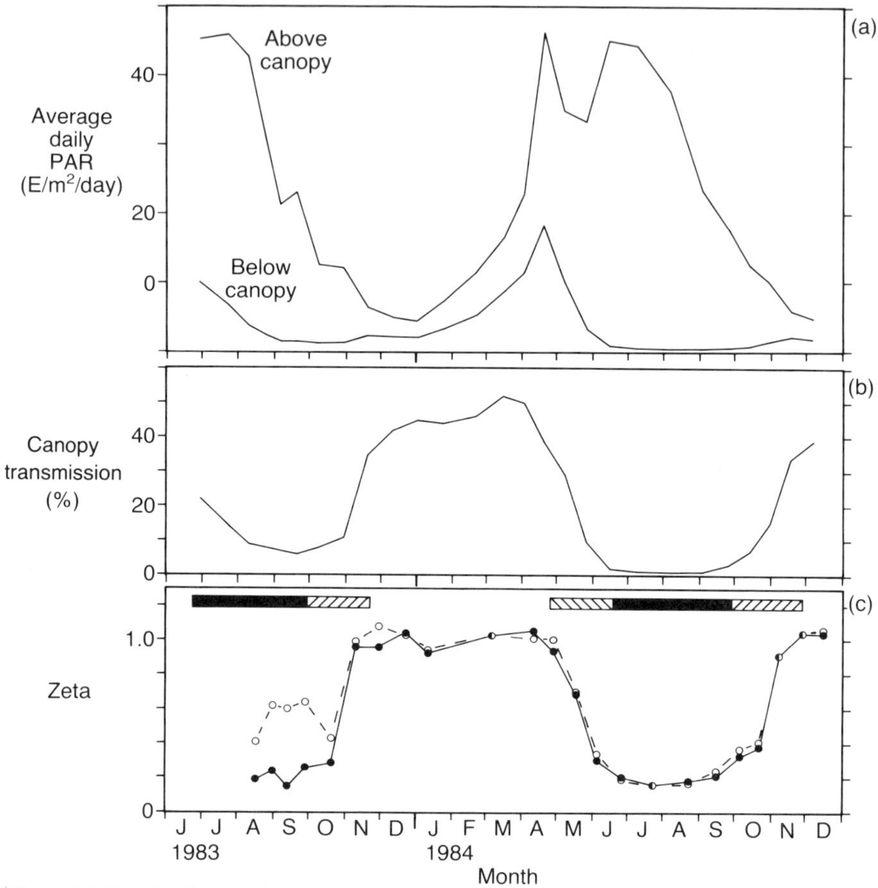

Figure 3.2 Results from a chestnut coppice during the second and third seasons of growth at Rogate, Sussex: (a) estimated incident PAR above canopy and measured PAR below canopy; (b) canopy transmission for PAR; (c) the R/FR ratio as zeta. Measured PAR was the average from integrated totals over 2–4 week periods and incident PAR and transmission were calculated for the same periods. Zeta was derived from spot measurements: close to a stool where there was always a canopy above, except for a short while after cutting, (●), and between the stools in the position last to be covered by the extending canopy (○). Coincident points are indicated by ◐. The black bar shows the duration of a full leaf canopy, the periods of leafing and leaf fall are hatched (from observations at the site).

ized conditions but they are still potentially misleading as the transmission recorded is valid only for those conditions at that particular time of both day and year.

The results for the coppice sites are summarized in Table 3.1 as the percentage transmission for the leafy and leafless canopy and the months when PAR was above or below two thresholds, chosen after experimental work on woodland plants (Mitchell and Woodward, 1988). Most woodland plants can grow and flower at 5 E/m^2/day. At 1 E/m^2/day shade-tolerant species can survive but few can flower. Future workers should refine these thresholds, but the important point is that they are expressed in absolute units.

Table 3.1 Summary of PAR measurements for coppice sites. The leafy period was taken as July to September, leafless as December to March. Brackets around the month indicate an incomplete estimate of the period because it was truncated by the start or end of measurements. Values for transmission are given as ranges for most periods because of changes in canopy extent or because even when the canopy was constant transmission varied (see text). Chestnut coppice at Rogate, Sussex, and hazel at Dickers Copse, near Basingstoke, Hampshire, were worked coppices. The willow and poplar plots were experimental plots at Long Ashton Research Station near Bristol (parts of Forestry Commission and Long Ashton Research Station experiments). For further details see Mitchell (1984)

Age of regrowth (years) and date	*Height at end of year (m)*	*Stools/ha*	*Canopy transmission (%)*		*Period when PAR (E/m²/day)*	
			leafy	*leafless*	*over 5*	*less than 1*
Chestnut (Castanea sativa)						
1(1984)	2.2	1680	35–60	50–65	(March)–September	none
1(1983)	1.9	1720	30–50	–	(June)–September	–
2(1984)	3.1	1720	3–10†	40–50	February–June	none
2(1983)	3.1	2030	6–20	–	(June)–July	–
3(1984)	4.4	2030	1–3*	–	February–May	June–October
6(1983)	6.3	1950	2–3*	40	–	July–October
7(1984)	7.2	1950	2–3*	–	March–May	July–October
9(1983)	8.6	1880	1–2*	30–40	–	June–October
10(1984)	9.3	1880	1–2*	–	March–May	July–October
Willow (Salix burjatica Nasarov '*Korso*')						
1(1983)	2.4	2500	10–20*	60–65	(June)–(September)	none
1(1984)	2.6	2500	10–15*	–	February–July	none
1(1984)	2.2	10 000	10–30*	25–45	(March)–June and September–October	none
1(1983)	2.3	10 000	2–10*	–	–	August
2(1984)	3.2	10 000	2–20*	–	February–April	June–July
1(1984)	2.1	40 800	3–10*	25–35	(March)–June	none
1(1983)	1.9	40 800	1–10*	–	–	July–August
2(1984)	2.8	40 800	4–10*	–	March–April	May–June
Poplar (Populus deltoides × P. trichocarpa '*Rap*')						
1(1983)	3.1	2500	15–40†	45–70	(June)–August	none
1(1984)	3.2	2500	10–40†	–	February–August	none
Hazel (Corylus avellana)						
1(1983)	1.3	2660	45–60	55–70	(June)–October	none
2(1984)	2.1	2660	–	–	February–(July)	none

*Closed canopy. For willow in the first year of regrowth canopy closure takes place from July (40 800 and 10 000 stools/ha) to August (2500 stools/ha).

†Canopy almost closed. For poplar this meant that from about August the crowns met within a row and across the rows but not along the diagonals.

It is clear from the chestnut sites that after canopy closure, normally in the third year, only a few percent of incident PAR passed through the leaf canopy and only 30–40% when the canopy was leafless. Reasonable levels of PAR were available only for a few months in late winter and early spring. Before the canopy closed, PAR over 5 E/m^2/day was available throughout the spring and summer.

The results for the series of densities in willow coppice were less clear cut because the canopy always closed during the first year and leaf fall started in August. At 2 m × 2 m spacing the stools had a lax mode of growth with many shoots nearly horizontal at the base but vertical at the tip; this produced a complete but thin canopy. Nevertheless there is evidence that with denser stands canopy transmission is lower in both the leafy and leafless seasons, and only at 10 000 stools/ha or above did a complete canopy reduce PAR to less than 1 E/m^2/day for part of the summer. The dense thickets of leafless stems at 10 000 or 40 800 stools/ha intercepted 55–75% of incident PAR. In the second year the period when PAR was above 5 E/m^2/day was cut back to a few months in early spring.

The poplar (*Populus deltoides* × *P. trichocarpa* 'Rap') coppice did not completely close the canopy in one year of growth, but at the end of the summer transmission was reduced to 10–15%. The hazel (*Corylus avellana*) coppice grew more slowly than the others and was still far from closing canopy after two years. In between the stools PAR remained high throughout the year.

All these measurements were made in simple coppice without the complication of standards, but coppice with standards accounts for 31% by area of all coppice in England (Crowther and Evans, 1984). It is safe to assume that standards will further reduce PAR received at ground level in proportion to their numbers per hectare, provided that they do not markedly restrict coppice regrowth, and they will probably make levels of PAR more variable from place to place.

Rackham (1975) estimated the amounts of solar radiation in four shade variants in Hayley Wood. By making various assumptions (Mitchell, 1984) it is possible to convert these estimates to PAR for comparison with the measurements made in coppices (Figure 3.3 and Table 3.2). The commonest type of coppice with standards at Hayley Wood is small coppice of hazel and hawthorn (*Crataegus oxycanthoides* and *C. monogyna*) together with larger coppice of ash (*Fraxinus excelsior*) and field maple (*Acer campestre*), and pedunculate oak (*Quercus robur*) standards with crowns covering a third to a half of the ground area. The top of the canopy is at about 15 m. The summer values for PAR and canopy transmission are similar to chestnut and willow coppice but the winter values are only half as great, suggesting that standards and two sizes of coppice cast much more shade

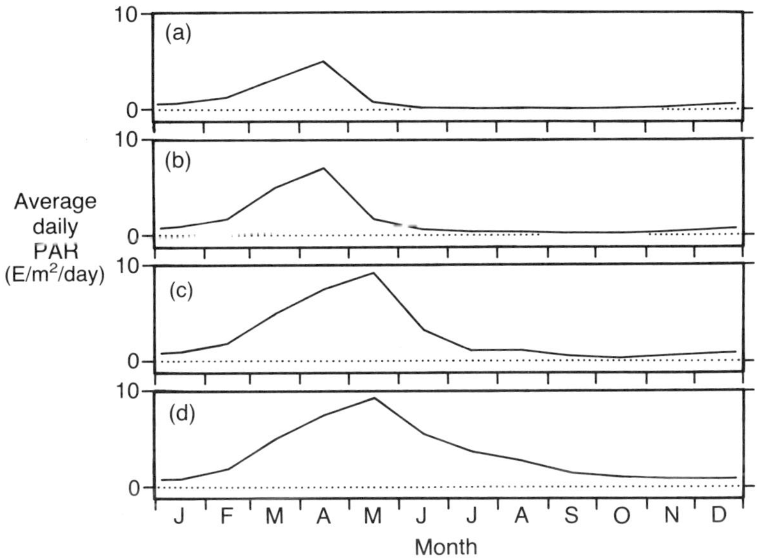

Figure 3.3 Annual change in PAR under the coppice shade variants of Hayley Wood, Cambridgeshire. The shade variants are: (a) dense shade under thick hazel or hawthorn coppice; (b) normal shade under coppice; (c) the shade of large coppice and standards but small coppice missing; and (d) glades, ride edges and under gaps in the canopy. (Adapted and redrawn from estimates of solar radiation given in Figure 31 of Rackham, 1975.)

than the leafless canopy of simple coppice 2–9 m tall. With a mixture of species as well as this more complex structure, the variation from place to place within a stand in Hayley Wood must also be greater.

Table 3.2 Summary of PAR estimates for Hayley Wood, Cambridgeshire. The leafy period was taken as July to September; leafless as December to March. (Adapted from Rackham (1975))

Shade variant	*Canopy transmission (%)*		*Period when PAR (E/m²/day)*	
	leafy	*leafless*	*over 5*	*less than 1*
A: thick coppice	1	10–20	none	May–January
B: normal coppice	1	10–30	March–April	June–January
C: small gap in coppice	2–3	15–30	March–May	July–January
D: glades and ride edges	7–10	15–30	March–June	October–January

3.4.2 High forest

There are few results for high forest that are directly comparable with those in coppice because measurements have rarely been made with similar instruments or at more than one point in the rotation or at different levels of stocking. However, it has been possible to convert three sets of measurements into the terms used for the coppice results (Table 3.3). The general pattern to emerge from Table 3.3 is the same as for coppices. Under a complete canopy PAR is less than 1 E/m²/day during the summer, and only above 5 E/m²/day for a few months in late winter and early spring. Where the canopy is not closed, as in the large clearing in Madingley Wood, PAR over 5 E/m²/day persists into the summer and levels below 1 E/m²/day do not occur.

Table 3.3 Summary of PAR estimates for high forest. The leafy period was taken as July to September, leafless as December to March. Estimates of PAR were made from the measurements of solar radiation reported

Site and/or year	*Canopy transmission (%)*		*Period when PAR (E/m²/day)*	
	leafy	*leafless*	*over 5*	*less than 1*
*Madingley Wood, near Cambridge, UK**				
Photosite, 1962	2	10–25	April	July–January
Photosite, 1963	1–2	10–25	April	July–January
Small clearing, 1962	6–7	20–35	March–May	none
Large clearing, 1962	10–12	20–35	March–June	none
Large clearing, 1963	11–16	25–35	March–July	none
Virelles-Blaimont, Belgium†				
April 1966–March 1967	2–3	30–50	March–April	June–September
Oak Ridge, Tennessee, USA‡				
1972–1973	1	10–30	March–April	June–September

*Madingley Wood consisted of mixed broadleaves about 20 m tall. The photosite was in a stand of smooth-leaved elm (*Ulmus carpinifolia*) with some pedunculate oak and a shrub layer of hazel. The small clearing was a similar area 5 m distant but beneath a small gap in the canopy. The large clearing was a gap about 20 m in diameter surrounded by oak, ash and hazel (Anderson, 1964a,b).

†The stand at Virelles-Blaimont (Grulois, 1967, 1969) had a dense canopy of pedunculate oak and beech (*Fagus sylvatica*) (219 stems/ha, 10–21 m tall) and hornbeam (*Carpinus betulus*) and field maple (917 stems/ha, 6–7 m tall). (It appears to be naturally regenerated high forest becoming two storeyed, not coppice with standards as the species and sizes might suggest.)

‡The stand at Oak Ridge (Hutchison and Matt, 1977) was almost pure tulip trees (*Liriodendron tulipifera*), 50 years old and 30 m tall; 810 stems/ha and basal area 26 m²/ha for trees and shrubs over 10 m. There were abundant shrubs up to 10 m tall.

Measurements through a rotation were made by Mitscherlich (summarized in Geiger, 1965). Visible light (measured with photographic exposure meters) as a percentage of that in the open was recorded in numerous stands of Norway spruce (*Picea abies*; mistranslated as fir in Geiger) from 15 to 130 years in age. The youngest stands transmitted 30–50% of incident light but at canopy closure (17–20 years) this fell to 5–10%; there then followed a steady rise to about 30% at 120 years.

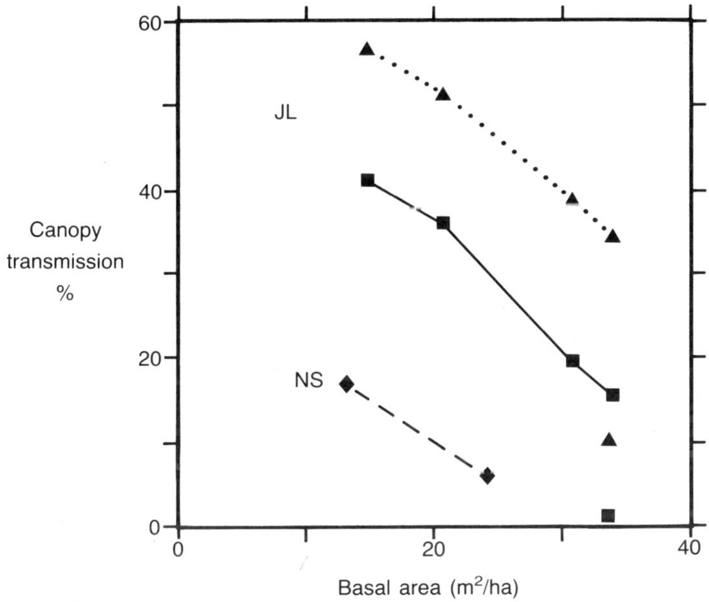

Figure 3.4 The effect of stocking (measured as basal area) on transmission of visible light in thinned stands of Japanese larch (JL) 46 years old, when in full leaf (■) or leafless (▲) (Sakura *et al.*, 1985) and of Norway spruce (NS), 29 years old (François *et al.*, 1985). The isolated points are for a stand of thicket-stage larch 17 years old.

Two examples of canopy transmission for visible light in pole stage stands thinned to different basal areas illustrate the large effect of differences in stocking (Figure 3.4). In stands of Japanese larch (*Larix kaempferi*) (Sakura *et al.*, 1985) the difference between the least and most densely stocked stands at any one time (20–25 percentage points) was larger than the difference between fully leaved and leafless canopies (15–20 percentage points). Much less light was transmitted by a thicket stand than by the pole stage stand with the same basal area. The two stands of Norway spruce, 29 years old and one year after the first heavy thinning (François *et al.*, 1985), show the same effect but at lower canopy transmission.

Ovington and Madgwick (1955) measured the percentage of visible light transmitted in plots of forest tree species at Bedgebury, Kent (Table 3.4). For a year or two after thinning transmission was higher, e.g. a change from 0.5–2% to 2–6% for Serbian spruce.

Table 3.4 Light transmitted by canopies of different tree species. Measurements were made with a selenium cell in 1953 and 1954 in plots of the species, mostly around 22 years old and early in the pole stage, at Bedgebury, Kent (Ovington and Madgwick, 1955)

	Visible light transmitted (%)	
	August–September	*April*
Heavily shading conifers*	0.5–2	0.5–2
Lightly shading conifers†	3–7	3–7
Hybrid larch‡	20	50
Broadleaves§	2–7	30–60

*Norway spruce, Serbian spruce (*Picea omorika*) (when unthinned), Lawson cypress (*Chamaecyparis lawsoniana*) and grand fir (*Abies grandis*).
†Corsican pine (*Pinus nigra* var. *maritima*) and Douglas fir (*Pseudotsuga menziesii*).
‡*Larix* × *eurolepis*, a fast-growing crop, heavily thinned.
§Sessile oak (*Quercus petraea*) (unthinned), red oak (*Quercus borealis*) (unthinned), Norway maple (*Acer platanoides*), roble (*Nothofagus obliqua*) and the shelterbelt of sessile oak standards (20 m tall) over chestnut coppice (11 m).

3.5 RED TO FAR-RED RATIO

3.5.1 Coppice

Measurements of the R/FR ratio were made at the coppice sites when the PAR integrators were read (Mitchell, 1984). They are given as zeta, the ratio of quantum irradiance in the 655–665 nm and 725–735 nm wavebands, which has a value around 1.15 in the open and down to 0.1 in deep shade of a leaf canopy (Smith, 1982). Results for chestnut coppice in the second and third season of growth are shown in Figure 3.2. The sharp changes in zeta when the leaves flush or fall are clear. There was little difference between the two measuring positions (at base of stool or between stools) except in the late summer of 1983 before the canopy closed.

These results are typical of all the coppices. Zeta is 0.2–0.4 for 3–5 months under a complete leaf canopy but 1.0–1.1 for about six months without leaves. A similar pattern was found in seven-year-old ash coppice in The Netherlands (Pons, 1983). Intermediate values occur during leafing and leaf fall but for only 4–6 weeks under complete canopies, except that with early leaf fall (as in the willow coppices) zeta can rise from August onwards, giving intermediate values for three months. But wherever there is an incomplete canopy and a large input of penetrating radiation (i.e. not scattered — transmitted or reflected — by a leaf) zeta has an intermediate value. For example, it reached only 0.6 between the crowns of almost closed poplar canopy (Table 3.1) and only 0.8 among the 2 m high bushes in the hazel coppice.

3.5.2 High forest

In high forest the R/FR ratio as zeta has been recorded for most of one particular year by Tasker and Smith (1977) under complete canopies of pedunculate oak, beech (*Fagus sylvatica*) and birch (*Betula pendula* and *B. pubescens*). Zeta was about 1.0 during the leafless period. It fell as the leaf canopy developed from early May to July to values of 0.4 (oak), 0.25 (beech) or 0.7 (birch). Under oak and beech zeta rose sharply at leaf fall to the winter value. The high summer value for zeta under birch may be related to its light canopy which allowed more solar radiation, with unaltered zeta, to penetrate than was the case for oak or beech.

3.6 DISCUSSION

3.6.1 Changes in PAR and R/FR ratio during the rotation

From the results presented above it is possible to produce a generalized time course of PAR and R/FR ratio for a year under a complete and dense canopy as in the thicket stage. Starting with the incident PAR and assuming canopy transmission of 50% when leafless and 2.5% when in full leaf, the values for December to March and July to September can be drawn in. The exact shape of the spring peak and the occurrence and shape of an autumn peak (always much lower and broader) depend on the phenology of leafing and leaf fall and hence on the tree species and the weather. Examples are given for an early leafer, e.g. hornbeam (*Carpinus betulus*), a late leafer, e.g. ash, and an early leafer with early leaf fall, e.g. willow coppice (Figure 3.5). For an evergreen species canopy transmission of 2.5% throughout the year is assumed.

Thus, with few assumptions, four seasonal patterns of PAR emerge. Further variety could be added by choosing different values for transmission to represent different species or stocking, and even more by varying the transmission during the leafless period to take account of the changes caused by solar geometry. But the basic pattern would remain because it is determined mostly by the yearly solar cycle and the phenology of the canopy: the leafy half-year lags behind the sunniest half-year. Differences between species depend not so much on the amount of PAR received at ground level but the time of year at which it is received. However, small changes in transmission can move the amount of PAR received across thresholds for survival or performance (e.g. flowering). For example, if the evergreen canopy transmits 5% instead of 2.5% (pine instead of spruce or a heavily thinned crop) then PAR is above 1 E/m^2/day from April to September, and reaches 2 E/m^2/day in June. Finally, to make the model match reality more closely, the variation from year to year has to be

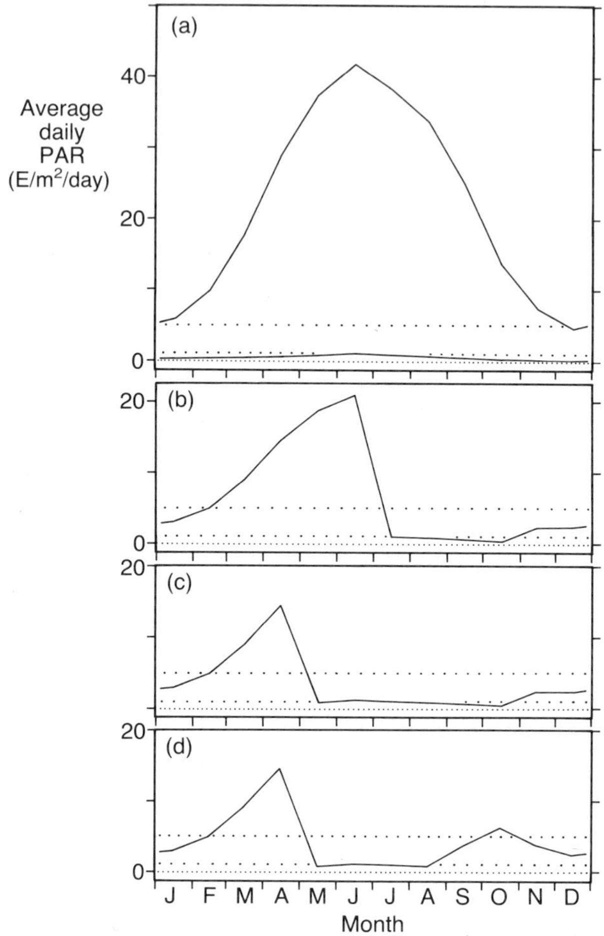

Figure 3.5 Estimated annual changes in PAR under four types of tree canopy in an average year in southern England: (a) incident PAR (upper line) and for an evergreen canopy (lower line); and for tree species that are (b) late in coming into leaf, (c) early and (d) early with early leaf fall. In all cases canopy transmission was set at 50% when leafless and 2.5% when in full leaf. The incident PAR is based on monthly averages for solar radiation for 10 years at Bracknell, Berkshire. The dotted lines are estimated thresholds for the growth and flowering of most woodland plants (5 E/m^2/day) and the survival but little or no flowering of shade-tolerant species (1 E/m^2/day).

considered. This means using the actual incident PAR and for shorter periods than months. Figure 3.2 shows the incident PAR estimated for Rogate, Sussex, from June 1983 to December 1984, averaged over periods of 2–4 weeks, and it is far from the smooth curve of ten-year monthly means shown in Figure 3.5. In addition, the effect of weather on the canopy has to be taken into account: the temperature of buds and expanding leaves for the rate of leafing; temperature and the occurrence of drought, storms

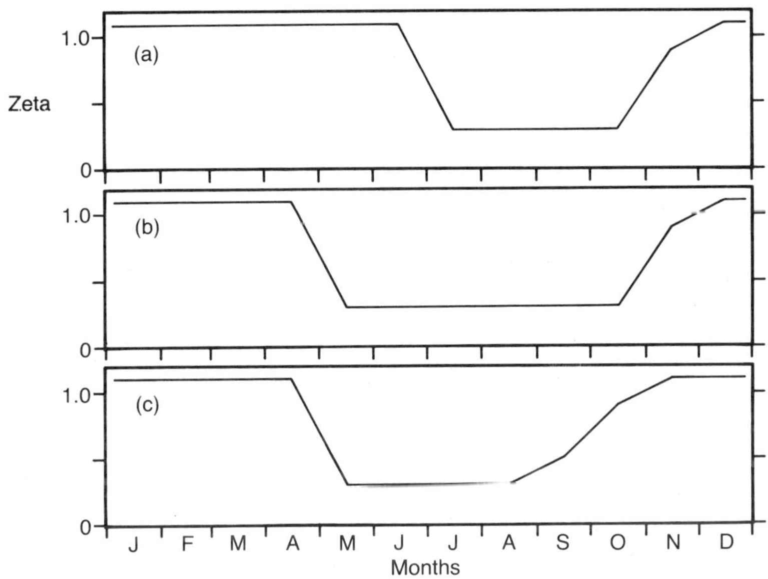

Figure 3.6 Estimated annual change in the R/FR ratio as zeta under the three deciduous canopies of Figure 3.5: (a) late leafer; (b) early leafer; and (c) early leafer with early leaf fall.

and frost for leaf senescence and abscission. Anderson (1964b) discussed in detail how the weather affected the incident radiation and the phenology of leafing and leaf fall to produce the observed patterns of PAR at ground level.

By a similar process Figure 3.6 has been constructed to show the variation in R/FR ratio under different canopies. The model is simpler because the incident R/FR ratio is constant (except around sunrise and sunset) but also less realistic because the R/FR ratio changes beneath apparently constant leaf canopies for poorly understood reasons (Tasker and Smith, 1977; Mitchell, 1984). Although whole-plant responses to a low R/FR ratio are known for a few species (Morgan and Smith, 1979), e.g. the flowering of *Lamiastrum galeobdolon* is reduced (Mitchell and Woodward, 1988), it is not yet possible to suggest threshold values for plant survival and performance, if indeed they exist for the R/FR ratio.

The analysis of PAR can now be extended from the changes within a year for the thicket stage to the differences between stages. In the young stage conditions change from those in the open to those under a complete canopy year by year as the canopy extends and rapidly at close spacings or if tree growth is vigorous. The peak value of PAR is reduced and under deciduous trees moves back from June to April. The PAR received in the summer decreases past the threshold for growth and flowering of light-demanding plants, for flowering of moderate shade tolerators and ultimately for flowering of deep shade tolerators. In the thicket stage PAR and R/FR ratio at ground level vary little from year to year or from place to place. In the pole stage spatial heterogeneity is introduced by thinning and there is perhaps slightly more change from year to year as the green canopy rises and thinning continues. In the veteran stage spatial heterogeneity increases further and the differences from place to place within the stand become permanent.

3.6.2 Differences between coppice and high forest

Stage for stage, the microclimate in coppice and high forest is the same. Coppice regrowth is more rapid than the establishment of young trees because there is a root system in place which contains reserves to fuel initial growth (Pontailler *et al.*, 1984) and the numerous shoots form a bush over the stool. For a given density of stools and vigour of growth, coppice will reach canopy closure more quickly than a high forest stand. The irregular spacing of many coppices may make this less obvious because larger gaps persist for longer and delay canopy closure, or at least the visual impression that it has occurred. High forest has a pole stage and potentially a veteran stage, during which the canopy height rises to two or three times that of coppice but the number of trees decreases from a few thousand to a few hundred. Comparing rotations, the difference between coppice and high forest is simply that the young stage recurs every 10–20 years in coppice but 50–200 years in high forest.

There is also a difference in the structure of the woodland. Worked coppices generally contain all age classes at one time since the areas cut are one or two hectares and the rotation is short. For high forest, only large woods under consistent management for some time are likely to have an even representation of age classes or stages.

3.6.3 Implications for ground vegetation

The seasonal pattern of PAR is associated with the phenology of the ground flora. Under deciduous trees plants able to take advantage of the spring peak are favoured and there is a characteristic vernal flora (Salisbury, 1921; Ellenberg, 1988; Persson, 1975). Under an evergreen canopy these plants are at no advantage so summer-green or evergreen plants do better (Beatty, 1984), e.g. wood sorrel (*Oxalis acetosella*) (Nihlgård, 1970).

The changing regime of PAR through the rotation is reflected in the ground vegetation, even if PAR is not the only environmental factor involved. In both coppice and deciduous high forest PAR is usually less than 1 E/m^2/day in summer under a complete canopy and only more than 5 E/m^2/day in late winter and early spring. Under incomplete canopies higher values of PAR occur in summer. As a coppice regrows the seasonal peaks in ground vegetation dry weight (Ford and Newbould, 1977) and number of species present (Ash and Barkham, 1976) occur earlier. The marginal species of the ground vegetation (those not found in the coppice before cutting and thought at one time to spread in from the edges) become progressively restricted to the spaces between the stools (Salisbury, 1916; Pons, 1976) and then disappear, many persisting in the seed bank (Brown and Oosterhuis, 1981; Chapter 8). The more shade-tolerant plants, which equally flourished in the young stage, persist (as green plants for at least part of each year) through the thicket stage although growth and flowering are less vigorous. The occurrence of the thicket stage protects them from being overwhelmed by the stronger growing marginal species (Rackham, 1975; Ellenberg, 1988).

The characteristic ground flora in coppice seems to come about because it benefits from the 'periodic impetus' of cutting (Salisbury, 1916, 1924); if cutting is delayed the ground flora is reduced. It appears that even shade-tolerant species do not persist in large numbers through a long thicket stage. Barkham and Hance (1982) were able to show, through a model of population biology for wild daffodil (*Narcissus pseudonarcissus*), that the cyclic changes under coppice produced a realistic and stable population, unlike that under constant conditions characteristic of either stage of the coppice rotation. Salisbury (1916) found that the ground flora under coppice with standards was intermediate between that of simple coppice and high forest, in proportion to the number of standards. Where there were many standards only felling with coppicing produced the typical response from the ground vegetation.

3.6.4 The influence of management

The growth stages and the order in which they occur cannot be altered. A thicket stage is inevitable, except in short-rotation coppice, and necessary for productive use of the site (Ford, 1979; Cannell *et al.*, 1987). It is the stage at which the stand exerts maximum influence on the site. In terms of ground vegetation the differences between tree species are then at their most pronounced (Hill, 1979; Mitchell and Kirby, 1989), whereas plantations of many species tend to converge on a common flora in the pole stage (Evans, 1984). The thicket is also the stage with the most uniform microclimate so that the edges and occasional small gaps assume greater importance in stand heterogeneity.

However, management does control, within limits, the length of stages and the timing of transitions. This is important because greater differences in microclimate arise from stage and stocking than from the characteristics of the tree species *per se* (except only that phenology, including whether deciduous or evergreen, is species-specific). The thicket stage is reached sooner (a) at closer spacings; (b) with tree species that grow vigorously, respond to weeding or fertilization or spread out laterally; and (c) if weed trees and shrubs are tolerated and allowed to form part of the canopy, at least for a while. The end of the thicket stage can be hastened by early thinning, or cutting of a coppice, and the stocking of the pole stage can be controlled by the thinning regime. Low stocking would bring forward the start of the veteran stage where gaps persist. In addition, arranging for the wood to consist of more, smaller and irregularly shaped stands would produce more edges where the microclimate is ameliorated compared with the interior of the stands. Nature conservation could benefit from an approach that planned the future structure of the woodland in terms of stages (Mitchell and Kirby, 1989).

3.7 CONCLUSION

Accepting the results discussed above for the young and thicket stages, albeit based on coppice, research is needed into the microclimate of the pole stage: its heterogeneity in time and space and how it varies with stocking. Evergreen species should not be neglected on the assumption that canopy transmission is constant throughout the year. Solar geometry and canopy changes that are less obvious than those of deciduous species will change the transmission, and thinning should have large effects.

It is axiomatic in physiological ecology that the occurrence and performance of plants and animals are related to the factors of microclimate, soil and other organisms, and not directly to silvicultural systems or tree species (except for monophagous herbivores). If the environmental factors relevant to the plants and animals of interest can be quantified, and if the responses of the species to these factors are understood and confirmed by field experiments, then management can be focused more closely on the nature conservation objectives and be made more flexible in incorporating other objectives. To some extent the microclimate can be manipulated by silvicultural treatments. Understanding the microclimate in terms of growth stages and stocking provides a usable framework for management and aids the planning of future research.

ACKNOWLEDGEMENTS

This work has been supported at various times by the Natural Environment Research Council, St John's College, Cambridge, the Cambridge Philosophical Society and the Nature Conservancy Council. I am grateful to K.G. Stott, R.I. Parfitt, M.L. Pearce, R.W. Sochacki, N. Austin and J. Hurle for the use of coppice sites in their care, and to M.D. Morecroft, Dr J.R. Packham and Dr M.G. Holmes for comments on a draft manuscript.

REFERENCES

Anderson, M.C. (1964a) Studies of the woodland light climate. I. The photographic computation of light conditions. *Journal of Ecology*, **52**, 27–41.

Anderson, M.C. (1964b) Studies of the woodland light climate. II. Seasonal variation in the light climate. *Journal of Ecology*, **52**, 643–63.

Ash, J.E. and Barkham, J.P. (1976) Changes and variability in the field layer of a coppiced woodland in Norfolk, England. *Journal of Ecology*, **64**, 697–712.

Barkham, J.P. and Hance, C.E. (1982) Population dynamics of the wild daffodil (*Narcissus pseudonarcissus*). III. Implications of a computer model of 1000 years of population change. *Journal of Ecology*, **70**, 323–44.

Beatty, S.W. (1984) Influence of microtopography and canopy species on spatial patterns of forest understory plants. *Ecology*, **65**, 1406–19.

Brown, A.H.F. and Oosterhuis, L. (1981) The role of buried seed in coppice-woods. *Biological Conservation*, **21**, 19–38.

Cannell, M.G.R., Milne, R., Sheppard, L.J. and Unsworth, M.H. (1987) Radiation interception and productivity of willow. *Journal of Applied Ecology*, **24**, 261–78.

Crowther, R.E. and Evans, J. (1984) *Coppice*, Forestry Commission Leaflet no. 83, HMSO, London.

Ellenberg, H. (1988) *Vegetation Ecology of Central Europe*, Cambridge University Press, Cambridge.

Evans, J. (1984) *Silviculture of Broadleaved Woodland*, Forestry Commission Bulletin, 62, HMSO, London.

Ford, E.D. (1979) An ecological basis for predicting the growth and stability of plantation forests, in *The Ecology of Even-aged Forest Plantations* (eds E.D. Ford, D.C. Malcolm and J. Atterson), Institute of Terrestrial Ecology, Cambridge, pp. 147–74.

Ford, E.D. and Newbould, P.J. (1970) Stand structure and dry weight production through the sweet chestnut (*Castanea sativa* Mill.) coppice cycle. *Journal of Ecology*, **58**, 275–96.

Ford, E.D. and Newbould, P.J. (1977) The biomass and production of ground vegetation and its relation to tree cover through a deciduous woodland cycle. *Journal of Ecology*, **65**, 201–12.

François, F., André, P. and Devillez, F. (1985) Effet de l'intensité de l'éclaircie sur l'extinction du rayonnement solaire en jeunes futaies de *Picea abies* (L.) Karst. *Annales des Sciences Forestières*, **42**, 323–38.

Geiger, R. (1965) *The Climate near the Ground*, Harvard University Press, Cambridge, Massachusetts.

Grulois, J. (1967) Extinction du rayonnement global, tropismes et paramètres foliaires. *Bulletin de la Société Royale Botanique de Belgique*, **100**, 315–34.

Grulois, J. (1969) Réflexion, interception et transmission du rayonnement de courtes longeurs d'onde: variation au cours d'une année. *Bulletin de la Société Royale Botanique de Belgique*, **102**, 13–26.

Hill, M.O. (1979) The development of a flora in even-aged plantations, in *The Ecology of Even-aged Forest Plantations* (eds E.D. Ford, D.C. Malcolm and J. Atterson), Institute of Terrestrial Ecology, Cambridge, pp. 175–92.

Holmes, M.G. and Smith, H. (1975) The function of phytochrome in plants growing in the natural environment. *Nature*, **254**, 512–14.

Hutchison, B.A. and Matt, D.R. (1977) The distribution of solar radiation within a deciduous forest. *Ecological Monographs*, **47**, 185–207.

Low, A.J. (1974) *Initial spacing in relation to establishment and early growth of conifer plantations*, Forestry Commission Research and Development Paper, 110, Forestry Commission, Wrecclesham, near Farnham, Surrey.

Mitchell, P.L. (1984) *Solar radiation under coppice and the physiological ecology of selected woodland plants*, PhD dissertation, University of Cambridge.

Mitchell, P.L. and Kirby, K.J. (1989) *Ecological Effects of Forestry Practices in Long-established Woodland and their Implications for Nature Conservation*, Oxford Forestry Institute Occasional Paper, 39, Department of Plant Sciences, Oxford.

Mitchell, P.L. and Woodward, F.I. (1987) Instrument for measuring temperature, photosynthetically active radiation and daylength, and its use in the measurement of daylength and temperature in coppice. *Journal of Applied Ecology*, **24**, 239–49.

Mitchell, P.L. and Woodward, F.I. (1988) Responses of three woodland herbs to reduced photosynthetically active radiation and low red to far-red ratio in shade. *Journal of Ecology*, **76**, 807–25.

Morgan, D.C. and Smith, H. (1979) A systematic relationship between phytochrome-controlled development and species habitat, for plants grown in simulated natural radiation. *Planta*, **145**, 253–8.

Moss, D. (1979) Even-aged plantations as a habitat for birds, in *The Ecology of Even-aged Forest Plantations* (eds E.D. Ford, D.C. Malcolm and J. Atterson), Institute of Terrestrial Ecology, Cambridge, pp. 413–27.

Nihlgård, B. (1970) Vegetation types of planted spruce forests in Scania, southern Sweden. *Botaniska Notiser*, **123**, 310–37.

Ovington, J.D. (1955) Studies of the development of woodland conditions under different trees. III. The ground flora. *Journal of Ecology*, **43**, 1–21.

Ovington, J.D. and Madgwick, H.A.I. (1955) A comparison of light in different woodlands. *Forestry*, **28**, 141–6.

Peace, W.J.H. and Grubb, P.J. (1982) Interaction of light and mineral nutrient supply in the growth of *Impatiens parviflora*. *New Phytologist*, **90**, 127–50.

Persson, H. (1975) Deciduous woodland at Andersby, eastern Sweden: field-layer and below-ground production. *Acta Phytogeographica Suecica*, **62**, 1–71.

Peterken, G.F. (1981) *Woodland Conservation and Management*, Chapman and Hall, London.

Pons, T.L. (1976) An ecophysiological study in the field layer of ash coppice. I. Field measurements. *Acta Botanica Neerlandica*, **25**, 401–16.

Pons, T.L. (1983) Significance of inhibition of seed germination under the leaf canopy in ash coppice. *Plant, Cell and Environment*, **6**, 385–92.

Pontailler, J.Y., Leroux, M. and Saugier, B. (1984) Évolution d'un taillis de châtaigniers après coupe: photosynthèse et croissance des rejets. *Acta Oecologia/Oecologia Plantarum*, **5**, 89–99.

Rackham, O. (1975) *Hayley Wood: its History and Ecology*, Cambridgeshire and Isle of Ely Naturalists' Trust, Cambridge.

Rackham, O. (1980) *Ancient Woodland*, Edward Arnold, London.

Ratcliffe, P.R., Hall, J. and Allen, J. (1986) Computer predictions of sequential growth changes in commercial forests as an aid to wildlife management, with reference to red deer. *Scottish Forestry*, **40**, 79–83.

Rose, C.I. (1979) Observations on the ecology and conservation value of native and introduced tree species. *Quarterly Journal of Forestry*, **73**, 219–29.

Sakura, T., Gimingham, C.H. and Millar, C.S. (1985) Effect of tree density on ground vegetation in a Japanese larch plantation. *Scottish Forestry*, **39**, 191–8.

Salisbury, E.J. (1916) The oak–hornbeam woods of Hertfordshire. I and II. The *Quercus robur–Carpinus* woods. *Journal of Ecology*, **4**, 88–117.

Salisbury, E.J. (1921) Phenology and habitat with special reference to the phenology of woodlands. *Quarterly Journal of the Royal Meterological Society*, **47**, 251–9.

Salisbury, E.J. (1924) The effects of coppicing as illustrated by the woods of Hertfordshire. *Transactions of the Hertfordshire Natural History Society*, **18**, 1–21.

Smith, H. (1982) Light quality, photoperception and plant strategy. *Annual Review of Plant Physiology*, **33**, 481–518.

Tasker, R. and Smith, H. (1977) The function of phytochrome in the natural environment. V. Seasonal changes in radiant energy quality in woodlands. *Photochemistry and Photobiology*, **26**, 487–91.

Whitmore, T.C. (1975) *Tropical Rain Forests of the Far East*, Oxford University Press, Oxford.

4

Soil-water relations in an ancient coppice woodland

IAN CUMMINGS and HADRIAN COOK

4.1 INTRODUCTION

Hydrologically the soil is the critical interface between the atmosphere and the land, across which water is exchanged. In woodland the heterogeneity in stand structure of both trees and ground flora means that soil-water regimes resulting from the interplay of transpiration, interception loss, throughfall and stemflow are more complex than those for agricultural crops. Different tree species tend to exploit the soil to differing extents due to the selective pressures of waterlogging and drought. For example, some arid zone forests of *Acacia* have a rooting depth of 15 m and only rarely experience water stress (Waring and Cleary, 1967; Hinckley *et al.*, 1983). However, in woodland ecosystems where precipitation is abundant, trees develop shallow root systems and are frequently water stressed (Waring and Schlesinger, 1985). Research linking soil hydrology and vegetation pattern is comparatively rare although Falinski (1986) has related the spatial pattern of vegetation in the Bialowieza forest (Poland) to five main soil-water types.

The ancient coppice woodlands of East Anglia are frequently associated with relatively high plateaux on heavy, waterlogged soils (Rackham, 1980; Ball and Stevens, 1981). Martin and Pigott (1975) have described the chalky boulder clay soils of Hayley Wood, Cambridgeshire, which are probably typical of many ancient woodland soils in East Anglia. These soils experience seasonal fluctuations in soil-water recharge and water-table level which give rise to alternate anaerobic and aerobic conditions in the upper metre or so of soil. Deeper soil layers are frequently anaerobic – except during severe droughts when the water-table is lowered. In Swanton Great Wood, Norfolk, the shallow water-table between late autumn and early spring may lead to restricted root development, so that during a relatively brief period without rain – or a longer period with reduced precipitation – the trees and ground flora may be stressed to a greater extent than under situations of frequent droughts and deeper water-tables (Waring and Schlesinger, 1985).

Soil water may be characterized by two variables – water content (commonly expressed volumetrically) and soil-water energy status. The energy status of soil-water may be defined (after Hillel, 1982) as:

$$\Phi_{tot} = \Psi_m + \Psi_g + \Psi_o + \ldots$$

where: Φ_{tot} is the total potential, Ψ_m is the matric potential, Ψ_g is the gravitational potential, and Ψ_o is the osmotic potential. In non-saline soils the osmotic potential term may be negligible, whereas the gravitational potential at each point is determined by the elevation of the point relative to a reference level (generally the soil surface) and is important in considerations of soil-water movement. The single largest component of total potential is usually the matric potential which represents the sum total of capillary and adsorptive forces retaining water in the soil matrix.

In soils drier than 'field capacity' (usually defined as water retained in the soil profile following a thorough re-wetting and two or three days free drainage in the absence of evapotranspiration) water movement may become restricted. Overall soil-water *retention* is greater in heavy textured soils than in light (Dent and Scammell, 1981) soils, but the available 'water capacity' (usually defined as water content held between matric potentials of −5 and −1500 kPa) of clay soils is frequently low. This results from soil-water being too tightly adsorbed on the surface of clay particles to be readily extracted by the root system. The water content of a clay soil declines very slowly over the higher potential range of available water content whereas that of a sandy soil declines rapidly. Although giving the appearance of being wet and retaining a significant total amount of water, heavy textured soils hold water under physical conditions that make uptake by plants difficult. For example, Leshem (1970) found that tree roots do not grow at matric potentials below about −700 kPa, a point at which absorption is so slow that trees are unable to maintain positive cell turgor.

Where recharge fails to keep pace with depletion the uptake of soil water by evapotranspiration results in the creation of a 'soil-water deficit' (SWD). For the root-zone this is defined as:

$$\mathrm{SWD} = \Theta_{fc} - \Theta_t$$

where: Θ_{fc} is defined as soil-water content set at a matric potential of −5 kPa (which is considered to be close to field capacity) and Θ_t is the profile water content at a specified time (t).

Soil-water relations and their effects on the ancient woodland perennial ground flora have received comparatively little attention from ecologists compared with the effects of light (e.g. Coombe, 1965). It has been known for some years, however, that the higher level in the soil of bulbs of *Hyacinthoides non-scripta* in coppiced areas is probably attributed to a corresponding rise in the water-table after coppicing (Salisbury, 1924). When coppicing coincides with either wet or dry growing seasons, the interaction between light conditions and soil-water regimes is found to produce colourful plant communities including marsh bedstraw (*Galium palustre*), ragged robin (*Lychnis flos-cuculi*) and lesser spearwort (*Ranunculus flammula*) (Rackham, 1975). It is generally assumed, however, that the performance of vegetation rooted in the soil is related to the sufficiency of the most limiting factor (e.g. light, soil water, temperature, nutrients and disturbance). Dent and Cook (1987) and Cook and Dent (1990) observed that for arable crops, soil-water-induced stress may result if the rate of water supplied to the crop falls below the demand in certain growth stages: performance is related not only to supply, but also to the *sufficiency* of supply (Melitz, 1986).

Early measurements of the effects of soil water on woodland ground flora (reviewed by Korstian and Coile, 1938) used the gravimetric method of measuring soil-water which, apart from disturbing the soil profile for each successive measurement, ignores the energy status of soil water. More recently, Barrett and Youngberg (1965), McKeague *et al.* (1984), and Moore and Vankatt (1986) gathered information on profile-available water during the growing season. However, the limitation of this approach is that a sufficient supply of water may not be available across the range of potentials under which water is held in soil.

Ecologists often underestimate the effects that fluctuations in water supply may have on the growth of the perennial ground flora in woodlands. For example, Collins *et al.* (1985) found no significant increase in soil water beneath canopy gaps and consequently discounted this as a limiting factor affecting the growth of groundflora. However, Inghe and Tamm (1985) have shown that a summer drought may indeed affect the flowering performance in the following year of the woodland perennial wood sanicle (*Sanicula europaea*). The demography of the wild daffodil (*Narcissus pseudonarcissus*) may also be affected by the previous spring and summer climate (Barkham, 1980). The same may be true for other perennial herbs which are confined to, or frequently associated with, ancient coppiced woodland (Peterken, 1974) due to poor seed dispersal, production or viability (Brown and Oosterhuis, 1981; Grime *et al.*, 1988), e.g. dogs mercury (*Mercurialis perennis*), bluebell (*Hyacinthoides non-scripta*), May lily (*Maianthemum bifolium*), wood brome (*Bromus ramosus*), ramsoms (*Allium ursinum*), yellow archangel (*Lamiastrum galeobdolon*) and wood melick (*Melica uniflora*).

In order to examine the significance of soil-water stress on one vernal species, wood anemone (*Anemone nemorosa*), an investigation was carried out using measurements of soil-water content and soil-water energy status (i.e. matric potential) in an ancient coppice woodland. This was complemented by a greenhouse experiment to determine threshold matric potentials for *A. nemorosa* and to estimate the frequency of water stress likely to be induced by various management practices.

4.2 METHODS

4.2.1 Study site

The study site was Swanton Great Wood in Norfolk, England, a national nature reserve designated by the Nature Conservancy Council. The reserve (52 ha) belongs to the Lime Province (Rackham, 1986) and contains eight of the ancient semi-natural communities of mixed coppice described by Peterken (1981). One area of about ten hectares is managed on a ten-year rotation for faggots, thatching hazel and firewood. The two most abundant tree species in the coppiced areas are small-leaved lime (*Tilia cordata*) and hazel (*Corylus avellana*). The reserve has a diverse flora (about 300 species of flowering plant) including many species frequently associated with ancient coppice woodland such as *Anemone nemorosa*, lily-of-the-valley (*Convallaria majalis*), *Mercurialis perennis* and *Lamiastrum galeobdolon*. In Norfolk, *Maianthemum bifolium* is only found in Swanton Great Wood.

The reserve occupies a plateau, around 76 m Ordnance Datum comprised of calcareous boulder clay in the south of the wood and glacial sands and gravels in the north. The surface run-off enters the headwaters of the River Stiffkey and a tributary of the River Wensum. Full descriptions of the reserve are given by Peterken (1970, 1981), Goodfellow and Peterken (1981), Elliott (1985) and Rackham (1986).

The soil textures for the study plots were broadly similar. The Soil Series for the three study compartments were mostly Wighill (Table 4.1) with some Honingham. Wighill is a decalcified till; Honingham a chalky boulder clay. However, in the observed soil profiles these are decalcified with no visible chalky granules, so strictly making the soils in the study site pseudo-Honingham (W.M. Corbett, personal communication). This decalcification is probably due to surface acidification caused by organic acids released from humus present on the soil surface. All soils in the study plots are stagnogleyic with a broad gley morphology.

4.2.2 Field experimental compartments

A rectangular plot of 2500 m^2 was marked out within each of three compartments representing a range of coppice ages and stool densities. The plot size was limited by the size of the smallest coppiced area. Within each compartment, the numbers, species, and diameter of coppiced stools, and the numbers of living stems on each stool were recorded. Stand characteristics for the three study compartments are summarized in Table 4.2 for

Table 4.1 Textural classification of soils in the study site

Compartment (replicate)	*Depth (cm)*	*Textural class*	*Soil Series*	*Colour**
I (a)	0–30	Sandy silt loam	Wighill	10YR 3/3
	30–40	Sandy silt loam	Wighill	10YR 5/4
	40–60	Sandy clay loam	Wighill	10YR 5/4
	60–80	Sandy clay loam	Wighill	10YR 5/8
I (b)	0–20	Sandy silt loam	Honingham	10YR 3/3
	20–40	Sandy silt loam	Honingham	10YR 5/4
	40–60	Clay	Honingham	10YR 6/6
	60–80	Clay	Honingham	10YR 5/8 (mottling) (10YR 6/2)
II (a)	0–30	Sandy loam	Wighill	10YR 4/4
	30–80	Clay	Wighill	10YR 5/8
II (b)	0–30	Sandy silt loam	Wighill	10YR 3/3
	30–40	Sandy silt loam	Wighill	10YR 4/4
	40–60	Sandy clay loam	Wighill	10YR 5/4
III (a)	0–30	Sandy silt loam	Honingham	10YR 4/4
	30–50	Sandy silt loam	Honingham	10YR 6/6
	50–60	Sandy clay	Honingham	10YR 6/8
	60–80	Sandy clay	Honingham	10YR 6/6
III (b)	0–50	Sandy silt loam	Wighill	10YR 4/4
	50–80	Sandy clay loam	Wighill	10YR 5/8 (mottling) (10YR 5/2)

**Source*: Munsell Soil Colour Chart

1987 when measurements began. Compartment I (CI) was composed of young, three-year-old coppice, Compartment II (CII) consisted of three-year-old coppice at a higher density and Compartment III (CIII) represented high-density coppice towards the end of the coppice cycle (nine years old in 1987).

Two randomly located circular plots were established within the rectangular compartments, in which soil-water content and matric potentials were measured. These areas were calculated to represent the stool density in the study site. A circle of radius greater than or equal to the minimum distance between the trees in a particular stand was considered adequately to represent the stand structure (Anuchin, 1970). The minimum distance between the centres of each stool (2.15 m) was calculated for Compartment I which had the lowest stool density, and hence widest spacing between stools, of all three compartments (2157 stools/ha). A radius of 2.15 m gives a circular area of 14.52 m^2 which is a compromise in area between producing homogeneity in vegetation structure on the one hand, and including a fair representation of vegetation heterogeneity on the other.

Table 4.2 Stand characteristics (1987) of the study compartments. CI, young low-density coppice; CII young high-density coppice; CIII, mature high-density coppice

Compartment	*CI*	*CII*	*CIII*
Year when coppiced (autumn/winter)	1984/85	1984/85	1978/79
Age of coppice in spring 1987 (years)	3	3	9
Stool density (number/ha)	2157	3014	3471
Average stool spacing (m)	2.15	1.82	1.70
Species	Number of stools (as % of total)		
Corylus avellana	85 (56.7)	78 (37.0)	31 (12.9)
Tilia cordata	39 (25.8)	39 (18.4)	116 (47.7)
Fraxinus excelsior	14 (9.4)	75 (35.6)	2 (0.1)
Betula spp.	9 (6.0)	11 (5.2)	58 (24.9)
Acer campestre	3 (2.0)	8 (3.8)	2 (0.1)
Salix spp.	0 (0)	0 (0)	34 (14.3)
Prunus padus	1 (0.1)	1 (0)	0 (0)
Total	151	211	243
Most common	*Corylus*	*Corylus*	*Tilia*
Ground area occupied by stools (m^2/ha)	169.86	692.43	250.57
% Area stools	1.7	6.9	2.5
% Bare ground	98.3	93.1	97.5
Number of live stems per compartment	3561±45	9393±107	2919±32

4.2.3 Soil-water measurements

Depletion of soil water is by the processes of evapotranspiration and drainage. However, there are problems in measuring these components of the water balance. Potential evapotranspiration may be obtained for an agricultural crop using the Penman equation modified by using a suitable crop coefficient to estimate actual crop evapotranspiration (e.g. Doorenbos and Kassam, 1979). For coppice woodland no such crop coefficients exist, and the heterogeneity of the canopy may render this approach unreliable. However, direct measurement of soil-water using the neutron probe and tensiometers (used in this study) is an established approach in forest hydrology (Cooper, 1980).

Bi-weekly measurements were made of soil-water content and soil matric potential in the three compartments over a period, spanning the growing season, of three years, 1987–1989. Soil-water deficits were calculated from these data assuming field capacity to be water content at a profile matric potential of −5 kPa in the profile (Hall *et al.*, 1977). Measurements of soil-water content were made using the Wallingford Neutron Probe (Bell, 1976).

Aluminium access tubes were randomly installed in the ground between coppiced trees within each of the two random circular plots per compartment. Six access tubes were inserted in one plot (a) and four in the second plot (b) within each of the three compartments. This arrangement was devised to accommodate banks of tensiometers. Neutron probe 'soundings' were made at four successive depths within each soil-water plot (at 20 cm, 35 cm, 50 cm and 70 cm). Probe readings spanned the period April 1987 to October 1989.

Soil matric potential measurements were made using tensiometers. Once installed the instruments remained in place throughout the experimental period. They were installed at four depths corresponding to the 'sounding depths' of the neutron probe. Soil-water deficits were calculated from soil-water contents using the neutron probe and matric potentials.

4.2.4 Rainfall data

Monthly rainfall data were obtained for the period 1961–1989, from the National Rivers Authority, for Hope House, Hindolveston, Norfolk – a site 1.5 km from Swanton Great Wood, Table 4.3). These data were used to calculate rainfall exceedance (Ward, 1975) over the period of soil-water measurements. The exceedance figure means that a given (monthly) rainfall figure is likely to occur, or be exceeded, in a given percentage of years. For example, the rainfall during May 1989 was low (14 mm) meaning that the probability of its exceedance in other years is high (94%).

Table 4.3 Rainfall patterns (mm) for March–October 1987, 1988 and 1989, and exceedance values (%) calculated from long-term rainfall (1961–1989)

	Year	*March*	*April*	*May*	*June*	*July*	*August*	*September*	*October*	*Total*
Rainfall	1987	55.0	52.6	72.5	79.5	91.9	139.5	48.1	120.4	659.5
Exceedance		43.5	55.6	13.3	23.3	16.7	3.3	47.6	13.3	3.3
Rainfall	1988	88.0	20.7	38.3	29.7	108.9	55.7	30.1	67.2	438.6
Exceedance		10.0	87.0	66.7	69.9	6.7	55.6	80.0	37.0	52.6
Rainfall	1989	45.4	65.5	14.1	96.0	40.5	37.1	42.3	60.0	400.9
Exceedance		62.5	26.3	93.5	20.0	62.5	76.9	58.8	50.0	76.9
Mean exceedance 1961–1989		51.6	52.2	50.5	55.6	61.8	61.2	60.7	67.1	

Mean rainfall March–October 1961 – 1989 = 452.3 mm

Source: National Rivers Authority

4.2.5 Pot experiment: controlled water additions to *A. nemorosa*

Fresh field topsoil (Wighill series) was collected from Compartment III and sieved into 20 plastic pots (each 25 cm in diameter). On 2 March 1989 about 250 non-branching *A. nemorosa* rhizome fragments were randomly excavated in Compartment III, from a 5 m radius circular plot. All rhizomes were packed in damp moss in the field and brought back to the greenhouse. Ten rhizomes were planted radially in each of the 20 pots at a depth of 3 cm. Each pot was kept well watered from 2 March until specific water treatments were introduced on 9 March 1989. Pots were weeded throughout the experiment which lasted from March to September 1989. Rhizomes were standardized prior to planting out in the pots in two ways. First, rhizome fragments consisting of two years' growth increments were selected. The new year's growth (spring 1988) of rhizomes were easily distinguished from the previous year's increment by their swollen bases and lighter colour. Secondly, only non-branching rhizomes were selected.

Four watering treatments, replicated five times, were chosen to represent the full range of water matric potentials experienced by *A. nemorosa* in the field study. The first treatment maintained pots with a water-table just below the surface (permanently waterlogged). This was achieved by standing pots in water baths filled with water to the appropriate height. The second treatment maintained pots around field capacity by watering the soil regularly. The third and fourth treatments represented dry and very dry conditions in the field. These treatments were maintained by varying the quantities of water added to maintain matric potentials below field capacity (Table 4.4).

Table 4.4 Monthly mean matric potential (−kPa) ± one SE recorded once a week in four *A. nemorosa* potting treatments (n = 5). P, permanently waterlogged; D, watered daily to maintain field capacity

Treatment	*Water added (dm^3)*	May	*Mean matric potential (5 pots)* −kPa June	July	August	September
Permanently waterlogged	P	–	–	–	–	–
Field capacity	D	4.9 ± 0.3	5.5 ± 0.4	5.0 ± 0.1	5.2 ± 0.2	5.4 ± 0.1
Dry	8.5	18.0 ± 1.6	28.5 ± 2.4	24.3 ± 3.9	22.0 ± 4.0	26.9 ± 4.3
Very dry	2.5	18.2 ± 2.4	21.6 ± 2.0	41.1 ± 2.5	40.1 ± 5.1	14.1 ± 2.5

Portable (or 'quick draw') tensiometer readings of the matric potential in the pots were taken at 10 cm depth, within the root-zone of *A. nemorosa*. Measurements of the length of the middle leaf were made each week from April–June 1989 after which all leaves in the pots had died. The rhizomes were eventually excavated on 28 September 1989 by which time the terminal bud had developed and horizontal growth was observed to have slowed down appreciably. Measurements were then made of the length and width of the two increments on each rhizome fragment from which the cylindrical volumes of the 1988 and 1989 increments were calculated.

4.3 RESULTS

4.3.1 Rainfall

The rainfall data for March, April and May (1987–1989) from Hope House, Hindolveston are lower than for the summer and autumn months. Rainfall and monthly exceedance values, however, show considerable variation for months within the three study years. For instance, April 1987 experienced a rainfall of 52.6 mm (exceedance value 55.6%) whereas April 1988 experienced only 20.7 mm (exceedance value 87%) which is considerably drier than average (Table 4.3).

Exceedance probabilities in the long term were also used to make deductions about rainfall during the period from March to October which relates to the phenology of *A. nemorosa* (Shirreffs, 1985). Table 4.3 shows that during this critical period, 1987 rainfall was well above average (659.5 mm); in 1988 it was only 14.2 mm below average and 1989 was considerably drier (51.9 mm below average). Three contrasting years are presented, which represent a range of rainfall, during which different coppice management treatments can be compared. Regional potential evapotranspiration averaged 514 mm for March–October (530 mm for the entire year) between 1941 and 1970 (Smith and Trafford, 1976).

4.3.2 Soil water

The two soil-water plots (replicates a and b) used to take neutron probe soundings in each treatment satisfied the statistical criterion for pooling data from each replicate (e.g. for Compartment I which had the greatest variation; t = 1.959, P=0.053, df* = 838). The data from the neutron probe (recorded as mean counts per second (mcps)) for each of the four depths (20 cm, 35 cm, 50 cm and 70 cm) were therefore averaged for each compartment, and plotted against time. Time-scales used were in serial days for the three years, commencing 1 April 1987 and ending on 26 October 1989.

Figure 4.1 shows the seasonal patterns of soil water for 20 cm and 35 cm depths for each of the three compartments between April 1987 and October 1989. Wetter soils during the year 1987 are shown by overall higher mcps.

* df = degrees of freedom

Analysis of variance showed that soil water differed significantly among the three treatments in the wet year, April 1987 to March 1988 and the dry period, April 1988 to October 1989. For a finer approximation to the differences in soil-water content means between treatments however, the Sheffe *a posteriori* Multiple Contrast Test (Sheffe, 1959) was used. During the very wet year, April 1987 to March 1988, significant differences in soil-water content were found between young low-density coppice (CI) and both young high-density (CII) and old high-density (CIII) coppice (Sheffe, $df = 558$, $P < 0.05$). There was no significant difference between young and old high-density coppice. Figure 4.2 shows these data graphically for April 1987–March 1988, and April 1988–October 1989. In the drier year (April 1988–October 1989) there were significant differences between all three compartments (Sheffe, $df = 837$, $P < 0.01$). Hence, both age and density of coppice affect soil-water content, but mainly in dry years.

The differences in the soil-water content for each depth for the three management treatments over the measurement period are shown in Figure 4.3. Significant differences in soil-water content within the root-zone of *A. nemorosa* occur between all three treatments at 20 cm and 35 cm depths (Sheffe, $df = 417$, $P < 0.05$). Below 35 cm there is some degree of overlap with differences mainly attributable to the density of coppice rather than its age. For instance, at 50 cm depth, there is a significant difference between young high and low-density coppice (Sheffe, $df = 417$, $P < 0.05$) but no difference between either of these treatments and old high-density coppice ($P > 0.05$). At 70 cm depth, young high-density coppice is significantly drier than either young low-density or old high-density coppice (Sheffe, $df = 417$, $P < 0.05$).

4.3.3 Pot experiment and soil-water stress thresholds

Measurements of mean leaf length per petiole for *A. nemorosa* are shown in Figure 4.4 for each of the four watering treatments. It can be seen that leaf length for most petioles in the permanently waterlogged treatment declined after 9 May. In pots kept at or lower than −5 kPa, leaf length only declined after 23 May at which time the field capacity treatment had a significantly greater leaf length than pots kept at either −15 kPa or −16 kPa. The very dry treatment, although similar in average root-zone matric potential, experienced greater fluctuations. Consequently, a matric potential of −16 kPa was considered to be the threshold at which soil-water-induced stress (measured in terms of root-zone matric potential) is likely to have a detrimental effect on leaf development. Leaf area and dry weight may be similarly affected since, for vegetative shoots, the length of the middle radical leaf was significantly correlated with leaf area ($r = 0.891$, $p < 0.01$) and leaf area with dry weight ($r = 0.746$, <0.05).

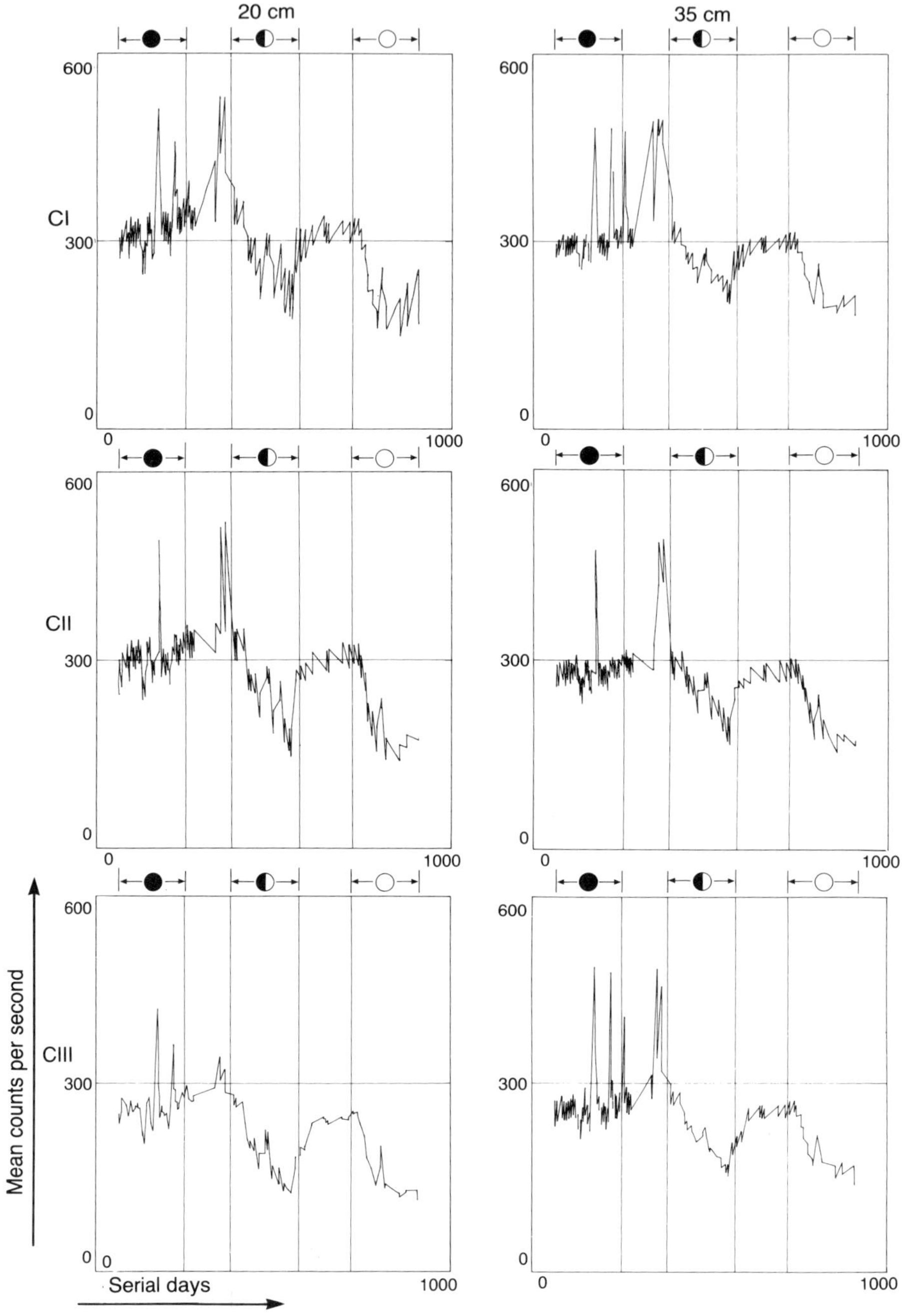

Figure 4.1 Soil-water content (as measured by the neutron probe in mean counts per second) at two depths in the three study compartments, from April to October, 1987–1989). CI, young low-density coppice; CII, young high-density coppice; CIII, old high-density coppice; ●, 1987, ◐, 1988; ○, 1989.

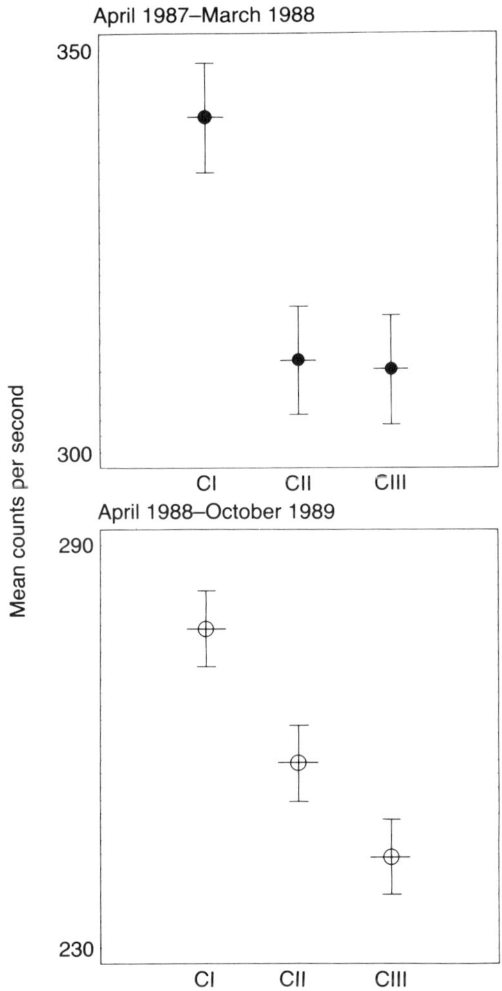

Figure 4.2 Soil-water content means and 95% confidence limits for the three study compartments in wet (•) and dry years (○). CI young low-density coppice, CII, young high-density coppice; CIII, old high-density coppice.

Rhizome performance was also measured. Figure 4.5 shows the relationship between percentage volume ratio in the rhizome increments between 1988 and 1989, and mean matric potential in the root-zone of the greenhouse pot experiment. The best-fit straight line predicts a zero change in volume ratio between the years at −40 kPa, and the spread of data points suggest negligible or negative changes in rhizome volume ratio between the years occurs below −28 kPa. Consequently, a matric potential in the root-

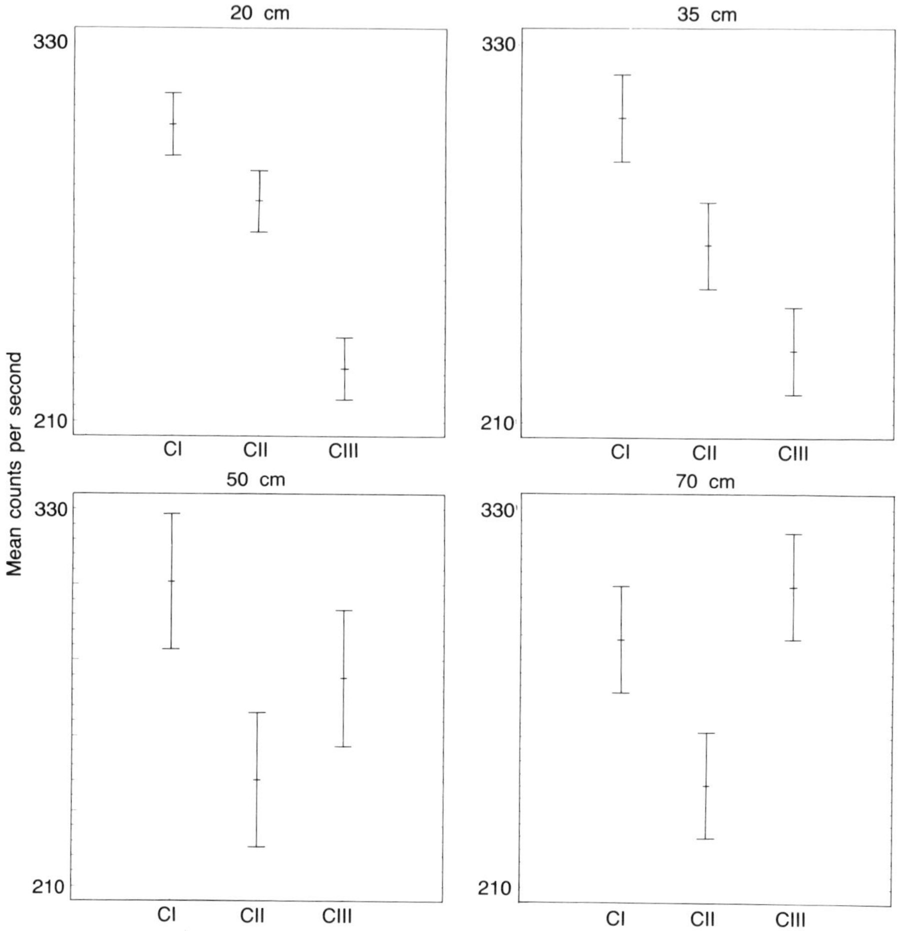

Figure 4.3 Soil-water content means (counts per second from the neutron probe) and 95% confidence limits for four depths in the three study compartments between April 1987 and October 1989. CI, young low-density coppice; CII, young high-density coppice; CIII, old high-density coppice.

zone of −30 kPa has been selected as the threshold below which soil-water-induced stress (measured in terms of root-zone matric potential in the pot) is likely to have a detrimental effect on rhizome development.

4.3.4 Critical thresholds in the field

Figure 4.6 shows the seasonal patterns of field matric potential for depths of 20 cm and 35 cm. For comparison the critical threshold potentials for *A. nemorosa* are shown at −16 kPa (leaf growth impaired) and −30 kPa (rhizome growth impaired). Field capacity (−5 kPa) is also included for

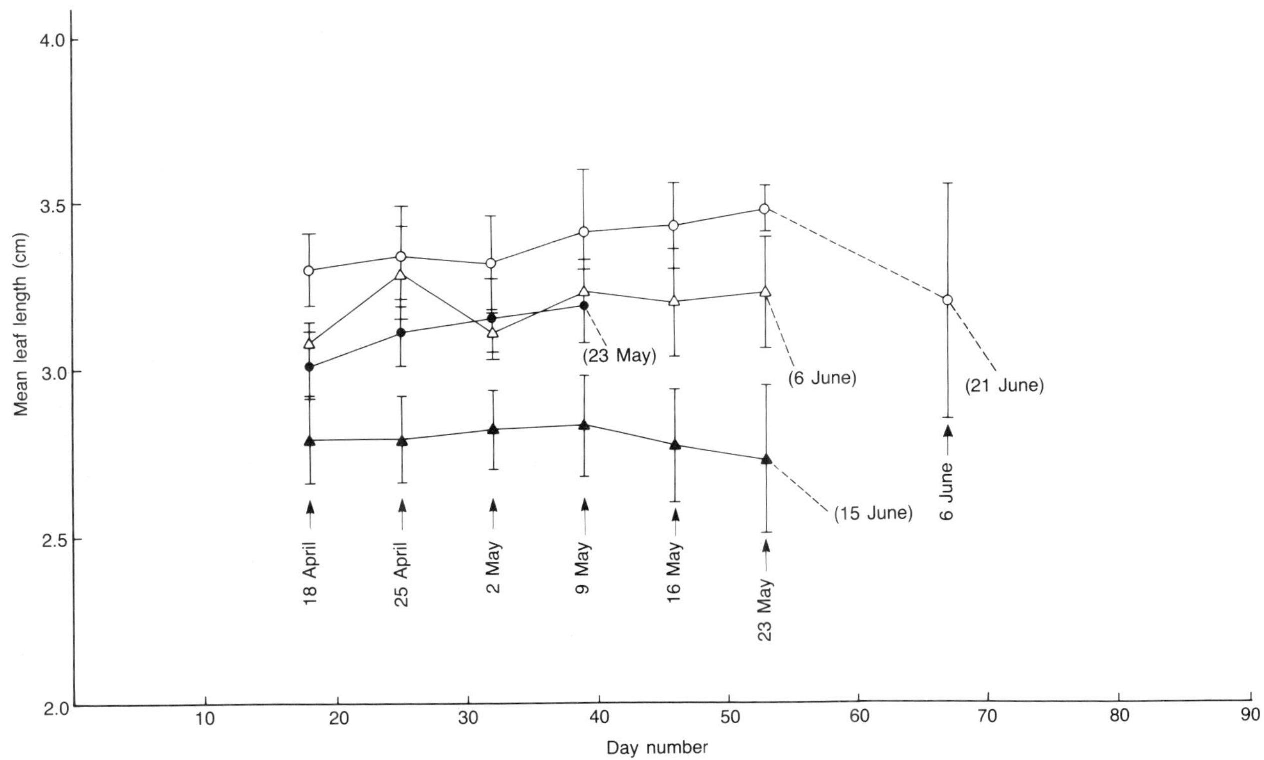

Figure 4.4 Mean leaf length per shoot and 95% confidence limits of *A. nemorosa* in four watering treatments (five replicates, ten petioles per pot) in the greenhouse pot experiment, April–June 1989. Dotted lines indicate when shoot mortality begins; dates in brackets refer to the time at which complete leaf senescence occurs in all pots. ●, permanently waterlogged ($\Psi_m = 0$); ○, field capacity (mean $\Psi_m = -4.68 \pm 0.38$ kPa); △, dry (mean $\Psi_m = -15.15 \pm 1.87$ kPa); ▲, very dry (mean $\Psi_m = -15.90 \pm 2.28$ kPa)

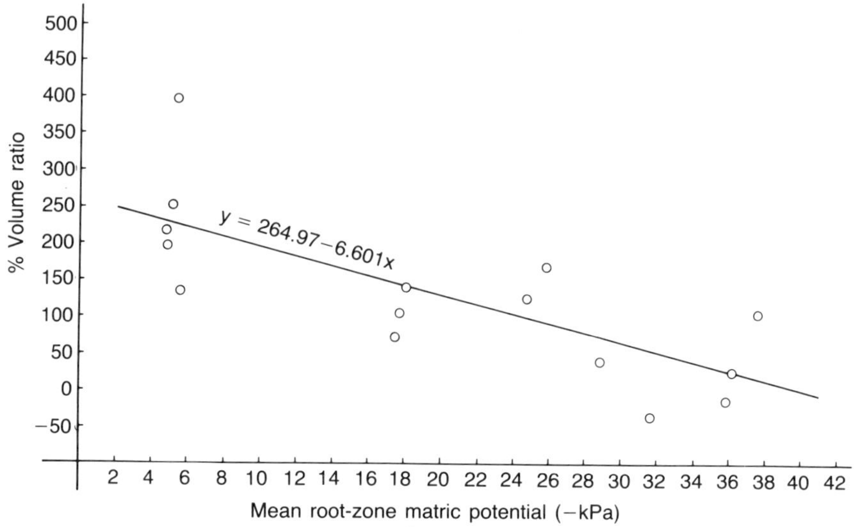

Figure 4.5 Percentage volume ratio for the mean 1988 and 1989 *A. nemorosa* rhizome increments (collected from the field in 1988 and transplanted to pots in 1989) and mean matric potentials in the root-zone of the greenhouse pot experiment (excludes waterlogged pots $\Psi_m = 0$).

comparison. Table 4.5 shows the percentage of 'stress' days derived from field data using the results of the greenhouse experiments. The critical growth period (April–October) for *A. nemorosa* includes the period during which both leaf and rhizome development will be affected. This is discussed more fully elsewhere (Cummings, 1991). Also shown are the maximum soil-water deficits achieved during the three summers. For the wet year (April 1987–March 1988) soil-water deficits were rarely observed under any management treatment, except for a brief period in July (Figure 4.7). This situation is changed in both the drier years (1988, 1989) during summer and autumn when the greatest soil-water deficits occurred (Table 4.5, Figure 4.7). The percentage of observed stress days (from the matric potential data) for both thresholds not only tends to increase with decreasing March–October rainfall but also with age of coppice. Furthermore, it can be seen in Table 4.5 that by increasing the density of coppice from 2157 to 3014 stools/ha, within the same age-class of coppice there is a doubling of the percentage of observed stress days (from 11.8% to 23.5%) in the dry period (March–October 1989) for a threshold potential likely to impair leaf growth in *A. nemorosa*. However, a proportion of stress days that is likely to affect rhizome development is only achieved in mature high-density coppice.

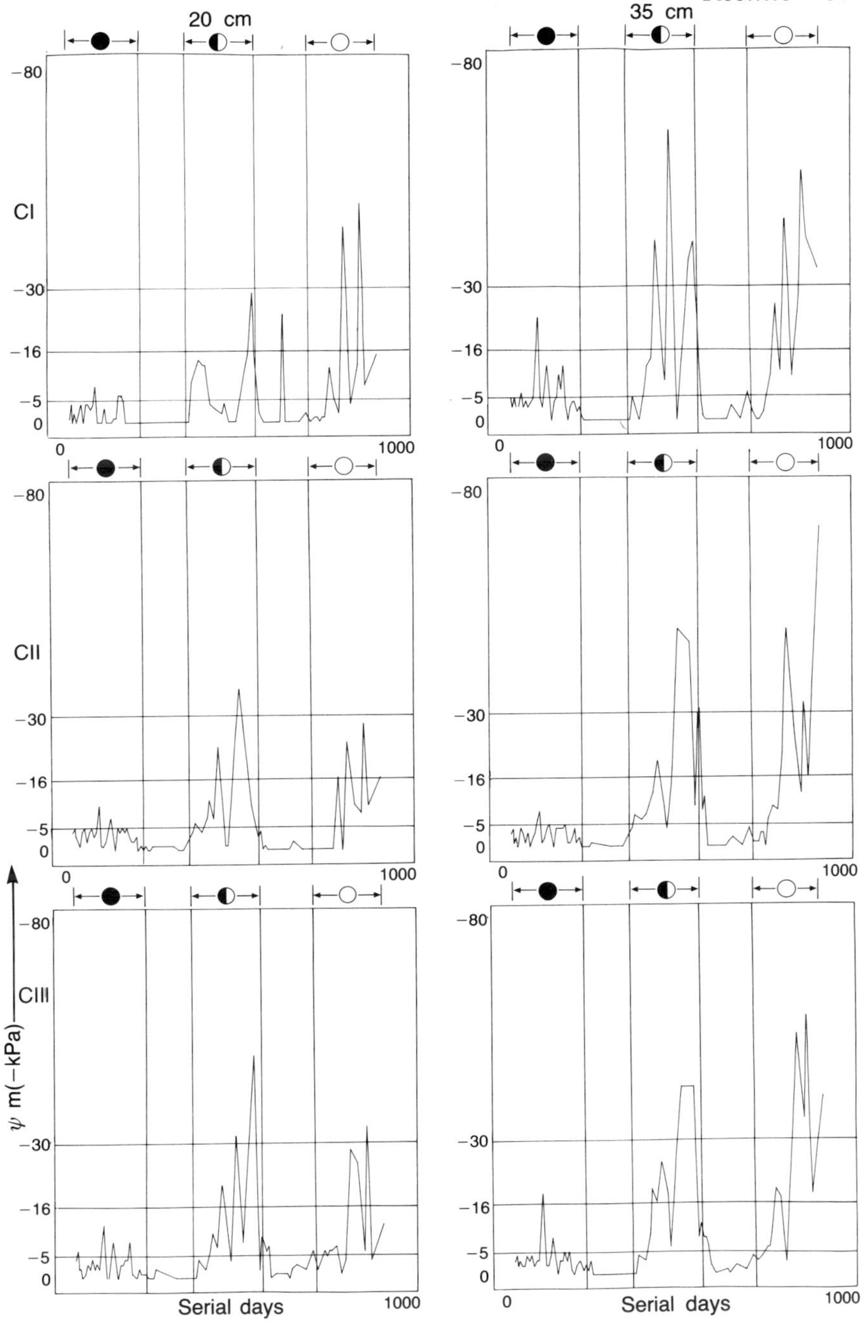

Figure 4.6 Soil-water matric potential (Ψ_m) measured by tensiometers (−kPa) at two depths in the three study compartments from April to October, 1987–1989. Threshold potentials for the onset of water stress in *A. nemorosa* leaves (−16 kPa) and rhizomes (−30 kPa) are shown, field capacity is at −5 kPa. CI, young low-density coppice; CII, young high-density coppice; CIII, old high-density coppice; ●, 1987, ◐, 1988; ○, 1989.

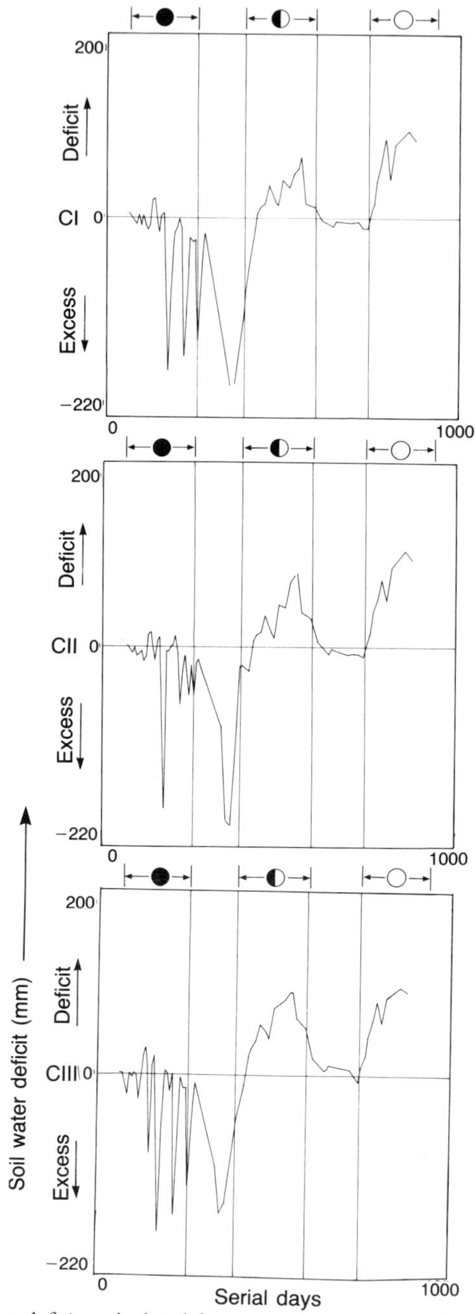

Figure 4.7 Soil-water deficits calculated from neutron probe soundings and matric potentials for the three study compartments, from April to October, 1987–1989. CI, young low-density coppice; CII, young high-density coppice; CIII, old high-density coppice; ●, 1987, ◐, 1988; ○, 1989.

Table 4.5 Percentage observed water-stress days and maximum soil-water deficits calculated for the critical growth period (April–October), for *A. nemorosa* leaves and rhizomes within three contrasting spring and summer rainfall periods (wet, dry, very dry) and three coppice management treatments. A, tensiometer field observations at or below −16 kPa; B, as for A but at or below −30 kPa; C, maximum soil-water deficit

Year	*March–October rainfall (mm)*	*Rainfall exceedance (%)*	*Low-density coppice (2157 stools/ha)*			*High-density coppice (3014–3471 stools/ha)*					
			3–5 year coppice			*3–5 year coppice*			*9–11 year coppice*		
			A (%)	*B (%)*	*C (mm)*	*A (%)*	*B (%)*	*C (mm)*	*A (%)*	*B (%)*	*C (mm)*
1987	659.5	3.3	0	0	24	0	0	16	0	0	29
1988	438.6	52.6	12.5	6.3	85	12.5	0	88	18.8	12.5	98
1989	400.9	76.9	11.8	0	95	23.5	0	105	23.5	17.6	105

4.4 DISCUSSION

In the study compartments at Swanton Great Wood, soil-water contents under younger coppice (3–5 years) were higher than beneath older coppice (9–11 years), but only in dry years were the differences between treatments significant. These findings confirm the observation made over sixty years ago by Salisbury (1924) that soil-water content increases after coppicing. Under older coppice, *A. nemorosa* may therefore be expected to experience greater physiological stress. Differences in soil-water were also observed between low and high-density coppice before canopy closure (5–6-year-old coppice), indicating that well-stocked coppice removes water from the soil at a faster rate than otherwise similar but poorly-stocked stands. These differences were especially important in the surface soil (0.35 cm) at the study site, but below 35 cm the effects of coppice density were insignificant. Soil-water deficits were correspondingly greater under high-density coppice, suggesting competition between roots for soil water.

At times of water shortage, differences in soil-water content between horizons will develop, since rooting systems preferentially deplete soil water in the surface horizons – providing there is no substantial recharge by rainfall. Even where recharge of the surface occurs, differences between the wetter surface horizons and the drier sub-surface layers will be marked. Transfer of soil water to a drier root-zone by upward movement of water is unlikely to be great in the clay soils of the study site (Cook, 1987) due to low unsaturated hydraulic conductivities.

The greenhouse pot experiment showed that the growth of *A. nemorosa* is clearly affected by root-zone matric potentials depressed to levels frequently encountered in the study site soils in dry growing seasons (April–October). *A. nemorosa* develops signs of physiological stress in leaf growth at or below a matric potential of −16 kPa and in rhizome growth at or below −30 kPa. At the study site the probability of matric potentials low enough to impair the growth of *A. nemorosa* increases in dry years. Inghe and Tamm (1985) and Barkham (1980) both observed that flowering and bulb depth in woodland perennials may be affected by summer droughts. It is suggested here that *A. nemorosa* may be a useful analogue for the response of other perennial herbs to soil-water stress, such as *Convallaria majalis*, *Mercurialis perennis*, *Allium ursinum*, *Maianthemum bifolium*, *Hyacinthoides non-scripta* and *Lamiastrum galeobdolon*.

The effects of waterlogging on the growth of *A. nemorosa* suggests that leaves and rhizomes were fairly tolerant to transient waterlogging, although complete senescence was observed after only three weeks in constantly waterlogged treatments (cf. Pigott, 1982). The shallow roots of perennial herbs may enable a certain degree of tolerance to waterlogging – an observation supported by Martin (1968) for *Mercurialis perennis*. The shallow root system may confer a competitive advantage over those species which, because of their deeper roots, are more prone to periodic waterlogging. By the same token, however, woodland perennials with a superficial root system may be more susceptible to drought. Because the ground flora is strongly influenced by soil-water, the effects of shade are not necessarily overriding (see Chapter 3). Coppice rotations therefore not only provide light conditions where some species flourish but they also affect soil-water regimes, favouring some species over others. The interaction may be particularly strong in some years where coppicing coincides with either wet or dry years.

Given the hydrological evidence presented in this paper for *A. nemorosa*, it is possible to speculate on the likely effects of coppicing on the spatial patterns of clonal perennial herbs in ancient woodland. Soil-water measurements and exceedance calculations of rainfall pattern suggest that soil-water supply to the root-zone is affected by a range of coppice age-classes and stool densities. Thus short coppice rotations (e.g. 5–10 years) may reduce the frequency of soil-water deficits and reduce water stress experienced by perennial herbs, so giving rise to intensive clonal growth which, after successive short rotations, may eventually result in locally homogeneous stands of flowering clonal perennial herbs. Where coppice trees have been removed to widen rides, *A. nemorosa* may respond to a combination of light and increased soil-water to dominate the vernal perennial ground flora – provided the rides are mown each summer. This effect can be clearly seen at Swanton Great Wood (Figure 4.8). Alternatively, a long

Figure 4.8 Flowering of *A. nemorosa* on the rides of Swanton Great Wood, Norfolk in April 1988, following a wet growing season (March–October 1987). This type of homogeneity in the vernal perennial ground flora probably reflects many years of interaction between high spring-light conditions and optimum soil-water supply. (Photograph: I.P.F. Cummings.)

coppice rotation of more than ten years will tend to increase water stress, causing extensive rhizome branching, and resulting in a patchy ground flora, the gaps reflecting areas of intensive root competition between the roots of trees and the ground flora (Figure 4.9). Under non-intervention management, water stress may continue to increase, resulting in fragmented clones and restricted flowering in isolated patches throughout the wood. Deeper rooted, shade-tolerant species of the perennial ground flora in non-intervention management woodland will probably have a competitive advantage over shallow rooting perennials.

In the event of a long-term climatic change causing a reduction in rainfall or an increase in transpirational stress, the implications for clonal perennial species analogous to *A. nemorosa* are clear. Drought stress will increase and population densities will decline. For clones already weakened by long periods of non-intervention (e.g. 100 years), there could be important conservation implications to be drawn from the results presented in this paper. Resumption of coppicing could strengthen clones and even prevent local extinctions, although coppicing would probably only be a serious conservation option where a coppice structure remains, and stool densities

Figure 4.9 Flowering of *A. nemorosa* beneath non-intervention hornbeam (*Carpinus betulus*) coppice, April 1988, in Hockley Woods, Essex. Gaps in the vernal perennial ground flora, as shown here, may result from intense root competition for soil water. The canopy allows in more light than in younger coppice. (Photograph: I.P.F. Cummings.)

are low (in the region of 2000 stools/ha). The distribution of clonal perennials in coppiced woodland will also be affected by soil texture since the ability of the soil to supply water to roots at a given rate will be influenced by the proportion of sand, silt and clay in the root-zone soil fraction.

ACKNOWLEDGEMENTS

We would like to acknowledge with thanks the interest and helpful criticism of a number of people during this work. In particular, we should like to thank Dr John Barkham and Dr Jonathan Mitchley for helpful comments on the manuscript; Bill Corbett and Don Saunders for help in the field; Phil Judge for assistance with the illustrations; the Nature Conservancy Council for permission to work at Swanton Great Wood National Nature Reserve; the National Rivers Authority and Meteorological Office, Bracknell, for rainfall and evapotranspiration data; and the School of Environmental Sciences at the University of East Anglia for grants and technical support.

REFERENCES

Anuchin, N.P. (1970) *Forest Mensuration*, Israel Prog. Sci. Transl., Jerusalem.

Ball, D.F. and Stevens, P.A. (1981) The role of ancient woodlands in conserving 'undisturbed' soils in Britain. *Biological Conservation*, **19**, 163–76.

Barkham, J.P. (1980) Population dynamics of the wild daffodil (*Narcissus pseudonarcissus*). II. Changes in number of shoots and flowers, and the effect of bulb depth on growth and reproduction. *Journal of Ecology*, **68**, 635–64.

Barrett, J.W. and Youngberg, C.T. (1965) Effect of tree spacing and understorey vegetation on water use in a pumice soil. *Soil Science Society of America Proceedings*, **29**, 472–5.

Bell, J.P. (1976) *Neutron Probe Practice*, Institute of Hydrology Report, 19, Natural Environment Research Council.

Brown, A.H.F. and Oosterhuis, L. (1981) The role of buried seed in coppicewoods. *Biological Conservation*, **21**, 19–38.

Collins, B.S., Dunne, K.P. and Pickett, S.T.A. (1985) Response of forest herbs to canopy gaps, in *The Ecology of Natural Disturbance and Patch Dynamics* (eds S.T.A. Pickett and P.S. White), Academic Press, London, pp. 218–34.

Cook, H.F. (1987) Cluster analysis applied to grouping soil series according to droughtiness. *Soil Survey and Land Evaluation*, 7, 177–85.

Cook, H.F. and Dent, D.L. (1990) Modelling soil water supply to crops. *Catena*, **17**, 25–39.

Coombe, D.E. (1965) The seasonal light climate and plant growth in a Cambridgeshire wood, in *Light as an Ecological Factor* (eds R. Bainbridge, G.C. Evans and O. Rackham), Blackwell, Oxford, pp. 148–66.

Cooper, J.D. (1980) *Measurements of Moisture Fluxes in Unsaturated Soil in Thetford Forest*, Institute of Hydrology Report, 66, Natural Environment Research Council.

Cummings, I.P.F. (1991) *The effects of root competition on* Anemone nemorosa L. *in ancient coppiced woodland*. PhD thesis, University of East Anglia, Norwich.

Dent, D.L. and Cook, H.F. (1987) Soil water sufficiency: data needs and error bars of a soil water potential model, in *Proceedings of the International Workshop on Quantified Land Evaluation Procedures* (eds K.J. Beek, P.A. Burrough and D.E. McCormack), ITC publication no. 6, Washington DC, pp. 111–18.

Dent, D.L. and Scammell, R.P. (1981) Assessment of long-term irrigation need by integration of data for soil and crop characteristics and climate. *Soil Survey and Land Evaluation*, **1**, 51–7.

Doorenbos, J. and Kassam, A.H. (1979) *Yield response to water*. FAO Irrigation and Drainage Paper, **33**, Food and Agriculture Organization, Rome.

Elliott, M.M. (1985) *Some aspects of invertebrate herbivores on deciduous woodland regeneration*, PhD thesis, University of East Anglia, Norwich.

Falinski, J.B. (1986) *Vegetation Dynamics in Temperate Lowland Primeval Forests*, W. Junk, Dordrecht.

Goodfellow, S. and Peterken, G.F. (1981) A method for survey and assessment of woodlands for nature conservation using maps and species lists: the example of Norfolk woodlands. *Biological Conservation*, **21**, 177–95.

Grime, J.P., Hodgson, J.G. and Hunt, R. (1988) *Comparative Plant Ecology*, Unwin Hyman, London.

Hall, D.G.M., Reeve, M.J., Thomasson, A.J. and Wright, V.F. (1977) *Water retention, porosity and density of field soils*, Soil Survey Technical Monograph, 9, Harpenden.

Hillel, D. (1982) *Introduction to Soil Physics*, Academic Press, London.

Hinckley, T.M., Duhme, F., Hinckley, A.R. and Richter, H. (1983) Drought relations of shrub species: assessment of the mechanisms of drought resistance. *Oecologia*, **59**, 344–50.

Inghe, O. and Tamm, C.O. (1985) Survival and flowering of perennial herbs. IV. The behaviour of *Hepatica nobilis* and *Sanicula europaea* on permanent plots during 1943–1981. *Oikos*, **45**, 400–20.

Korstian, C.F. and Coile, T.S. (1938) *Plant Competition in Forest Stands*, Duke University School of Forestry Bulletin, 3, Duke University Press.

Leshem, B. (1970) Resting roots of *Pinus halapensis*: Structure, function and reaction to water stress. *Botanical Gazette* (*Chicago*), **131**, 99–104.

Martin, M.H. (1968) Conditions affecting the distribution of *Mercurialis perennis* L. in certain Cambridgeshire woodlands. *Journal of Ecology*, **56**, 777–93.

Martin, M.H. and Pigott, C.D. (1975), Chapter III: *Modern Ecology*, in *Hayley Wood: its history and ecology*, (Ed. O. Rackham), Cambridgeshire and Isle of Ely Naturalists' Trust, Cambridge, pp. 61–71.

McKeague, J.A., Eilers, R.G., Thomasson, A.J. *et al.* (1984) Tentative assessment of soil survey approaches to the characterization and interpretation of air-water properties of soils. *Geoderma*, **3**, 69–100.

Melitz, P.J. (1986) The sufficiency concept in land evaluation. *Soil Survey and Land Evaluation*, **6**, 8–17.

Moore, M.C. and Vankatt, J.L. (1986) Responses of the herb layer to the gap dynamics of a mature beech–maple forest. *American Midland Naturalist*, **115**, 336–47.

Peterken, G.F. (1970) *Swanton Novers Woods, Norfolk*. Report for the Nature Conservancy, no. EA/450. Monks Wood Experimental Station, Abbots Ripton.

Peterken, G.F. (1974) A method for assessing woodland flora for conservation using indicator species. *Biological Conservation*, **6**, 239–45.

Peterken, G.F. (1981) *Woodland Conservation and Management*, Chapman and Hall, London.

Pigott, C.D. (1982) The experimental study of vegetation. *New Phytologist*, **90**, 389–404.

Rackham, O. (1975) *Hayley Wood: its history and ecology*. Cambridgeshire and Isle of Ely Naturalists' Trust, Cambridge.

Rackham, O. (1980) *Ancient Woodland*, Edward Arnold, London.

Rackham, O. (1986) The Ancient Woods of Norfolk. *Transactions of the Norfolk and Norwich Naturalists' Society*, **27**, 161–77.

Salisbury, E.J. (1924) The effects of coppicing as illustrated by the woods of Hertfordshire. *Transactions of the Hertfordshire Natural History Society*, **18**, 1–21.

Sheffe, H. (1959) *The Analysis of Variance*, John Wiley, New York.

Shirreffs, D.A. (1985) *Anemone nemorosa* L. Biological flora of the British Isles. *Journal of Ecology*, **73**, 1005–20.

Smith, L.P. and Trafford, B.D. (1976) *Climate and Drainage*, MAFF Technical Bulletin, 34, HMSO, London.

Ward, R. (1975) *Principles of Hydrology* (2nd edn), McGraw-Hill, London.

Waring, R.H. and Cleary, B.D. (1967) Plant moisture stress: evaluation by pressure bomb. *Science*, **155**, 1248–54.

Waring, R.H. and Schlesinger, W.H. (1985) *Forest Ecosystems: Concepts and Management*, Academic Press, London.

Part Three

CONTRASTING PROCESSES IN COPPICE AND FOREST ECOSYSTEMS

– 5

Coppicing and natural disturbance in temperate woodlands – a review

MARTIN N. EVANS and JOHN P. BARKHAM

5.1 INTRODUCTION

Disturbance is a key factor in the dynamics of all plant communities (White, 1979). In forests the destruction of trees results in obvious structural and environmental changes making a *prima facie* case for studying disturbance processes. As a result of such studies, the concept of 'climax' woodland has been more or less rejected, having been questioned nearly half a century ago by Jones (1945). In its place ideas of non-equilibrium and stochasticity have become fashionable. Techniques of woodland management, whether coppicing, selection thinning, clear-cutting or prescribed burning, are now described as 'artificial disturbance regimes' (Runkle, 1985). Comparisons are drawn between these and natural disturbance regimes, and indeed they may deliberately mimic them to maintain a certain species composition (Runkle, 1985) and structure (Lorimer, 1989).

For purposes of discussion, the definition of disturbance given by White and Pickett (1985) will be adopted here:

> *A disturbance is any relatively discrete event in time that disrupts ecosystem, community, or population structure and changes resources, substrate availability, or the physical environment.*

This definition clearly covers a great range of phenomena that are encountered in temperate woodlands (some of which are shown in Table 5.1). Also, an event such as a volcanic eruption may cause a number of different kinds of disturbance (Antos and Zobel, 1985). However, disturbance does not necessarily involve the destruction of biomass, although it does under some other definitions, notably that of Grime (1977). For example, if a tree falls in a forest, it may not die but instead resprout along the length of the trunk, resulting in the *reorganization* of biomass. Similarly, a ground flora community may experience a sudden edaphic change: for instance, slopes of some river valleys are occasionally flushed by more base-rich water; such an event may alter the competitive balance between species, and in this respect it is a disturbance, but biomass is not destroyed.

Table 5.1 Examples of canopy and non-canopy disturbances with references for their effects on the ground flora of temperate woods

(a) Canopy disturbance

Cause/type	*Reference*
Treefall gaps	Watt (1925); Davison and Forman (1982); Collins *et al.* (1985); Moore and Vankat (1986); Hutton (1989); Lamb (1989); Whitbread (1989)
Canopy defoliation by gypsy moth (*Lymantria dispar*)	Ehrenfeld (1980)
Forest fire	Boerner (1981)
Artificial canopy gaps	Collins and Pickett (1987, 1988)
Coppicing	Ash and Barkham (1976); Rackham (1980); Peterken (1981); Chapter 7
Clear-cutting and selection thinning	Metzger and Schultz (1981, 1984)
Prescribed burning	Boerner (1981)

(b) Non-canopy disturbance

Cause/type	*Reference*
Deer (various species) grazing, trampling and soil disturbance	Hough (1965); Rackham (1975); Barkham (1980); Falinski (1986)
European bison (*Bison bonasus*) grazing, trampling and soil disturbance	Falinski (1986)
Native wild boar (*Sus scrofa*) grazing and rooting	Falinski (1986)
Introduced wild boar (*Sus scrofa*) grazing and rooting	Bratton (1974, 1975)
Mole (*Talpa europaea*) burrowing	Mellanby (1971); Barkham (1980)
American red squirrel (*Tamiasciurus hudsonicus*) digging	Bratton (1974)
Chipmunk (*Tamias striatus*) litter removal and soil disturbance	Collins and Pickett (1988)
Snowshoe hare (*Lepus americanus*) grazing	Antos and Zobel (1985)
Bank vole (*Clethrionomys glareolus*) grazing and burrowing	Barkham (1980); Falinski (1986)
Other rodents grazing	Falinski (1986)
Pheasant (*Phasianus colchicus*) trampling, grazing and defecation	Ludolf *et al.* (1988)
Invertebrate (various species) grazing and soil disturbance	Barkham (1980)
Burial under volcanic tephra	Antos and Zobel (1985)
Anthropogenic disturbances (camping, trails, etc.)	Hoehne (1981)
Artificial soil disturbance	Sauer and Struik (1964)

The response of the ground flora to disturbance is basically made up of two components (Marks, 1974; Ash and Barkham, 1976):

1. Reorganization of the vegetation already present prior to the disturbance;
2. Establishment of new plants, from seeds lying dormant in the soil or immigration of propagules.

The relative importance of reorganization or establishment depends on the size, severity, timing and qualitative nature of the disturbance (Connell, 1978; Runkle, 1985; Hughes and Fahey, 1988; Foster, 1988b; Spies and Franklin, 1989). These factors in turn interact with the sensitivity of individual species (Collins and Pickett, 1988) or even species genotypes (Thompson, 1985) to produce a complex pattern of response.

In this review, three main issues will be addressed. Firstly, the features of coppicing and natural canopy gap formation that affect the ground flora will be compared, including environmental and biotic changes and the distribution of disturbance events in space and time. Secondly, the significance to the ground flora of disturbances other than canopy removal (termed here 'non-canopy disturbances') will be discussed. Finally, the roles of all types of disturbance in maintaining ground flora diversity will be considered together with the implications for coppice management.

The effects of natural canopy gaps on the ground flora of temperate woodlands are little known (Collins, Dunne and Pickett, 1985), and although the effects of coppicing are rather better documented, they are also poorly understood. For purposes of comparison, therefore, possible **mechanisms** of change will be emphasized rather than the actual ground flora changes themselves. This approach is also necessary because work on the two types of disturbance is geographically separated – research on coppicing having been carried out largely in lowland Britain, that on natural canopy gaps in the eastern United States. However, this dichotomy has recently been broken down somewhat by studies of the effects of the Great Storm of October 1987 in south-east England (e.g. Whitbread, 1988, 1989; Hutton, 1989; Lamb, 1989); the more widespread storms of early 1990 may also promote study of natural disturbance in British woods. Fire will not be discussed as a natural agent of canopy disturbance: though important in some temperate forests (Jones, 1945; White, 1979), it is generally not a feature of British semi-natural woodlands due to the well-distributed pattern of rainfall and extremely low flammability of most native tree species (Rackham, 1986).

5.2 COMPARISON BETWEEN COPPICING AND NATURAL CANOPY DISTURBANCE

5.2.1 Macroenvironmental changes

While coppicing an area produces a gap in the woodland structure, in the case of natural disturbance there is some debate over what constitutes a true canopy gap. Brokaw (1982) points out that a hole may form at any level in the forest structure without necessarily having any ecological significance for the levels below it: for example, in some Japanese forests the microclimate of the forest floor is changed very little by canopy gap formation due to shading by dwarf bamboo (Nakashizuka, 1989). A true gap is therefore 'a "hole" in the forest extending through all levels down to an average height of two [metres]' (Brokaw, 1982). However, Lieberman *et al.* (1989) question this definition, pointing out that the reality is actually a highly heterogeneous canopy riddled with holes of all sizes.

As the size of a gap is important in determining the degree of environmental change, another approach is to define the minimum critical gap size which invokes an ecological response. However, actual figures for this vary (Collins and Pickett, 1987; Nakashizuka, 1989). Collins *et al.* (1985) point out that the height and structure of the surrounding canopy are crucial: a more meaningful concept is the ratio of the diameter of the gap, D (for a circular gap), to the mean height of the surrounding trees, H. This ratio has been shown to be of some environmental and ecological significance (Geiger, 1965; Minckler and Woerheide, 1965). Recently, a similar concept has been used to define gaps in terms of solar penetration to the woodland floor (Smith *et al.*, 1989).

The subject of the canopy gap environment has been well reviewed by Geiger (1965), Collins *et al.* (1985), Lamb (1989) and Barkham (Chapter 7). There are obvious similarities in the changes associated with coppicing and natural canopy gap formation. In both, the calm, shaded, microclimatically stable woodland floor (Geiger, 1965; Ash and Barkham, 1976) experiences a sudden change: light, soil moisture and air movement are increased, while air and surface temperatures fluctuate more widely between higher day-time and lower night-time values (see Table 7.1, Chapter 7). The principal factors determining the magnitude of these changes are the physical dimensions and orientation of the gap (Geiger, 1965; Minckler and Woerheide, 1965; Grubb, 1977; Collins *et al.*, 1985), and its relationship to the surrounding forest structure (Collins *et al.*, 1985).

Canopy gap formation may also induce changes in soil chemistry, especially nutrients, although these are largely uninvestigated, most work concentrating on the effects of large-scale clear-cuts (Vitousek, 1985). Theoretically, nutrient availability should increase due to greater insolation and reduced transpiration, accelerating the rate of decomposition and mineralization (Peterken, 1981; Canham and Marks, 1985); however, there is little direct evidence for this, and the situation may be complicated by changes in the litter input and in the invertebrate fauna (Shure and Phillips, 1987).

Although conditions are generally more favourable and resources more available for plant growth in canopy-gap areas (Canham and Marks, 1985), gaps may sometimes have adverse environmental features. For example, lower night-time temperatures increase the risk of frost damage (Geiger, 1965), and snow persistence in gaps may be longer, resulting in a shorter growing season (Veblen, 1985). Also, although the water-table may rise as a result of gap formation the surface layers, in which much of the ground flora is rooted, may dry out due to increased evaporation (Peterken, 1981; Collins *et al.*, 1985). Severe disturbance can result in considerable loss of soil and nutrients by erosion and runoff, badly degrading the site for plant growth (Canham and Marks, 1985; Runkle, 1985). The gap environment is highly heterogeneous (Minckler and Woerheide, 1965; White and Pickett, 1985), its effects extending well into the surrounding canopied area (Geiger, 1965).

This multitude of environmental changes makes it difficult to determine the proximal causes of plant responses to canopy gap formation, especially when they act in combination (e.g. Anderson *et al.* (1969) suggest that light and soil moisture may act synergistically). The situation is further complicated by the fact that one mosaic, the altered woodland canopy, is being superimposed on another, the patchy field layer (Ehrenfeld, 1980). The response of the ground flora to canopy gaps of any kind is therefore determined by a complex interaction of environmental, physiological and phenological factors. Some of the changes are summarized in Table 7.2 and have also been reviewed by Collins *et al.* (1985).

5.2.2 Soil disturbance and microtopographic changes

In unmanaged woodlands, the fall of a tree may involve its uprooting (Stephens, 1956), although its actual frequency compared with other types of damage is equivocal (Foster, 1988b). Different tree species vary in their propensity to different kinds of damage, whether uprooting, snapping of the trunk, or loss of limbs (Falinski, 1978; Runkle, 1982; Rackham, 1986; Foster, 1988b; Whitbread, 1988). The incidence of uprooting also depends on soil conditions (Nakashizuka, 1989) and the effects these have on the depth of tree root penetration (Mueller and Cline, 1959), the weather conditions in which the treefall occurs, as well as other local, historical and chance factors (Foster, 1988b). Where it does occur, uprooting causes soil disturbance and microtopographic change in the form of a 'pit', in which the tree roots and surrounding soil used to reside and a 'mound' comprising the uplifted mass, with the difference in microrelief between them varying from a few centimetres to over a metre (Beke and McKeague, 1984). Pit and mound microtopography is widespread in unmanaged temperate woodlands (Stephens, 1956) and is characteristic of over 90% of the area of the Appalachian forests (Beke and McKeague, 1984), covering 14–50% of the forest floor (Stephens, 1956; Beatty, 1984).

The formation of a pit and mound produces a site that is largely free from above and below ground competition and from litter (Thompson, 1980; Putz, 1983); nowhere else in the canopy-gap area is this necessarily the case (Putz, 1983). Pits and mounds are therefore prime sites for the establishment of new individuals, which may be of species not already growing in the vicinity (Thompson, 1980). Cook and Lyons (1983) found 17 species of tree and herb germinating on 'uproots', the latter all apparently germinating from the seedbank. Thompson (1980) noted that herb species show differential abilities to colonize different parts of an uproot, by vegetative spread or by seed, probably because the microclimatic and soil variation provided by uproots is considerable (Beke and McKeague, 1984; Foster, 1988b), as summarized in Table 5.2. For instance, pits are wetter, and so may support aquatic and wetland species (Falinski, 1978), while the leached tops of root plates may allow the establishment of calcifuge species, such as foxglove (*Digitalis purpurea*), in woods growing on chalk (Whitbread, 1989). Dispersal of seed to different parts of the uproot, particularly by animals (see below) may also be an important differentiating factor. Pits and mounds are long-term features in temperate woodlands (Stephens, 1956): their continued erosion leads to the exposure of more buried seeds (Cook and Lyons, 1983) and provides new colonization sites for plants (Henry and Swan, 1974; Foster, 1988a). Their influence on the microenvironment and hence on ground flora patterns continues for some considerable time after canopy repair (Beatty, 1984; Bratton, 1976).

Uprooting also has profound implications for the development of woodland soils, bringing up unweathered material from a greater depth in the soil and thereby allowing soil 'rejuvenation' (Armson and Fessenden, 1973; Beke and McKeague, 1984; Brown, 1977; Foster, 1988b). In a Michigan beech–maple forest, Brewer and Merritt (1978) calculated that at the observed rates, if uniformly distributed, the entire soil surface would be turned over in 3500–7500 years. The thicknesses of some soil horizons may reflect such turnover (Armson and Fessenden, 1973), and soil profiles disturbed by uprooting may be recognizable 500 years later (Stephens, 1956).

Although coppice woods usually lack large, over-mature trees capable of creating pits and mounds (Peterken, 1981), soil disturbance is also an important feature of coppicing, particularly as it is carried out in the winter

Table 5.2 A comparison of the characteristics of pits and mounds with references for their effects on the ground flora of temperate woodlands

Characteristic	*Pits*	*Mounds*	*Reference*
Soil moisture	Higher (may contain standing water)	Lower	Falinski (1978); Beatty (1984); Beke and McKeague (1984)
Soil nutrients	Higher	Lower	Beatty (1984)
Soil pH	Higher	Lower	Beatty (1984); Whitbread (1989)
Soil organic matter	Higher	Lower	Beatty (1984)
Soil aeration	Lower	Higher	Putz (1983)*
Soil microflora	?	?	Putz (1983)*
Soil erosion	Lower, shorter term	Higher, longer term	Cook and Lyons (1983); Foster (1988a)
Litter cover	Higher	Lower	Thompson (1980); Putz (1983)*; Beatty (1984)
Light			
quantity	Lower	Higher	Putz (1983)*
spectral composition	Lower red component	Higher red component	Putz (1983)*
Summer temperature	Colder	Warmer	Beatty (1984)
Winter temperature	Warmer	Colder	Beatty (1984)
Frost heave	Lower	Higher	Beatty (1984)
Period frozen	Longer	Shorter	Beke and McKeague (1984)
Snow accumulation	Higher	Lower	Beatty (1984)
Buried seed	Less	More	Thompson (1980); Cook and Lyons (1983); Foster (1988b)
Ability to catch wind-dispersed seed	Increased	Increased	Foster (1988b); Matlack (1989); Nakashizuka (1989)
Animal defecation	Increased	Increased	Thompson (1980)
Myrmecochory (seed dispersal by ants)	Increased	Increased	Thompson (1980)
Burrowing animals	Present	Present	Putz (1983)*; Whitbread (1988)

*Information from tropical forests, but may apply to uproots in temperate regions.

when the ground is often wet and soft. Felling, dragging the poles, and trampling can damage shoots and underground parts of ground flora plants (Rackham, 1980; Ash and Barkham, 1976), e.g. fragmenting the rhizomes of wood anemone (*Anemone nemorosa*) (N.R. Cowie, personal communication). The bonfires on which the tops are burned may also kill ground flora (Ash and Barkham, 1976), providing colonization sites (Cummings, 1982). In contrast with pit and mound formation in natural woods, soil disturbance in coppice woods is not to as great a depth (usually only a few centimetres) but it may be more widespread rather than concentrated in the area immediately around the bases of fallen trees.

In both unmanaged and coppiced woodlands, soil disturbance is important in allowing establishment responses, breaking up the local dominance characteristic of some perennial woodland herbs and letting other species gain a foothold (Thompson, 1980; Chapter 7). However, the mechanism by which soil disturbance alone seems to promote germination from the seed bank is unclear, and probably varies between species (Sauer and Struik, 1964; Collins and Pickett, 1988). It may be exposure of seed to warmth (Ash and Barkham, 1976), or a light response (Sauer and Struik, 1964). The spectral differences between light in gap and canopied areas (Geiger, 1965; Chapter 3) may also be of significance. Microtopographic factors at the smallest textural scale may also influence seed germination (Harper, Williams and Sagar, 1965).

A component seen in natural treefall gaps but often lacking from coppicing is the remaining tree biomass in the form of decaying wood. Dead wood may cover a significant area in natural gaps: for instance, Falinski (1978) noted that in one area of Bialowieza Forest in Poland, logs covered 12–15% of the forest floor. Logs are long-term features (Foster, 1988a; Handel, 1978), outlasting the canopy gaps from which they originated. For the ground flora, they represent an important diversifying agent on the forest floor (Bratton, 1976), providing refugia from competition (Handel, 1978) and from other forms of disturbance such as burial by volcanic tephra (Antos and Zobel, 1985). Herb species vary in their ability to colonize logs, thereby affecting ground flora patterns (Falinski, 1978; Thompson, 1980). As logs rot down the nutrient changes affect the ground flora (Falinski, 1978; Whitbread, 1989). The implications of the lack of dead wood in coppiced woods are discussed more widely in Chapter 6.

5.2.3 Canopy gap fauna

Mammal activity in gap areas is likely to be greater than where there are no gaps (Collins and Pickett, 1987): for example, Falinski (1978) noted that fallen trees were preferred pathways for forest rodents at night. Barkham (1980) observed that the intensity of animal activity was related to woodland management and the time since last management disturbance. The effects of coppicing on small mammal and bird communities are described respectively in Chapters 9 and 11. Invertebrates, too, respond to canopy gap formation, especially after coppicing (Chapters 13 and 14), while in Appalachian forests Shure and Phillips (1987) suggest that the changes in litter quantity and quality resulting from canopy gap formation affect detritivore populations.

While it is intuitive that grazing, digging, seed dispersal, pollination and comminution by animals attracted to canopy gap areas must affect the ground flora, directly or indirectly, hard evidence for this is fairly scarce (some possible effects are suggested in Table 7.2). Thompson (1985) concluded that disturbance affects not only the species themselves, but also the interactions between them, and that such effects vary within a disturbed patch. For example, Thompson and Wilson (1978) found that birds were more likely to find, and consequently remove, fleshy fruits in gaps than under closed canopy, while Falinski (1978) noted that rooting of the ground flora by wild boar (*Sus scrofa*) was obstructed by fallen trees. Thompson (1980) observed that logs and pits were often used as defecation sites by birds and mammals: in this way, animals may act as important agents of seed dispersal. Ant dispersal is potentially important following a treefall, especially for species colonizing logs (Handel, 1978). In this respect, the relative faunal poverty of British woods may be of some significance; since a number of British woodland ground flora species are apparently adapted for ant dispersal (myrmecochory), but the necessary ant species are generally unavailable (Barkham, 1992). On a smaller spatial scale, another possible problem is lack of appropriate pollinators in canopy gap areas due to their relative scarcity and isolation (Cook and Lyons, 1983), emphasizing the importance to the ground flora of interconnecting compartments to maintain invertebrate populations in coppice woodlands (Whitbread, 1989; Chapters 13 and 14).

5.2.4 Persistence of canopy gap changes

Many of the changes resulting from canopy gap formation described above are short lived. In unmanaged woodlands gaps are rapidly filled by the growth of saplings or by extension growth from surrounding canopy trees, the size and severity of the disturbance determining the rate of recovery (Runkle, 1985). As a result, gaps in some temperate forests may be difficult to recognize on the ground after about five years (Moore and Vankat, 1986). Similarly in a coppiced area, canopy closure occurs after 4–10 years (Ash and Barkham, 1976).

Increasing shade during canopy closure reduces light levels (Peterken, 1981; Moore and Vankat, 1986), and a newly closed coppice canopy, especially of hornbeam (*Carpinus betulus*) may actually cast a denser shade than it did before coppicing (Rackham, 1980). Soil moisture is reduced (Moore and Vankat, 1986), and nutrient cycling stabilizes (Vitousek, 1985). 'Feedback' effects may operate as a result of the within-patch variation in plant responses noted by Thompson (1985): for example, Poulson and Platt (1989) found that the north–south gradient of decreasing light, characteristic of gaps in northern temperate forests, was reversed over a 13-year period by the rapid growth of vegetation in the northern part of the gap but not in the southern part.

The initial conditions following disturbance are of prime importance in determining subsequent ground flora changes, and early establishment is of great benefit. Cook and Lyons (1983), looking at the establishment of *Viola fimbriatula* on soil disturbed by uprooting, found that germination declined rapidly after the first year following treefall. Recruitment to the population was limited to the seedlings emerging immediately following the disturbance, and of these a few individuals grew larger, lived longer and produced nearly all of the seed; cohorts germinating subsequently failed to establish. Seedling establishment also declines soon after coppicing, and most species are already present from the start (Ash and Barkham, 1976). A caveat to this, however, is that the response time of some perennial ground flora species may be long (Rackham, 1980). Long response times to natural canopy gaps are also postulated (Moore and Vankat, 1986; Collins and Pickett, 1987, 1988; Whitney and Foster, 1988), and the patchiness resulting from a single disturbance event may persist for many years (Watt, 1947).

5.2.5 Spatial and temporal distribution of gaps

Coppicing and natural canopy gap formation differ in their distributions in space and in time. In unmanaged forests, the disturbance regime is usually expressed as a 'gap formation rate' or 'canopy turnover rate', given as percentage of land area affected per year (Runkle, 1985). In most forests the canopy disturbance regime falls into a relatively narrow range of 0.5–2.0% per year, giving a return interval of 50–200 years (Runkle, 1985), although this has been questioned by Barden (1989) and Fox (1989). However, a gap formation rate is not particularly meaningful by itself: disturbances may be few, rare and large or alternatively many, frequent and small; but usually they combine elements of both (Runkle, 1985; Spies and Franklin, 1989). The length of a coppice rotation is generally 5–20 years (Rackham, 1986); as a canopy disturbance coppicing is therefore a whole order of magnitude more frequent than natural canopy gap formation. A coppice rotation is also much more spatially and temporally rigid: after a fixed period of growth from the stools a fixed area is cut. It therefore lacks the element of stochasticity found in natural disturbance regimes, and it is probably at this point that comparison between the two breaks down most severely: there are no natural analogues for a system where simultaneous canopy disturbance over a known area is guaranteed after a known period.

When a tree is removed from the canopy in an unmanaged woodland it may be replaced by another tree of the same or a different species – there being an element of chance involved in replacement. This stochasticity characterizes the non-equilibrium condition so well demonstrated by Henry and Swan (1974). On a small plot in an old-growth New Hampshire forest, they found, in effect, three quite different forests, all initiated by major disturbances (fire or windthrow). After coppicing, however, stochasticity in species replacement is limited because regeneration of the woodland occurs principally from the cut stools (Peterken and Jones, 1989). Coppicing also greatly extends the life of the woody species involved and as a result some coppice stools in ancient woodlands are of great age (Rackham, 1986). Together, these two factors give coppice woodland a structure that 'changes repetitively and within narrow limits', and a composition that 'could remain stable . . . even at the smallest scale' (Peterken and Jones, 1989).

The implications of this for the ground flora are unclear. Canopy trees are known to affect the environment beneath them, both in terms of the light environment and the soil conditions (Geiger, 1965; Crozier and Boerner, 1984), and they compete with the field layer for moisture and nutrients (Peterken, 1981; Chapter 4). Also, in their early years, trees grow in the same woodland floor environment as the ground flora (Peterken, 1981). Consequently, correlations are sometimes observed between canopy tree species and ground flora composition (Peterken, 1981; Crozier and Boerner, 1984), but as Whitney and Foster (1988) point out, the dynamic nature of the overstorey and the slow response time of the ground flora mean that the latter is in a perpetual state of flux, and correlations between them may therefore be rare. In coppice woodland, on the other hand, the compositional stability of the overstorey imposed by management may result in a similar, albeit dynamic, stability of the ground flora.

5.3 NON-CANOPY DISTURBANCE

White and Pickett (1985) point out that disturbance is relative to the spatial and temporal scale of the system in question. Different considerations apply to woodland herbs from those that apply to trees (Collins *et al.*, 1985), and the two vary at different scales, the ground flora responding as a community to environmental variation that trees can only respond to as individuals (Peterken, 1981). Despite this, many studies have tended to regard the disturbance of woodlands purely in terms of the canopy, with the forest floor being thought of as 'undisturbed' until a canopy gap forms above it. However, the ground flora can experience 'non-canopy' disturbance at spatial and temporal scales that are totally different and often irrelevant to the canopy trees. Such disturbance may have implications for ground flora dynamics just as important as the formation of canopy gaps – natural or anthropogenic. Some examples are shown in Table 5.1(b). The range of activities is very wide indeed; a woodland herb may be grazed by deer (Rackham, 1975; Chapter 12), trampled by pheasants (Ludolf *et al.*, 1988), moved in the soil by moles (Barkham, 1980) or dug up by wild boar (Falinski, 1986).

Grazing in woods has been studied most extensively where the wild animals involved are rather more visible, such as European bison (*Bison bonasus*) in Bialowieza Forest in Poland (Falinski, 1986); but in British woods it only tends to be noticed where it is particularly severe and a management problem, such as the browsing of coppice regrowth by deer (Rackham, 1980) or where tree regeneration is inhibited as a result of it (Pigott, 1985). However, in some cases grazing can profoundly affect the ground flora: e.g. Rackham (1975) found that in Hayley Wood, Cambridgeshire, roe deer (*Capreolus capreolus*) reduced the population of oxlips (*Primula elatior*) fourfold in under 30 years.

Animals are also potentially important agents of soil disturbance in woodlands (Falinski, 1978). In a few forests of continental Europe wild boar is important not only because of its surface grazing activities, but also for its digging for the underground parts of ground flora plants – called 'rooting' (Falinski, 1986). Selectivity by wild boar and differences in the sensitivity of ground flora species to rooting bring about changes in plant community structure, with the less sensitive spring species becoming relatively more important than the more sensitive summer ones (Bratton, 1974; Falinski, 1986). Wild boar may have more drastic effects, though: their introduction to the southern Appalachians reduced ground flora cover in certain places to less than 5%, resulting in the local extinction of some species (Bratton, 1974, 1975). In British woodlands moles are probably the most comparable agents of soil disturbance. As with tree uprooting, they help maintain the base status of the surface layers by bringing up material from deeper down the soil profile (Rackham, 1980), thereby reversing the effects of leaching; there are also direct localized effects on the ground flora (Mellanby, 1971). Animal disturbances may also interact: e.g. Barkham (1980) observed that bulbs of wild daffodil (*Narcissus pseudonarcissus*) exposed or brought close to the surface by animal activity were more prone to slug attack.

Revegetation of the ground flora may result from vegetative growth of surviving plants and not necessarily from seed germination. It appears that non-canopy disturbance alone is poor at initiating establishment responses. However, exceptions do occur, such as the observation by Collins and Pickett (1988) that the establishment of *Aster acuminatus* and *Rubus* spp. was associated more with soil disturbance by chipmunks (*Tamias striatus*) than with canopy removal; and the removal of the strongly competitive dogs mercury (*Mercurialis perennis*) by picking or trampling allowed the growth of oxlip (*Primula elatior*) and bluebell (*Hyacinthoides non-scripta*) (Rackham, 1980). Further studies are clearly needed on the effects on the ground flora of non-canopy disturbances, both by animals (grazing, trampling and soil disturbance) and by abiotic causes, such as erosion by water or localized topographic effects.

5.4 IMPLICATIONS FOR DIVERSITY AND WOODLAND MANAGEMENT

It is generally accepted that disturbance is important in maintaining diversity within lower and upper limits. Disturbances may be too mild or infrequent to prevent competitive exclusion from occurring, but more dramatic ones may reduce diversity, as fewer species are adapted to the more extreme conditions (Connell, 1978). Small disturbances promoting diversity may be an argument for coppicing, while the effects of drastic disturbance are seen in the excessive rooting and loss of diversity caused by wild boar (Bratton, 1974).

There are numerous mechanisms through which natural canopy-gap formation may act to enhance ground flora diversity. 'Environmental release' is undoubtedly one, making conditions more favourable for the germination establishment and growth of plants. The soil disturbance and microtopographic changes associated with natural canopy gaps are also of great significance in breaking up large monospecific patches and facilitating the establishment of 'new' species. In Illinois, Thompson (1980) found that up to 26% of species colonizing logs, pits and mounds were not present nearby, while Moore and Vankat (1986) found that herb species richness was similar beneath gaps of different ages (undisturbed by pits and mounds) and under intact canopy. Seeds may be 'funnelled' to canopy gap areas, where stochastic factors in propagule availability may be important in enhancing diversity on a larger scale. There are instances, however, where ground flora diversification is severely limited in canopy gap areas, e.g. by the luxuriant growth of sugar maple (*Acer saccharum*) seedlings noted by Brewer (1980) in a Michigan woodland.

Coppicing, while broadly similar to natural canopy gap formation, lacks a number of features which may diversify the ground flora of unmanaged woodlands, notably pit and mound formation and the accumulation of dead wood. The enhanced order and stability that coppicing imposes on the structure, composition and disturbance regime of a woodland may limit the range of opportunities for ground flora development over space and time. However, with coppicing the greater frequency of canopy disturbance may compensate for some of these negative factors in terms of diversity. Also, stochastic factors do play a role in varying ground flora response to coppicing, particularly variability in the weather of the following spring (Rackham, 1980), and also propagule availability.

Much British ancient woodland is no longer coppiced and is described variously as 'neglected', 'stored', 'derelict', 'non-intervention', or simply 'unmanaged'. These woods are presently in a 'transition phase', in which the absence of management is allowing the onset of natural disturbance regimes and the consequent breakdown of the coppice structure. Characterization of these regimes will need further long-term study (Peterken and Jones, 1987; Peterken and Backmeroff, 1988) but the little long-term monitoring that has been carried out in unmanaged ancient woodlands has demonstrated that natural canopy disturbance can be a powerful agent of change and diversification in woodland structure and composition (Peterken and Jones, 1987, 1989). Unfortunately, there have been very few concomitant studies looking at the implications of these changes for the ground flora. Consequently, the evidence of ground flora change is largely anecdotal: the loss of certain species is attributed to a lack of coppicing (Rackham, 1980, 1986), although re-coppicing after up to a 100 years of neglect may see no apparent loss of ground flora diversity (Rackham, 1986; I.P.F. Cummings and R. Hobbs, personal communication). Nonetheless, coppicing is often regarded by conservationists as a *sine qua non* for maintaining ground flora diversity and a useful management tool enhanced by the predictability of the distribution and degree of the disturbance that it generates.

Protagonists of 'non-intervention' management often argue that coppiced ancient woodlands are simply 'reverting to their natural state', and as the ground flora species seen today must have survived in primary woodland, they will continue to do so in unmanaged woodlands. However, patchiness from past management and other human activities may persist for many decades (Peterken and Jones, 1989), and so the true expression of the ground flora in relation to natural features of a site (Chapter 7) may therefore be a protracted process. Furthermore, many centuries of continuous coppicing may have selected not only for a certain ground flora species but also for particular ecotypes. There are also considerations of species–area relations and island effects: Pickett and Thompson (1978), discussing the implications of disturbance studies for nature reserve design, point out that knowledge of the disturbance regime of a habitat is crucial, to determine its 'minimum dynamic area', or the smallest area maintaining internal recolonization sources, thereby minimizing the chance of local extinction. An illustration of this was given by Scheiner (1988) who, in a study of Michigan woods, attributed the long-term decline of *Danthonia spicata*, a perennial grass, to the decrease in suitable open habitat. Nature reserves should therefore be larger than the largest disturbed patch size: coppicing regimes, which maintain a mosaic of patches at different stages of regeneration following disturbance, appear to fit the rule. However, the minimum dynamic areas of British woodlands under natural disturbance regimes are unknown, but may be larger than the size of many ancient woodlands. Care should therefore be taken in extrapolating from studies of natural disturbance carried out in the large forests of the eastern United States, to the small, isolated fragments of ancient woodland set in the agricultural landscape of Britain.

Another important consideration is that of seed bank viability. Certain species, though largely confined to woodlands, germinate only in canopy gap areas, surviving under closed canopy as buried seed (Rackham, 1980; Chapter 8). In a given patch, the chance of survival of such a species in the long term depends on both the gap formation rate and the longevity of the seed bank. If the longest interval between gap formation is shorter than the longevity of viable seed in the soil the species will survive; if it is longer the species will become extinct in that patch. This sort of concept lends itself to predictive computer modelling, which could encompass the stochastic element in canopy gap formation.

Under the longer periods of closed canopy that occur in unmanaged woodlands, non-canopy disturbance may play an increasingly important role in maintaining ground flora diversity. Its effects on the ground flora under a closed canopy are poorly known but its chief importance may lie simply in breaking up monospecific clumps and preventing competitive exclusion from occurring. In this respect, the absence from British woods of certain large mammals, such as wild boar (extinct in the wild since 1260 (Rackham, 1986)), may mean that unless other animals such as moles generate sufficient soil disturbance and disruption to the ground flora, diversity in unmanaged woods may not be maintained in the long term. The faunal impoverishment of British woods, by geographical isolation and by millenia of human activity, may therefore prevent them from ever returning to a truly primeval state.

Ignorance about the consequences of disturbance for the long-term dynamics of the ground flora makes it difficult for managers to make confident conservation management prescriptions. Further studies are needed which should attempt to understand the whole spectrum of disturbances experienced by the ground flora, how they interact with each other and with the plants themselves. Finally, it should be remembered that the question of disturbance and its effects on diversity is only one, perhaps rather esoteric and parochial aspect of the problem. Much of the aesthetic and amenity value of many ancient woodlands is totally unrelated to their diversity: most visitors actually come to enjoy monospecific stands of flowers such as bluebell and wood anemone – not to count the species.

ACKNOWLEDGEMENTS

We are grateful to Ian Cummings, Neil Cowie and Richard Hobbs for their useful comments. Martin Evans wishes to thank the School of Environmental Sciences, University of East Anglia for supporting his studentship.

REFERENCES

Anderson, R.C., Loucks, O.L. and Swain, A.M. (1969) Herbaceous response to canopy cover, light intensity, and throughfall precipitation in coniferous forests. *Ecology*, **50**, 255–63.

Antos, J.A. and Zobel, D.B. (1985) Recovery of forest understories buried by tephra from Mount St Helens. *Vegetatio*, **64**, 103–11.

Armson, K.A. and Fessenden, R.J. (1973) Forest windthrows and their influence on soil morphology. *Proceedings of the Soil Science Society of America*, **37**, 781–3.

Ash, J.E. and Barkham, J.P. (1976) Changes and variability in the field layer of a coppiced woodland in Norfolk, England. *Journal of Ecology*, **64**, 697–712.

Barden, L.S. (1989) Repeatability in forest gap research: studies in the Great Smoky Mountains. *Ecology*, **70**, 558–9.

Barkham, J.P. (1980) Population dynamics of the wild daffodil (*Narcissus pseudonarcissus*). II. Changes in number of shoots and flowers, and the effect of bulb depth on growth and reproduction. *Journal of Ecology*, **68**, 635–64.

Barkham, J.P. (1992) Population dynamics of the wild daffodil (*Narcissus pseudonarcissus*). IV. Clumps and gaps. (In preparation).

Beatty, S.W. (1984) Influence of microtopography and canopy species on spatial patterns of forest understorey plants. *Ecology*, **65**, 1406–19.

Beke, G.J. and McKeague, J.A. (1984) Influence of tree windthrow on the properties and classification of selected forested soils from Nova Scotia. *Canadian Journal of Soil Science*, **64**, 195–207.

Boerner, R.E.J. (1981) Forest structure dynamics following wildfire and prescribed burning in the New Jersey pine barrens. *American Midland Naturalist*, **105**, 321–33.

Bratton, S.P. (1974) The effect of the European wild boar (*Sus scrofa*) on the high-elevation vernal flora in Great Smoky Mountains National Park. *Bulletin of the Torrey Botanical Club*, **101**, 198–206.

Bratton, S.P. (1975) The effect of the European wild boar, *Sus scrofa*, on gray beech forest in the Great Smoky Mountains. *Ecology*, **56**, 1356–66.

Bratton, S.P. (1976) Resource division in an understorey herb community: responses to temporal and microtopographic gradients. *American Naturalist*, **110**, 679–93.

Brewer, R. (1980) A half-century of changes in the herb layer of a climax deciduous forest in Michigan. *Journal of Ecology*, **68**, 823–32.

Brewer, R. and Merritt, P.G. (1978) Windthrow and tree replacement in a climax beech–maple forest. *Oikos*, **30**, 149–52.

Brokaw, N.V.L. (1982) The definition of treefall gap and its effect on measures of forest dynamics. *Biotropica*, **14**, 158–60.

Brown, J.L. (1977) Etude de la perturbation des horizons du sol par un arbre qui se renverse et de son impact sur la pedogenese. *Canadian Journal of Soil Science*, **57**, 173–86.

Canham, C.D. and Marks, P.L. (1985) The response of woody plants to disturbance: patterns of establishment and growth, in *The Ecology of Natural Disturbance and Patch Dynamics* (eds S.T.A. Pickett and P.S. White), Academic Press, London, pp. 197–216.

Collins, B.S., Dunne, K.P. and Pickett, S.T.A. (1985) Responses of forest herbs to canopy gaps, in *The Ecology of Natural Disturbance and Patch Dynamics* (eds S.T.A. Pickett and P.S. White), Academic Press, London, pp. 218–34.

Collins, B.S. and Pickett, S.T.A. (1987) Influence of canopy opening on the environment and herb layer in a northern hardwoods forest. *Vegetatio*, **70**, 3–10.

Collins, B.S. and Pickett, S.T.A. (1988) Demographic responses of herb layer species to experimental canopy gaps in a northern hardwoods forest. *Journal of Ecology*, **76**, 437–50.

Connell, J.H. (1978) Diversity in tropical rainforests and coral reefs. *Science*, **199**, 1302–10.

Cook, R.E. and Lyons, E.E. (1983) The biology of *Viola fimbriatula* in a natural disturbance. *Ecology*, **64**, 654–60.

Crozier, C.R. and Boerner, R.E.J. (1984) Correlations of understorey herb distribution patterns with microhabitats under different tree species in a mixed mesophytic forest. *Oecologia*, **62**, 337–43.

Cummings, I.P.F. (1982) *Bradfield Woods: bonfires and their effects on the flora.* MSc. thesis, University of Salford.

Davison, S.E. and Forman, R.T.T. (1982) Herb and shrub dynamics in a mature oak forest: a 30-year study. *Bulletin of the Torrey Botanical Club*, **109**, 64–73.

Ehrenfeld, J.G. (1980) Understorey response to canopy gaps of varying sizes in a mature oak forest. *Bulletin of the Torrey Botanical Club*, **107**, 29–41.

Falinski, J.B. (1978) Uprooted trees, their distribution and influence in the primeval forest biotope. *Vegetatio*, **38**, 175–83.

Falinski, J.B. (1986) *Vegetation Dynamics in Temperate Lowland Primeval Forests*, W. Junk, Dordrecht.

Foster, D.R. (1988a) Disturbance history, community organization and vegetation dynamics of the old-growth Pisgah forest, south-western New Hampshire, USA. *Journal of Ecology*, **76**, 105–34.

Foster, D.R. (1988b) Species and stand response to catastrophic wind in central New England, USA. *Journal of Ecology*, **76**, 135–51.

Fox, J.F. (1989) Bias in estimating forest disturbance rates and tree lifetimes. *Ecology*, **70**, 1267–72.

Geiger, R. (1965) *The Climate Near the Ground*. Harvard University Press, Cambridge, Massachusetts.

Grime, J.P. (1977) Evidence for the existence of three primary strategies in plants and its relevance to ecological and evolutionary theory. *American Naturalist*. **111**, 1169–94.

Grubb, P.J. (1977) The maintenance of species-richness in plant communities: the importance of the regeneration niche. *Biological Reviews*, **52**, 107–45.

Handel, S.N. (1978) The competitive relationship of three woodland sedges and its bearing on the evolution of ant-dispersal of *Carex pedunculata*. *Evolution*, **32**, 151–63.

Harper, J.L., Williams, J.T. and Sagar, G.R. (1965) The behaviour of seeds in the soil. I. The heterogeneity of soil surfaces and its role in determining the establishment of plants from seed. *Journal of Ecology*, **53**, 273–86.

Henry, J.D. and Swan, J.M.A. (1974) Reconstructing forest history from live and dead plant material – an approach to the study of forest succession in southwest New Hampshire. *Ecology*, **55**, 772–83.

Hoehne, L.M. (1981) The groundlayer vegetation of forest islands in a suburban matrix, in *Forest Island Dynamics in Man-dominated Landscapes* (eds R.L. Burgess and D.M. Sharpe), Springer-Verlag, New York, pp. 41–54.

Hough, A.F. (1965) A 20-year record of understorey vegetational change in a virgin Pennsylvania forest. *Ecology*, **46**, 370–3.

Hughes, J.W. and Fahey, T.J. (1988) Seed dispersal and colonization in a disturbed northern hardwood forest. *Bulletin of the Torrey Botanical Club*, 115, 89–99.

Hutton, D.N. (1989) *Permanent transect records of selected storm damaged woodland in Kent. Distribution of storm damage in relation to topography at Farningham Woods, Kent*, Chief Science Directorate Report, 922, Nature Conservancy Council, Peterborough.

Jones, E.W. (1945) The structure and reproduction of the virgin forest of the north temperate zone. *New Phytologist*, **44**, 130–48.

Lamb, J. (1989) *The ecological effects of the Great Storm of October 1987*, Chief Science Directorate Report, 937, Nature Conservancy Council, Peterborough.

Lieberman, M., Lieberman, D. and Peralta, R. (1989) Forests are not just Swiss cheese: canopy stereogeometry of non-gaps in tropical forest. *Ecology*, **70**, 550–2.

Lorimer, C.G. (1989) Relative effects of small and large disturbances on temperate hardwood forest structure. *Ecology*, **70**, 565–7.

Ludolf, C., Payne, S. and Robertson, P. (1988) The effect of pheasant release pens and strawed rides on ancient woodland ground flora, in *The Game Conservancy Review of 1988*, The Game Conservancy, Fordingbridge, pp. 42–6.

Marks, P.L. (1974) The role of pin cherry (*Prunus pensylvanica* L.) in the maintenance of stability in northern hardwood ecosystems. *Ecological Monographs*, **44**, 73–88.

Matlack, G.R. (1989) Secondary dispersal of seed across snow in *Betula lenta*, a gap-colonizing tree species. *Journal of Ecology*, 77, 853–69.

Mellanby, K. (1971) *The Mole*, Collins, London.

Metzger, F. and Schultz, J. (1981) Spring ground layer vegetation 50 years after harvesting in northern hardwoods forests. *American Midland Naturalist*, **105**, 44–50.

Metzger, F. and Schultz, J. (1984) Understorey response to 50 years of management of a northern hardwood forest in Upper Michigan. *American Midland Naturalist*, **112**, 209–23.

Minckler, L.S. and Woerheide, J.D. (1965) Reproduction of hardwoods 10 years after cutting as affected by site and opening size. *Journal of Forestry*, **63**, 103–7.

Moore, M.C. and Vankat, J.L. (1986) Responses of the herb layer to the gap dynamics of a mature beech–maple forest. *American Midland Naturalist*, **115**, 336–47.

Mueller, O.P. and Cline, M.G. (1959) Effects of mechanical soil barriers and soil wetness on rooting of trees and soil-mixing by blow-down in central New York. *Soil Science*, **88**, 107–11.

Nakashizuka, T. (1989) Role of uprooting in composition and dynamics of an old-growth forest in Japan. *Ecology*, **70**, 1273–8.

Peterken, G.F. (1981) *Woodland Conservation and Management*, Chapman and Hall, London.

Peterken G.F. and Backmeroff, C.E. (1988) *Long-term monitoring in unmanaged woodland nature reserves*, Research and Survey in Nature Conservation, 9, Nature Conservancy Council, Peterborough.

Peterken, G.F. and Jones, E.W. (1987) Forty years of change in Lady Park Wood: the old-growth stands. *Journal of Ecology*, **75**, 477–512.

Peterken, G.F. and Jones, E.W. (1989) Forty years of change in Lady Park Wood: the young-growth stands. *Journal of Ecology*, **77**, 401–29.

Pickett, S.T.A. and Thompson, J.N. (1978) Patch dynamics and the design of nature reserves. *Biological Conservation*, **13**, 27–37.

Pigott, C.D. (1985) Selective damage to tree-seedlings by bank voles (*Clethrionomys glareolus*). *Oecologia*, **67**, 367–71.

Poulson, T.L. and Platt, W.J. (1989) Gap light regimes influence canopy tree diversity. *Ecology*, **70**, 553–5.

Putz, F.E. (1983) Treefall pits and mounds, buried seeds, and the importance of soil disturbance to pioneer trees on Barro Colorado Islands, Panama. *Ecology*, **64**, 1069–74.

Rackham, O. (1975) *Hayley Wood: its history and ecology*, Cambridgeshire and Isle of Ely Naturalists Trust, Cambridge.

Rackham, O. (1980) *Ancient Woodland*, Edward Arnold, London.

Rackham, O. (1986) *The History of the Countryside*, Dent, London.

Runkle, J.R. (1982) Patterns of disturbance in some old-growth mesic forests of eastern North America. *Ecology*, **63**, 1533–46.

Runkle, J.R. (1985) Disturbance regimes in temperate forests, in *The Ecology of Natural Disturbance and Patch Dynamics* (eds S.T.A. Pickett and P.S. White), Academic Press, London, pp. 17–33.

Sauer, J. and Struik, G. (1964) A possible ecological relation between soil disturbance, light-flash and seed germination. *Ecology*, **45**, 884–6.

Scheiner, S.M. (1988) Population dynamics of an herbaceous perennial *Danthonia spicata* during secondary forest succession. *American Midland Naturalist*, **119**, 268–81.

Shure, D.J. and Phillips, D.L. (1987) Litter fall patterns within different-sized disturbance patches in a southern Appalachian mountain forest. *American Midland Naturalist*, **118**, 348–57.

Smith, W.K., Knapp, A.K. and Reiners, W.A. (1989) Penumbral effects on sunlight penetration in plant communities. *Ecology*, **70**, 1603–9.

Spies, T.A. and Franklin, J.F. (1989) Gap characteristics and vegetation response in coniferous forests of the Pacific Northwest. *Ecology*, **70**, 543–5.

Stephens, E.P. (1956) The uprooting of trees: a forest process. *Soil Science Society of American Proceedings*, **20**, 113–16.

Thompson, J.N. (1980) Treefalls and colonization patterns of temperate forest herbs. *American Midland Naturalist*, **104**, 176–84.

Thompson, J.N. (1985) Within-patch dynamics of life histories, populations and interactions: selection over time in small spaces, in *The Ecology of Natural Disturbance and Patch Dynamics* (eds S.T.A. Pickett and P.S. White), Academic Press, London, pp. 253–64.

Thompson, J.N. and Willson, M.F. (1978) Disturbance and the dispersal of fleshy fruits. *Science*, **200**, 1161–3.

Veblen, T.T. (1985) Stand dynamics in Chilean *Nothofagus* forests, in *The Ecology of Natural Disturbance and Patch Dynamics* (eds S.T.A. Pickett and P.S. White), Academic Press, London, pp. 35–51.

Vitousek, P.M. (1985) Community turnover and ecosystem nutrient dynamics. *The Ecology of Natural Disturbance and Patch Dynamics* (eds S.T.A. Pickett and P.S. White), Academic Press, London, pp. 325–33.

Watt, A.S. (1925) On the ecology of British beechwoods with special reference to their regeneration. Part II, Sections II and III. The development and structure of beech communities on the Sussex Downs. *Journal of Ecology*, **13**, 27–73.

Watt, A.S. (1947) Pattern and process in the plant community. *Journal of Ecology*, **35**, 1–22.

Whitbread, A. (1988) *Natural disturbance in the woodlands of south-east England: the October 1987 gale*, Nature Conservancy Council, Peterborough.

Whitbread, A. (1989) Nature's stormy cycle. *Natural World*, **26**, 29–30.

White, P.S. (1979) Pattern, process and natural disturbance in vegetation. *Botanical Reviews*, **45**, 229–99.

White, P.S. and Pickett, S.T.A. (1985) Natural disturbance and patch dynamics: an introduction, in *The Ecology of Natural Disturbance and Patch Dynamics* (eds S.T.A. Pickett and P.S. White), Academic Press, London, pp. 3–13.

Whitney, G.G. and Foster, D.R. (1988) Overstorey composition and age as determinants of the understorey flora of woods of central New England. *Journal of Ecology*, **76**, 867–76.

6

Accumulation of dead wood – a missing ingredient in coppicing?

K.J. KIRBY

6.1 INTRODUCTION

Forestry practices have caused major changes to the flora and fauna of our woods (Mitchell and Kirby, 1989). When the main objective of management was the production of wood and timber such changes were of little consequence to the forest manager. However, restoration of coppice is frequently undertaken for other reasons, including nature conservation. Therefore the effects of coppicing must be reassessed in terms of these new objectives.

Species of young-growth stages have benefited from the coppice system, but what about those at the opposite end of the trees' life cycle – the stages associated with dead and dying wood? Dead and dying wood is one of the two or three greatest resources for animal species in a natural forest. If fallen timber and slightly decayed trees are removed the whole system is gravely impoverished of perhaps a fifth of its fauna (Elton, 1966). In Wytham Woods, Oxfordshire, 18 of the 60 species of breeding birds depend wholly or partly on dead or dying trees for their nest sites. It seems likely that large rotting logs and stumps are an important resource for fungi also, but there have been few attempts to link the occurrence or abundance of fungi to the evaluation of sites for nature conservation and their subsequent management (Kirby, 1988).

6.1.1 Dead wood in natural forests

The former natural forests in Britain, by analogy with surviving near-natural stands elsewhere in the north temperate zone, were probably largely high forest with substantial amounts of dead wood (Elton, 1966; Falinski, 1986; Harmon *et al.*, 1986). The dead wood would have been present as dying or dead branches in the canopy, heart rot in the centres of trees developing up from the base and down from rot holes where branches had broken away from the trunk, stumps, dead standing trees or boles, and fallen trees and logs of a range of sizes.

Each type of dead wood provides habitat for a different range of organisms and species. Elton (1966) found *Hylesinus crenatus* (a bark beetle on ash (*Fraxinus excelsior*) in the rather thick bark of the trunk; *H. fraxini* was the main colonizer of fallen trunks and boughs on the tree, *H. orni* overlapped with the last but preferred smaller branches, while *H. oleiperda* went into the twigs. Standing dead trees and rot holes provide roost sites for birds and mammals, and dead branches in the canopy provide feeding sites for woodpeckers (Smith, 1987). Fallen trees and logs provide safe sites for tree seedling establishment (McKee *et al.*, 1982), distinctive microhabitats for bryophytes and lichens and sometimes higher plants (Falinski, 1986; Gustafsson and Hallingbäck, 1988; Söderström, 1988), and pathways for small mammals (Olszewski, 1968; Maser *et al.*, 1979). The fallen log physically suppresses plants beneath it and the subsequent decay locally alters soil nutrient conditions, so further influencing patterns in the ground flora (Falinski, 1986). Thus, changing the abundance or distribution of the dead-wood resource has far-reaching implications for the woodland ecosystem.

6.1.2 The change to coppice

No natural forests remain in Britain, but there are many fragments left of ancient semi-natural woodland (Kirby *et al.*, 1984; Walker and Kirby, 1989), some of which may have a continuity of woodland cover going back to the original natural forests (Peterken, 1977). In much of the country, particularly in the lowlands, these woods have at some time been managed as coppice or coppice with standards (Peterken, 1981; Rackham, 1990). Some of the forms in which dead wood formerly occurred were eliminated on many sites, while others were substantially reduced in abundance. Standing dead trees and fallen wood were removed as a valuable resource; any coppice poles that were left were of small diameter and so likely to decay quickly even if not removed from the site (Boddy and Swift, 1983; Harmon *et al.*, 1986). Most standards were harvested while young (Rackham, 1990) so that little dead branch wood was produced in the canopy. Pollards might be maintained on boundaries (and in adjacent hedgerows) but otherwise there were few trees which could develop a central rotting stem. Coppice stools could be centuries old and rotten in their heart, but these would be unlikely to substitute in amount or quality for other sources of dead wood that were lost.

Invertebrates associated with dead wood appear to have declined (Warren and Key, 1991) judging by their frequency in archaeological remains (Girling, 1982) and their limited distributions now (Crowson, 1962; Hammond, 1974; Hunter, 1977). They appear to be poor colonists compared with other groups (Wormell, 1977) and are considered to be vulnerable on a European scale (Speight, 1989). The loss of large diameter dead wood is particularly significant because different species may make use of large, rather than small logs, and also because a large log may provide a food resource for a much longer period. In a similar way, specialist bryophytes and beetles are declining as areas of primeval boreal forests are converted to managed stands (Heliövaara and Väisänen, 1984; Söderström, 1988). Some wood-rotting fungi may also have become rarer but I know of no studies to support this view.

Had all woods in Britain been managed regularly and intensively as coppice the flora and fauna associated with over-mature trees and dead wood would have been even more impoverished than it is. However, periods of neglect occurred when the coppice was not cut or trees might be cut and not harvested. Wood pasture systems encouraged the retention of old pollards (Harding and Rose, 1986) and in a few areas, patches of over-mature high forest survived – most notably in the native pinewoods of Scotland (Hunter, 1977).

6.1.3 Recent changes in coppice woods

Many ancient semi-natural woods that were once coppices have not been cut for 30–40 years and, in the north and west, not for 100 years or more (Peterken, 1981). Some have been singled and extraction of wood has continued; others, probably the majority, have been neglected so that dead wood has started to accumulate in them. Some do have a reasonably well-developed assemblage of dead-wood invertebrates, although they may lack the species most dependent on a long continuity of dead-wood habitats; Monks Wood in Cambridgeshire has been included on a list of sites important for specialist mature timber beetles.

There is now renewed interest in coppice management, particularly for nature conservation reasons. On some sites coppice restoration could eliminate the dead-wood accumulation of the last 50 years, but elsewhere, where there is no market for the coppice products, large amounts of material are left on site. Is either of these situations desirable? If dead wood is to be encouraged on coppice sites how should it be done? More information is needed on the amount of dead wood present in both neglected and worked coppice.

6.2 MEASUREMENT OF DEAD WOOD IN BRITISH SITES

Detailed measurements of amounts of dead wood have been made, e.g. in mixed coppice (Swift *et al.*, 1976) and in a neglected beech stand (Phillipson, 1983). Permanent transects showing forest structure may allow estimates of the turnover of at least the major trunks (Peterken and Backmeroff, 1988; Koop, 1989). However, these methods may not be appropriate for extensive surveys because they are time-consuming.

Measuring dead wood within sample quadrats is complicated by the need to measure length as well as diameter. Instead, a simpler method that may be used to assess one of the major elements – fallen logs – is described in the next section, with results from worked and neglected coppices.

6.2.1 Line intersect sampling

Dead logs, branches or trunks are counted if they are more than 5 cm in diameter where they cross the central line of a transect. Using this line intersect sampling method (Van Wagner, 1968; Warren and Olson, 1964), the total length (L) of fallen timber in area (A) is related to the number of intersections (N) with a transect line of length (l) by the formula:

$$L = \frac{\pi NA}{2l}$$

This simplifies to:

$$L = \frac{\pi N}{2l}$$

if the length per unit area is required.

In the surveys described below, the logs were assigned to 10 cm diameter classes (except that the lowest class was only 6–10 cm) and one of four decay classes where: 1 = recently fallen and still with bark on; 2 = bark loose, sometimes lost but wood still sound; 3 = rot starting; and 4 = well rotted. A more detailed system of decay classes may be useful in some instances (e.g. Maser and Trappe, 1984). Stand structure was also recorded along the transect to provide an indication of likely future inputs. All stems of trees and shrubs more than 1.5 m high within 2 m either side of the central line (i.e. within 100 m^2) were assigned by species to 10 cm diameter (breast height) classes. The lowest class was split into two sub-classes: 5 cm or less and 6–10 cm. Any snags, i.e. dead standing trees or stumps more than 1.5 m high, within 2 m of the central line were noted. It was found in practice that a minimum of ten transects from relatively homogeneous areas were required to provide a reasonable sample.

6.2.2 Trial surveys in coppice woods

Table 6.1 and Figure 6.1 summarize results from a trial of the method in three ages of coppice in Ham Street and Blean Woods National Nature Reserves, Kent, and in Lady Park Wood, Gwent.

Table 6.1 Line intersect sampling results from Blean and Ham Street woods (Kent) and Lady Park Wood (Gwent)

Site	*Blean*			*Ham Street*		*Lady Park*		
Grid reference	*TR 115 610*			*TR 010 340*		*SO 546145*		
Main tree species	*Oak* *Birch* *Chestnut* *Beech*			*Hornbeam* *Oak* *Birch*		*Beech* *Birch* *Oak* *Ash* *Lime*		
Stand age (years)	0–1	30	60–70	0–1	20	50–60	46	80–115
Area (ha)	2	3	2	1.2	1.6	1.0	–	–
Mean number of intersections per 25 m transect	0.5	0.4	1.8	0.6	0.8	3.0	4.8	5.8
(SE*)	(0.2)	(0.2)	(0.3)	(0.5)	(0.4)	(0.7)	(0.8)	(1.8)
Number of intersections in ten transects stem diameter (cm)								
6–10	5	3	14	4	8	23	38	35
11–20	–	1	4	2	–	6	9	19
20+	–	–	–		–	1	1	4
Estimated length of dead wood (m/ha)	314	251	1133	377	503	1882	3021	3650
Estimated volume of dead wood (m^3/ha)	1	4	16	7	2	31	36	46
Number of snags† in ten transects	0	3	24	0	3	8	10	6

*Standard error
†Dead standing stems >1.5 m high

The much greater levels of dead wood in the neglected coppice compartments compared with worked coppice are apparent. Little dead wood is left after cutting, whereas young stands that develop following natural catastrophes do so amidst high levels of dead wood (Tritton, 1980). Initially the coppice stems are too small to contribute much to the dead-wood pool, but after about 30 years there is a major reduction in stem numbers and a consequent rise in standing dead stems ('snags') and fallen dead wood. At Lady Park Wood (Peterken and Jones, 1987, 1989) there was a decline in stem numbers between *c.* 46 and *c.* 90 years' growth but relatively little difference in the total length of dead wood present. There was a greater difference in dead-wood volume because there are more large fallen logs in the old growth. Other factors such as tree species' growth rate, resistance of logs to decay, site type and catastrophic events (wind, disease) affect the amount of dead wood in a stand.

This type of survey could be extended to a wider range of sites and management conditions to provide a quantitative basis for discussions

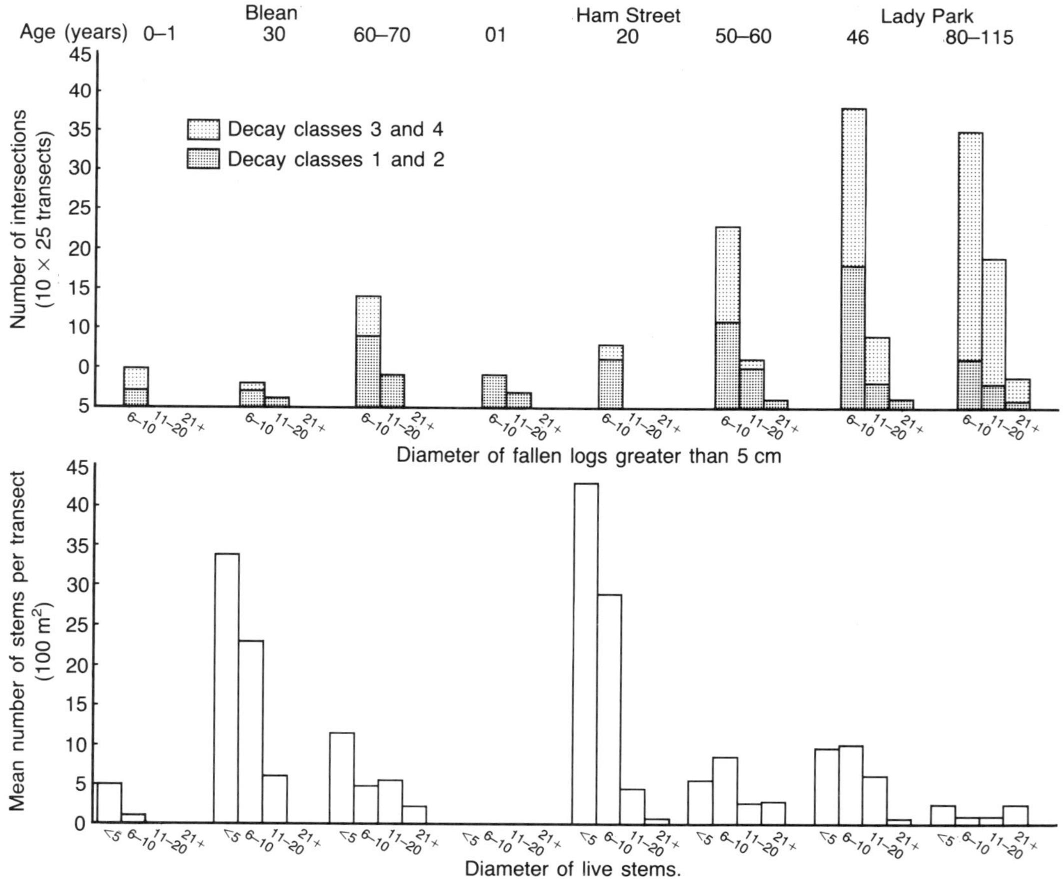

Figure 6.1 Live stems, dead wood frequency and decay class frequency in different stands, at Ham Street and Blean Woods, Kent, and Lady Park Wood, Gwent.

about how much dead wood should be retained. A more extensive survey based on this approach has been carried out by F. B. Goldsmith and M. King for the Wildlife Research Branch of the Forestry Commission.

6.3 MANAGEMENT OF DEAD WOOD

Sometimes the prime reason for coppice management on a site is to mimic as closely as possible 'traditional' practice – a form of industrial archaeology. For this, not only should all fallen wood be removed, rather than burnt on site, but, as suggested in detailed historical records for Blean Wood, perhaps even litter and small twigs should be swept up and taken away (D. Maylam, personal communication). Such practices would be likely to deplete soil nutrients over time (Rackham, 1980). Consequently nutrient levels may have increased this century because of more recycling and less export of nutrients in neglected coppice and high forest compared with worked coppice. This may have contributed to the apparent increase in nitrophilous species such as bramble (*Rubus fruticosus* agg.).

There are a few sites, however, where maintaining traditional practices is an end in itself. Here coppicing is usually carried out to achieve a set of conditions that will benefit particular species – mostly those of the open phase in a rotation. Modifying traditional practice is acceptable provided it does not compromise the survival of open-phase species, and is desirable if it increases the scope for other groups of species to survive or colonize the site, or makes part of the woodland more natural.

Several different strategies to increase the dead wood are possible: setting aside whole stands; leaving individual trees to senesce and rot on site; maintaining and encouraging pollards and old stools and deliberately damaging a few trees to encourage rotting; leaving some of the coppice produce spread out to rot, and creating dead-wood piles.

6.3.1 Minimal-intervention stands

Stands and whole woods may be designated as minimal-intervention areas so that natural ecological processes can be followed. Such areas need to be as large as possible to reduce edge effects and encompass the full range of gap-dynamic processes (Pickett and Thompson, 1978). There are other roles, however, that minimal-intervention areas can fulfil – providing zones where the soil will not be compacted or disturbed by timber extraction machinery (e.g. Moehring, 1970) and for the accumulation of dead wood in various forms. The latter can be met even by single compartments in a coppice, as the results from the Blean Wood and Ham Street Wood nature reserves illustrate. In such compartments, singling stems to promote a high forest structure and more rapid development of large-diameter trees could be considered when the area is first set aside. The cut logs should be left to rot in the stand to provide some dead wood earlier than would occur through natural tree death.

6.3.2 Retaining individual trees to over-maturity

In minimal-intervention compartments the open-phase species characteristic of coppice are likely to be lost. Individual trees can, however, be retained to over-maturity (in high forest 4–5/ha is recommended (Mitchell and Kirby, 1989) but fewer could be retained in coppice) to provide dead wood in the canopy and eventually rotting trunks, while the bulk of the wood is coppiced. Since the adults of dead-wood beetles and flies may require flowers as nectar sources, the presence of open areas (cut coppice and rides) near to the dead-wood source may be advantageous. The large crowns of over-mature trees shade out the coppice regrowth, however, so it is better to locate such trees as groups in the corner of coppice blocks rather than scattered through the coupe. This also ensures that the trunks are at least partly shaded, reducing the risk of heat sterilization of wood. Fast-growing species such as birch (*Betula* spp.) and aspen (*Populus tremula*) may be left to increase the dead wood in the short term, but trees such as oak (*Quercus* spp.), beech (*Fagus sylvatica*) or lime (*Tilia cordata*) that grow to larger dimensions before dying are also needed.

Both minimal-intervention stands and groups of over-mature trees could pose problems for the site manager or owner because they may be seen as a safety risk. They should be located in the least accessible parts of the wood.

6.3.3 Pollards, old stools and increasing dead-wood formation

Regularly pollarded trees (as opposed to the long-neglected pollards common in wood pasture) do not provide dead branchwood in the canopy, but can provide a greater potential volume of rotten heartwood than most old coppice stools. Their branches (poles) do not cast much shade over the coppice beneath and they are less of a safety hazard than very old maiden trees. Thus, a greater number of pollards than maidens can be incorporated within a worked coppice wood. Pollards were often located on external boundaries where their regrowth would be out of reach of cattle in neighbouring fields, but now there may be more value in establishing new pollards on internal edges. Good nectar sources for adult dead-wood species are more likely to be found within the wood because surrounding old meadows or hedgerows have now been replaced by fields of arable or improved grass. New pollards are easily created by cutting stems 15–30 cm in diameter and should include a range of species. Large, overgrown pollards are less of a feature of coppice than of wood pasture, but where they occur there may be benefits in trying to re-pollard them to reduce risks both to pollards and people, although the success of this procedure can be uncertain (Mitchell, 1989).

Old coppice stools, particularly high-cut ones, should receive careful attention during cutting. If necessary, the younger growth around them should be thinned to reduce competition with the regrowth from the old stool.

Where trees are to be left to provide dead wood the commercial value of the wood in their trunks is unimportant. Therefore there may be some value in deliberately damaging them to speed up rot development (Warren and Key, 1991). Fortunately most species that feed on dead wood do not attack whole maiden stems so this is not a threat to timber production on most broadleaf sites. However, care will be needed to ensure that this does not create an unsafe tree.

6.3.4 Using the cut coppice stems as a source of dead wood

Where labour is very limited or only a small part of the crop is saleable the cut coppice may be left spread over the site. The disadvantages of this are that the material is all small and rots at about the same rate, so providing little long-term continuity of dead wood. The effects on the ground flora through direct suppression and a flush of nutrient enrichment are widespread over the coupe; bramble may scramble over the cut material, adding to the general difficulty of moving through the site and possibly increasing the suppression of other ground flora species (Mason and Long, 1987). At the least, twiggy material should be collected up and burnt, even if the main stems are left spread out.

Building piles of coppice material is a common strategy. Much effort may be involved in creating these and, after a series of cuts, if most of the wood has been put into piles the area which they occupy can be extensive. The piles are frequently left exposed to sun and heat and, because of their height, may take a long time to be sheltered by vegetation. This greatly reduces their value for dead-wood invertebrates. Long thin piles would be a better mimic of natural dead wood than the common cord stack. If log piles are used the manager should allocate a fixed area for them beyond which timber will be removed or burnt. Priority should be given to retaining wood that is already dead or shows some sign of decay. By analogy with the area under dead wood in natural forests it is suggested that a maximum of 5–10% of the total coppice area should be allowed for existing and future log piles and all fire sites. This is based on the projected surface area of logs for a series of undisturbed hardwood stands (Falinski, 1986; Harmon *et al.*, 1986). A minimum decay time of ten years for log piles is likely and for many 20–30 years (i.e. up to two coppice rotations) may be more realistic.

6.3.5 Limits to what we can expect from an increase in dead wood in coppices

Sites which are currently valued for their dead-wood fauna and flora have generally had a long continuity of abundant dead wood. Some specialist dead-wood species appear to have quite specific requirements as to the size, location or state of decay of the dead wood that they use and are very poor colonizers. Such species must be conserved where they occur now (often wood pasture sites, e.g. Windsor Great Park, Figure 6.2) and are unlikely to benefit from increased dead wood in coppices where it was previously lacking. It is the commoner dead-wood species that are likely to benefit from an increase in dead wood, just as it is the common open-phase species that come in when overgrown rides are widened.

In the long term it may be possible to consider judicious introduction of dead-wood species, just as there have been a few introductions of butterflies to woodland. At present our knowledge is too incomplete to consider this; we need to know more about the requirements of such species and our ability to meet them.

6.4 CONCLUSIONS

Past management has led to major reductions in the dead-wood resource of coppice woods compared with natural forests with consequent reductions in associated species. Where coppicing is re-introduced for nature conservation, encouraging more dead wood and maintaining some of that which has accumulated during periods of neglect should be considered – although this is a secondary objective to promoting the open-phase species.

'Habitat' piles created by stocking surplus poles, provide one type of dead-wood environment but their location and extent should be limited. Surplus sound timber should be removed or burnt. A wider range of dead-wood conditions can be provided by designating minimal-intervention areas, pollarding trees and retaining some trees until their natural death (Figure 6.3).

However, there is a need for much more research into the amount of dead wood that accumulates under different regimes of management and neglect and the particular species associated with such dead wood in all its various shapes and sizes.

Figure 6.2 Decaying timber in Windsor Great Park, Berkshire. (Reproduced with permission from the Nature Conservancy Council.)

ACKNOWLEDGEMENTS

This work is part of the Nature Conservancy Council's research programme. Stefa Birkenhead, Damian Hughes, Roger Key, Mundi Smith and Rachel Thomas helped with the data recording and the preparation of this chapter.

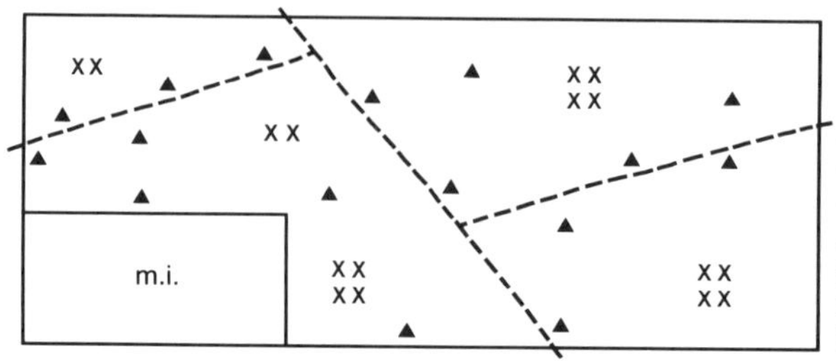

Figure 6.3 A hypothetical dead-wood strategy map for a coppice.

REFERENCES

Boddy, L. and Swift, M.J. (1983) Wood decomposition in an abandoned beech and oak coppiced woodland in south-east England: patterns of wood-litter fall. *Holarctic Ecology*, **6**, 320–32.

Crowson, R.A. (1962) Observations on the Coleoptera in Scottish oak woods. *The Glasgow Naturalist*, **18**, 177–95.

Elton, C.S. (1966) *The Pattern of Animal Communities*, Chapman and Hall, London.

Falinski, J.B. (1986) *Vegetation Dynamics in Temperate Lowland Primeval Forests*, W. Junk, Dordrecht.

Girling, M.A. (1982) Fossil insect faunas from forest sites, in *Archaeological Aspects of Woodland Ecology* (eds M. Bell and S. Limbrey), British Archaeological Report 146, Association for Environmental Archaeology, Oxford, pp. 129–46.

Gustafsson, L. and Hallingbäck, T. (1988) Bryophyte flora and vegetation of managed and virgin coniferous forests in south-west Sweden. *Biological Conservation*, **44**, 283–300.

Hammond, P.M. (1974) Changes in the British coleopterous fauna, in *The Changing Flora and Fauna of Britain* (Ed. D.L. Hawksworth), Academic Press, London, pp. 323–69.

Harding, P.J. and Rose, F. (1986) *Pasture-woodlands in Lowland Britain. A review of their importance for wildlife conservation*, Institute of Terrestrial Ecology, Abbots Ripton.

Harmon, M.E., Franklin, J.F., Swanson, F.J. *et al.* (1986) Ecology of coarse woody debris in temperate ecosystems. *Advances in Ecological Research*, **15**, 133–302.

Hunter, F.A. (1977) The ecology of pinewood beetles. In *The Native Pinewoods of Scotland* (eds R.G.H. Bunce and J.N.R. Jeffers), Institute of Terrestrial Ecology, Cambridge, pp. 42–55.

Heliövaara, K. and Väisänen, R. (1984) Effects of modern forestry on north-western European forest invertebrates: a synthesis. *Acta Forestalia Fennica*, **189**.

Kirby, K.J. (1988) The conservation of fungi in Britain. *Mycologist*, **2**, 5–7.

Kirby, K.J., Peterken, G.F., Spencer, J.W. and Walker, G.J. (1984) *Inventories of ancient semi-natural woodland*, Focus on nature conservation, 6, Nature Conservancy Council, Peterborough.

Koop, H. (1989) *Forest Dynamics*, Springer-Verlag, Berlin.

Maser, C., Anderson, R., Cromack, K. Jr, Williams, J.T. and Martin, R.E. (1979) Dead and downed woody material, in *Wildlife Habitats in Managed Forests, the Blue Mountains of Oregon and Washington* (Ed. J. Thomas), USDA Forest Service Agriculture Handbook 553, United States Department of Agriculture, Washington DC.

Maser, C. and Trappe, J.M. (1984) *The Seen and Unseen World of the Fallen Tree*, USDA General Technical Report PNW–164, United States Department of Agriculture, Washington DC.

Mason, C. and Long, S. (1987) Management of lowland broadleaved woodland, Bovingdon Hall, Essex, in *Conservation Monitoring and Management* (Ed. R. Matthews), Countryside Commission, Cheltenham, pp. 37–42.

McKee, A., LeRoi, G. and Franklin, J.F. (1982) Structure, composition and reproductive behavior of terrace forests, South Fork Hoh River, Olympic National Park, in *Ecological Research in National Parks of the Pacific North-west* (eds E.E. Starkey, J.F. Franklin and J.W. Matthews), National Park Service Cooperative Studies Unit, Corvallis, Oregon, pp. 22–9.

Mitchell, P.L. (1989) Repollarding large neglected pollards: a review of current practice and results. *Arboricultural Journal*, **13**, 125–42.

Mitchell, P.L. and Kirby, K.J. (1989). *Ecological Effects of Forestry Practices in Long-established Woodland and their Implications for Nature Conservation.* Oxford Forestry Institute Occasional Paper, 39, Department of Plant Sciences, University of Oxford.

Moehring, D.H. (1970) Forest soil improvement through cultivation. *Journal of Forestry*, **68**, 328–331.

Olszewski, J.L. (1968) The role of uprooted trees in the movements of rodents in forests. *Oikos*, **19**, 99–104.

Peterken, G.F. (1977) Habitat conservation priorities in British and European woodland. *Biological Conservation*, **11**, 223–36.

Peterken, G.F. (1981) *Woodland Conservation and Management*, Chapman and Hall, London.

Peterken, G.F. and Backmeroff, C. (1988) *Long-term monitoring in unmanaged woodland nature reserves*, Research and Survey in Nature Conservation **9** Nature Conservancy Council, Peterborough.

Peterken, G.F. and Jones, E.W. (1987) Forty years of change in Lady Park Wood: the old-growth stands. *Journal of Ecology*, **75**, 477–512.

Peterken, G.F. and Jones, E.W. (1989) Forty years of change in Lady Park Wood: the young-growth stands. *Journal of Ecology*, **77**, 401–29.

Phillipson, J. (1983) Wood biomass, production and litter fall in an English Beech wood. *Oecologia*, **57**, 189–95.

Pickett, S.T.A. and Thompson, J.N. (1978) Patch dynamics and the design of nature reserves. *Biological Conservation*, **13**, 27–37.

Rackham, O. (1980) *Ancient Woodland*, Edward Arnold, London.

Rackham, O. (1990) *Trees and Woodlands in the British Landscape*, 2nd edn, Dent, London.

Smith, K.W. (1987) The ecology of the great spotted woodpecker. *RSPB Conservation Review*, **1**, 74–7.

Söderström, L. (1988) Sequence of bryophytes and lichens in relation to substrate variables of decaying coniferous wood in northern Sweden. *Nordic Journal of Botany*, **8**, 89–97.

Speight, M.C.D. (1989) *Saproxylic Invertebrates and their Conservation*, Nature and Environment Series 42, Council of Europe, Gland, Switzerland.

Swift, M.J., Healey, I.N., Hibberd, J.K. *et al.* (1976) The decomposition of branch wood in the canopy and floor of a mixed deciduous woodland. *Oecologia*, **26**, 139–49.

Tritton, L.M. (1980) *Dead wood in the northern hardwood forest ecosystem*, PhD thesis, Yale University, Connecticut, USA.

Van Wagner, C.E. (1968) The line-intersect method in forest fuel sampling. *Forest Science*, **14**, 20–6.

Walker, G.J. and Kirby, K.J. (1989) *Inventories of ancient, long-established and semi-natural woodland in Scotland*, Research and Survey in Nature Conservation, 22, Nature Conservancy Council, Peterborough.

Warren, M.S. and Key, R.S. (1991) Woodlands: past, present and future potential for insects, in *The Conservation of Insects and their Habitats*, Royal Entomological Society Symposium, 15 (in press).

Warren, W.G. and Olsen, P.E. (1964) A line intersect technique for assessing logging waste. *Forest Science*, **10**, 267–76.

Wormell, P. (1977) Woodland insect population changes on the Isle of Rhum in relation to forest history and woodland restoration. *Scottish Forestry*, **31**, 13–36.

Part Four

EFFECTS OF COPPICING ON THE GROUND FLORA

– 7

The effects of coppicing and neglect on the performance of the perennial ground flora

J.P. BARKHAM

7.1 INTRODUCTION

Disturbance is important in determining the structure of biological communities and in the maintenance of species diversity (Lamb, 1989). The members of many communities are clearly space limited and, in the absence of external disturbance, such communities may move towards a homogeneous association of the competitively dominant species (Levin and Paine, 1974) and towards a lower species density. The latter is also characteristic of areas of frequent major disturbance, whereas the 'intermediate disturbance hypothesis' (Connell, 1978) suggests that the highest species density will occur between the two extremes. Conservationists generally like high species diversity. A number of woodland studies indicate that plant species diversity is related to the type and intensity of disturbance (Metzger and Schultz, 1981).

Woodland management involves disturbance. Conservation managers of ancient woodland in lowland Britain are usually confronted with a choice from three options:

1. non-intervention or neglect
2. high forest
3. coppicing

In terms of relative frequency of disturbance, the first is low and the last high. Does coppicing result in a scale and frequency of disturbance that maximizes or minimizes plant species density?

Coppicing is a very variable process – varying according to both locality and time. Under a coppice rotation there is a regular cycle of cutting and clearing which was traditionally 5–8 years long in medieval times (Rackham, 1976), but which has been as long as 20 years in more recent times as markets for coppice products have declined. As an approach to conservation management coppicing is an attempt to re-establish a regime similar to that practised in most lowland woods from the middle of the 13th century to the 20th century, although the system was already in decline in the early 19th century (Salisbury, 1924).

Under a non-intervention system, management would hope to intervene to prevent or at least minimize human influence, thereby allowing, albeit on a very small scale, natural processes of woodland development.

The ground flora of woodland is adapted in one way or another to the dynamics of the tree and shrub canopy overhead (Chapter 5). The most obvious groupings are (a) the shade-avoiding annuals, biennials and short-lived perennials that are commonly represented in the seed bank (Brown and Oosterhuis, 1981) and (b) species that are tolerant of varying degrees of canopy shade. Although a few of the latter are annuals, it is the shade-tolerant perennials that are considered here. They are most important from a nature conservation point of view because they include showy, popular plants such as bluebell (*Hyacinthoides non-scripta*), wood anemone (*Anemone nemorosa*), and primrose (*Primula vulgaris*) and others such as dog's mercury (*Mercurialis perennis*), ivy (*Hedera helix*) and bramble (*Rubus fruticosus* sens. lat.) which, although lacking aesthetic appeal, are abundant species capable of covering large areas of ground. This group also includes a number of species which are rare or local and which are characteristic only of ancient woodlands (Peterken, 1974).

Because the environment experienced by the shade-tolerant ground flora is different under regimes of coppicing and non-intervention, it is reasonable to expect the plants to respond in different ways. This chapter explores the possible consequences of the two management strategies. Is the ground flora impoverished by neglect? Is its diversity sustained or reduced by coppicing?

A speculative summary of the environmental characteristics of ancient woodland under the two different management regimes considered here is given in Table 7.1. The likely consequences of these characteristics for the shade-tolerant woodland flora are given in Table 7.2.

7.2 METHODS

Data were drawn from a number of studies carried out in Brigsteer Park Wood, near Kendal, Cumbria, England (Barkham, 1980a). The wood lies on a west-facing carboniferous limestone escarpment, with pockets of acid glacial drift soils overlying limestone pavement in more level areas. Its history as woodland is probably continuous, going back to prehistoric times. It was traditionally managed as a coppice. Part of the wood was cleared in the 1960s and replanted with a mixture of coniferous and deciduous tree species. The remainder has not been managed for more

Table 7.1 Environmental characteristics of ancient woodland resulting from coppice and non-intervention forms of management

Characteristic	*Coppice*	*Non-intervention*
Light/shade	Regular phases; frequency depending on length of rotation; length of high light phase depends on coppice regrowth rate and density of coppice stools (Chapter 3); increased spring light 2-fold, summer light 20-fold in high light phase (Rackham, 1975)	Irregular and relatively infrequent high light phase; length of high light phase dependent on size of gap and on rate of canopy closure (Runkle, 1985); in the absence of rapid regrowth of woody species, ground layer competition important
Soil/air temperature	Earlier heating of soil in spring during high light phase (Salisbury, 1924); higher soil surface temperatures in summer (Ash and Barkham, 1976); plants in sun 6–9°C warmer than in shade (Rackham, 1980)	Slow heating in spring; slow cooling at the end of summer (Moore and Vankat, 1986); fluctuations damped until natural gap creation (Geiger, 1965; Veblen, 1985)
Soil disturbance	Trampling and compaction: if coppicing carried out in wet weather, significant damage to surface (top 5 cm) of soil and rhizomes and roots of perennials; likely to be greatest around stools, stacking areas and bonfire sites (Ash and Barkham, 1976; Cummings, 1982); effect on mole activity unknown. Fewer or no tip-ups caused by windthrow	Tip-ups due to windthrow; alters drainage characteristics of soil and creates hummocks and hollows (Chapter 5; Lamb, 1989); increasing amounts of dead wood on the surface (Falinski, 1986). Moles active in surface layers of all suitable soils, causing mixing of horizons and frequent disruption of root and rhizome systems (Rackham, 1980). Little trampling and compaction
Soil water	May depend on soil type; in young coppice, more soil water throughout the growing season in all years, less in old coppice (Chapter 4), but Rackham (1980) comes to opposite conclusion; old coppice presumed similar to non-intervention, but may be less soil water	Less available soil water, unless and until a natural canopy gap is created (Moore and Vankat, 1986; Minckler and Woerheide, 1965) compared with young coppice (but compare Rackham, 1980)
Soil nutrients	Depletion of phosphorus (Rackham, 1975) and, presumably, other nutrients due to long period of cropping (Peterken, 1981); pulse release of nutrients,	Different effects of different tree species on availability of phosphorus (Harrison, 1970); build-up of organic matter at soil surface and in soil; steady nutrient

Table 7.1 *Continued*

Characteristic	*Coppice*	*Non-intervention*
Soil nutrients (*contd.*)	high in years immediately following cutting, low in canopy regrowth phase with low organic matter input and cooler summer temperatures	release rate depends on soil type; differences between pits and mounds; (see Table 5.2), other gap effects poorly known (Vitousek, 1985).
Leaf litter	Rapid disappearance during open phase; rising rate of accumulation during shade phase; maximum wind-sorting and redistribution during open phase; mean annual litter production probably relatively low	Build-up as canopy develops in balance with decomposition rate in all sites except waterlogged or very acid ones where steady accumulation; minimal wind sorting; mean annual litter production relatively high; gaps same as open phase in coppice (Shure and Phillips, 1987)
Mammals		
deer	Disturbed, then encouraged by availability of browse and dense cover of regrowth (Rackham, 1975)	Least attractive
rabbits	Disturbed, then encouraged by availability of browse and dense cover of regrowth (Rackham, 1975)	Ubiquitous
mice and voles	Disturbed, then encouraged by availability of browse and dense cover of regrowth (Rackham, 1975), then ubiquitous (Chapter 11)	Ubiquitous (Falinski, 1986)
moles	On suitable soils, less active immediately after coppicing due to compaction, then ubiquitous	Ubiquitous on all suitable soils (Mellanby, 1971)
Birds		
pheasants	Consume herbivorous invertebrates in open phase	Unimportant
perching fructivores	Unimportant in open phase; may increase in importance as regrowth develops	Defecation of viable seed particularly bramble (*Rubus fruticosus*) and holly (*Ilex aquifolium*; more fruit taken from gaps (Thompson and Willson, 1978)
Invertebrates	Few herbivores of ground-layer plants; those of tree and shrub species may be significant in shade phase	Same as shade phase under coppice; canopy gaps cause change in litter patterns which may affect detritivore populations (Shure and Phillips, 1987)

Table 7.2 Perennial herbaceous plant response to coppice and non-intervention environments

Characteristic	*Coppice*	*Non-intervention*
Light/shade	Light phase: regular flowering and seed production; increased growth rates of those species able to respond positively to high light; diminished assimilation rate of some species (Salisbury, 1924); increased growth of light-demanding competing species with higher relative growth rates. Shade phase: flowering and seed production irregular and infrequent; disappearance of shade-intolerant species (Rackham, 1980); seedlings susceptible to fungal attack	Gradual increase of competitive dominants of the shade phase at the expense of lesser competitors (Brewer, 1980), flowering and seed production relatively infrequent though always present in high light phase of early spring; resources put into clonal growth; occasional treefall gaps favour species already present in them; small ground flora gaps favour clonally-growing species adjacent to them (Moore and Vankat, 1986; Ehrenfeld, 1980)
Soil/air temperature	Light phase: increased growth rates of roots and rhizomes; germination of seeds of competing heliophytes; earlier growth in warm springs; enhanced pollination, fruit production and dispersal (Thompson and Willson, 1978; Collins *et al*., 1985)	Favours early growth in cold late springs; later growth than open ground in warm early springs
Soil disturbance	Damage to roots and below-ground perennating parts, particularly of near-surface rooted species (Ash and Barkham, 1976); spatially non-uniform damage leads to gap creation in most damaged areas; differences between soil types in terms of liability to damage and rapidity of recovery	Tip-ups expose roots, bulbs and rhizomes of perennials, seeds of seedbank species; species redistribution and recolonization of pits and mounds (Beatty, 1984; Thompson, 1980); effect of moles and other animals little known (see Table 5.1; Falinski, 1978; Cook and Lyons, 1983; Lamb, 1989; Whitbread, 1989)
Soil water	Favours rapid growth rates in light phase; favours species with relatively high relative growth rates (RGR); in shade phase same or more severe conditions than in non-intervention system (Chapter 4)	Favours slow growth rates; relatively high probability of summer drought in surface soil layers affecting growth, maintenance, and interspecific competition (Moore and Vankat, 1986)
Soil nutrients	Favours stress-tolerant species with low relative growth rates (these species are unable to use temporary nutrient flushes as effectively as high RGR species,	Favours shade-tolerant competitors, i.e. those species in guild of low RGR plants that have a relatively high RGR in low light conditions

Table 7.2 *Continued*

Characteristic	*Coppice*	*Non-intervention*
Soil nutrients *contd.*	and therefore suffer in competition with them if the light phase is prolonged)	
Leaf litter	Lack of accumulation at surface favours smaller, slow-growing species with minimal tolerance of overlying litter deposition, particularly grasses (Sydes and Grime, 1981a)	Favours species with physiognomic adaptations for piercing thick layers of overlying litter, particularly herbs (Sydes and Grime, 1981b); accumulating dead wood favours particular species (Falinski, 1978; Bratton, 1976; Thompson, 1980; Handel, 1978), while smothering others
Mammals		
deer	Reduce vigour of bramble (*Rubus fruticosus*) in high light and scrub phase of regrowth; browse ground flora (Rackham, 1975)	As for coppice, but less significant as populations lower
rabbits	Unpalatable species favoured in competition with palatable; where abundant, ground vegetation consists largely of less palatable species	As for coppice
mice and voles	As for rabbits, but less significant	As for coppice (Falinski, 1986)
moles	Less surface soil disturbance after coppicing because of compaction, results in consolidation of root and rhizome systems; presumably affects competitive interactions of plant species	Continuous soil disturbance affects root and rhizome systems (Barkham, 1980b; Mellanby, 1971) presumably affects competitive interactions of ground-layer species
Birds	Probably minimal effects except localized damage by pheasants kept at high densities (Ludolf *et al.*, 1988)	Promote growth of woody species in competition with ground-layer herbs
Invertebrates	Ground-layer herbs seem to have few invertebrate herbivores; competition from develop ing tree and shrub seedlings and saplings probably much reduced in shade phase by invertebrate herbivores	Very little known

than 30 years (non-intervention sites), apart from one compartment last coppiced in the traditional way in 1972 (coppiced site).

7.2.1 Permanent plots

Fifty permanent plots, 2 m × 2 m in size, were established in 1969 to monitor changes in the population size of the wild daffodil (*Narcissus pseudonarcissus*) (Barkham, 1980b). In 1969 and 1987 the percentage cover of all other species present in these quadrats was visually estimated (Barkham, 1992, in preparation). In this chapter, data from three groups of plots in non-intervention areas are examined in relation to those from the coppiced site. Overall changes in the vegetation are described in terms of the change in position of the plots on a graph with the first two axes derived from DECORANA (Hill, 1979). This was achieved by carrying out a series of DECORANA analyses, adding 1987 data for each plot, one plot at a time, to the complete set of 1969 data. By doing this, the overall distribution of 1969 points was minimally affected by the addition of 1987 data. Losses and gains of species are identified and particular attention is given to changes in the cover of the most abundant ground flora species, which includes *N. pseudonarcissus* – changes in the shoot numbers of which were detailed for most years from 1969 to 1989.

7.2.2 Demography of N. pseudonarcissus

For a detailed description of the demography of *N. pseudonarcissus* see Barkham (1989a). Here, some of the results of a subsequent ten-year experiment to investigate the significance of 'gaps' are described (Barkham, 1992b, in preparation). The clumped distribution of ground layer plants of ancient coppice woods is a familiar feature. In the case of *N. pseudonarcissus*, dense pure clumps are often separated from one another by spaces (gaps) unutilized by other species. To investigate the significance of these gaps in relation to the way in which the ground layer vegetation develops under coppice and non-intervention systems, mature bulbs, seedlings and seeds were planted into a series of such gaps, replicated, and their performance compared with controls.

Twenty years is a short time over which to investigate the consequences of environmental change for the ground flora of woodland. Therefore demographic data for *N. pseudonarcissus*, collected in the early 1970s (Barkham, 1980a), were used to produce a simulation model in order to predict population changes over several hundred years (Barkham and Hance, 1982).

Figure 7.1 The effects of coppicing: this time sequence of photographs was taken from the same position during April of each year in Brigsteer Wood, Cumbria, and shows the changes which take place in vegetation through the coppice cycle. (a) 1974: one growing season after coppicing, spring ephemerals flower prolifically (*Narcissus pseudonarcissus* and *Anemone nemorosa*) and there are early signs of bramble growth. (b) 1977: bramble (*Rubus fruticosus*) and raspberry (*R. idaeus*) begin to form a dense thicket before the re-growing coppice shoots form a shading canopy. At this stage, the ground layer herbs are experiencing a fully shaded

environment once again. (c) 1981: re-establishment of a full coppice canopy is complete and most of the bramble had died back. (d) 1990: this happened to be the most spectacular flowering season for *N. pseudonarcissus* in 20 years, and growth was about three weeks in advance of normal after an exceptionally warm early spring. Apart from differences caused by these factors, there has been little change over the previous six years. (Photographs: J.P. Barkham.)

7.2.3 Other long-term studies

There are few published long-term studies of ground flora changes in managed or unmanaged woodlands, and none for Britain (Peterken and Backmeroff, 1988; Lamb, 1989). Data from other countries are referred to here in support of the argument presented (Hough, 1965; Brewer, 1980; Metzger and Schultz, 1981; Davison and Forman, 1982; Inghe and Tamm, 1985).

7.3 RESULTS

7.3.1 Vegetation development of coppice and non-intervention sites

Overall changes

Figure 7.1 illustrates some of the vegetation changes that take place through a coppice cycle. The changes over the 18-year period in the vegetation of each of 25 permanent plots are represented in Figure 7.2. Each site is represented by a pair of points, one for each of the two sampling dates (1969 and 1987). The amount of change has been small in most sites. For most plots it has been 'centrifugal', i.e. the plots have moved away from the centre, indicating that the vegetation as a whole has become less uniform, although the vegetation within plot-groups in most cases has remained similar. For both the coppice and oak high forest sites the points have drawn closer together, as have the two groups to each other, indicating the increased similarity of the ground flora of the plots in each of those sites. Change is not obviously greater or lesser in the non-intervention sites than in the coppice site. In the former, individual plots show changes ranging from almost none to very large.

The reasons for the large changes in plots 1–4 are as follows:

1. increases in *Anemone nemorosa* and *Hyacinthoides non-scripta* from (19% to 30% and from 21% to 56% respectively);
2. increase in *Hyacinthoides non-scripta* (from 28% to 64%) and decrease in *Mercurialis perennis* and *Rubus fruticosus* (from 10% to 2% and from 50% to 15% respectively);
3. increase in *Hedera helix* (from zero to 39%) and decrease in *Rubus fruticosus* (from 49% to 6%);
4. increase in *Ctenidium molluscum* (from 8% to 36%) and decrease in *Eurhynchium striatum* (from 53% to 8%), little change in cover of herbs.

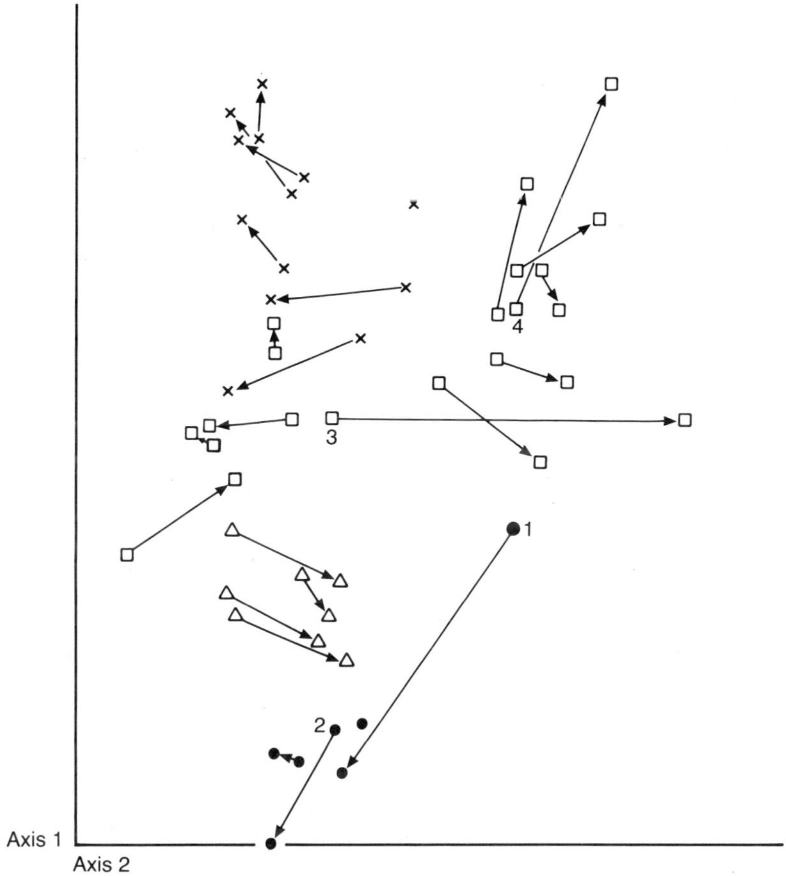

Figure 7.2 Vegetation changes in plots between 1969 and 1987 in coppiced and undisturbed sites. The positions the plots were determined for each of the two years on the first two axes of detrended correspondence analysis (DECORANA). Arrows link each pair of positions. Longer arrows indicate greater change, shorter ones less. Where there is an isolated plot, there was no change. •, Coppice sites; undisturbed sites; □, old coppice on limestone pavement or rubble; x, yew wood with some old coppice on scree; △, oak high forest with scattered hazel.

Changes in species number and cover

The mean number of vascular plant species in the permanent plots declined in each of the four sites but significantly so only in the coppice site (Table 7.3). The losses were largely of species occurring only in very small amounts (1% cover) in 1969. Bryophytes were not included in this analysis because of the possibility, on both sampling occasions, of failing to observe particular species present in very small quantities.

Table 7.3 The mean number of vascular plant species per quadrat for three undisturbed old-coppice sites and for one actively-coppiced site at Brigsteer Wood, Cumbria, in 1969 and 1987

	Mean number of vascular plant species		*Significance of difference*
	1969	*1987*	
Coppice site	10.0	8.0	$p<0.05$
Undisturbed sites			
old coppice on limestone pavement or rubble	10.6	10.3	–
yew wood with some old coppice on scree	6.8	6.2	–
oak high forest with scattered hazel	7.5	6.8	–

This picture is amplified in Figure 7.3 which shows species losses and gains for the coppice and undisturbed sites. Most losses and gains were of tree and shrub seedlings. These, especially sycamore (*Acer pseudoplatanus*) and ash (*Fraxinus excelsior*), germinate freely but survive for only a few years in full canopy shade. There were no gains for this group in the coppice site (the recording times being at the closed-canopy stage). Herb gains and losses, though variable, were broadly similar in the two sorts of site. Edge and rideside species were absent from the coppice plots and oak

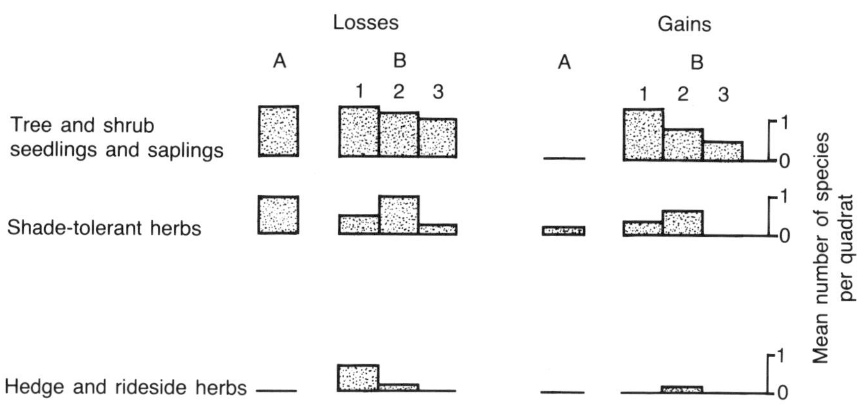

Figure 7.3 The mean number of species lost from permanent plots between 1969 and 1987 and the number gained by them. A, coppice site; B, undisturbed sites: 1, old coppice on limestone pavement or rubble; 2, yew wood with some old coppice on scree; 3, oak high forest with scattered hazel.

high forest showed losses and no gains on limestone pavement, and were of negligible importance under yew.

Significant changes in percentage vegetation cover in all sites (Table 7.4) were in contrast to the few significant species' gains and losses (Table 7.3). However, in the coppice site the only significant change was a decrease in bryophyte cover although there was a corresponding (but not significant) increase in vascular plant cover. In the undisturbed sites there were decreases in cover in old coppice on limestone pavement, but increases under yew and oak.

Table 7.4 Changes in mean percentage cover between 1969 and 1987 of the total ground vegetation and its vascular plant and bryophyte components in three undisturbed old coppice sites and one actively coppiced site at Brigsteer Wood, Cumbria (0, no change; +, increase; −, decrease)

		Mean percentage cover 1969	1987	*Significance of difference*	
Coppice site	total	150	144		0
	vascular plants	105	110		0
	bryophytes	45	34	$p<0.01$	−
Undisturbed sites	total	129	112	$p<0.01$	−
old coppice on limestone pavement rubble	vascular plants	72	63	$p<0.01$	−
	bryophytes	57	49	$p<0.01$	−
yew wood with some old coppice on scree	total	36	47	$p<0.01$	+
	vascular plants	12	17	$p<0.05$	+
	bryophytes	24	30	$p<0.05$	+
oak high forest with scattered hazel	total	86	115	$p<0.01$	+
	vascular plants	69	100	$p<0.01$	+
	bryophytes	17	15	-	0

Changes in quantities of key species

Certain widespread and abundant perennials in Brigsteer Wood are also characteristic of a wide range of shaded woodland habitats in Britain as a whole. Changes in their cover under coppice and non-intervention management are therefore of particular interest. These changes are recorded in Table 7.5. They indicate that although different species have responded in different ways over the 18-year study period, few have changed significantly in percentage cover. Notable exceptions to this are *Anemone nemorosa* which in general has increased, and *Rubus fruticosus* which has decreased. The latter may have been due to generally increased shading throughout, but was more likely the product of the increased density of roe deer (*Capreolus capreolus*) – bramble is their principal winter food (Hearney, 1983).

Table 7.5 Changes in the percentage cover of key species of woodland-shaded habitats between 1969 and 1987 in Brigsteer Wood, Cumbria (0, no change; +, increase; −, decrease; brackets indicate change is not significant ($p>0.05$)). (Data from Barkham, 1991b)

Key species	*Coppice site*	*Old coppice on limestone pavement or rubble*	*Yew wood with some old coppice on scree*	*Oak high forest with scattered hazel*
Herbs				
Anemone nemorosa	+	+	(+)	+
Mercurialis perennis	(−)	0	(+)	+
Hyacinthoides non-scripta	+	(−)	0	(−)
Hedera helix	(+)	+	(+)	absent
Narcissus pseudonarcissus	+	−	(+)	(+)
Rubus fruticosus	−	−	−	−
Mosses				
Ctenidium molluscum	0	(+)	+	(−)
Eurhynchium striatum	(−)	(−)	0	(−)
Thuidium tamariscinum	(−)	(−)	0	0

7.3.2 Changes in the number of *N. pseudonarcissus*

In the field

Shoot numbers of *N. pseudonarcissus* were recorded in most years for all permanent plots. Such counts yield much more precise data about change in status than estimates of cover, and are particularly useful for this species whose linear, spreading leaves make accurate cover recording difficult.

The data show (Figure 7.4) that there has been a more or less exponential increase in shoot numbers in the coppice site (A1), and a rise of 98% in shoot numbers over a 20-year period. In contrast, in the limestone pavement site (B1) the trend has been an almost exponential decline of similar magnitude and an overall loss of 51%. In the two other undisturbed sites there have been smaller fluctuations and little change overall.

Simulated changes

The effects of time on the field data for rates of mortality, clonal growth and reproduction from seed was investigated by the simulation exercise shown in Figure 7.5. From this it can be seen that the result of long-continuing shade-site-rates of mortality, clonal growth and seed reproduction is extinction (1 and 2). Rates for the same items recorded for open sites resulted either in a steady rate of increase after about 500 years (3), or in complete ground cover after only 100 years (4). Rates that were characteristic of shaded and open sites alternating for five-year periods, so

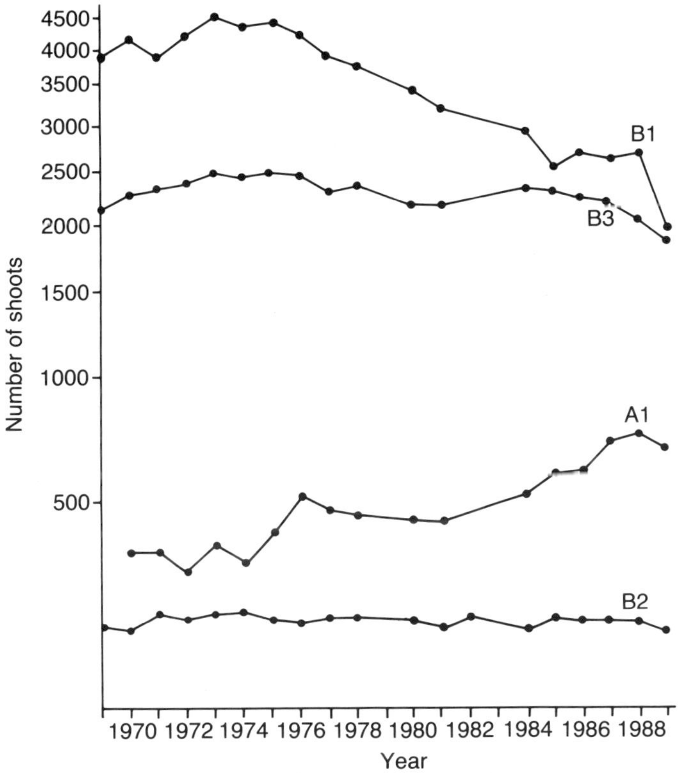

Figure 7.4 Changes in the number of *N. pseudonarcissus* shoots in permanent plots 1969–1989 in a coppice site and in three undisturbed sites in Brigsteer Wood, Cumbria. Data are drawn from Barkham (1991b). A, Coppiced site with time of coppicing indicated (four 4 m^2 plots); B1, undisturbed site with old coppice on limestone pavement or rubble (eleven 4 m^2 plots); B2, undisturbed site with yew wood and some old coppice on scree (six 4 m^2 plots); B3, undisturbed site with oak high forest and some scattered hazel (four 4 m^2 plots).

simulating a coppice cycle in a simple way, resulted in a nearly stable population over a 1000-year period (5).

Crucial to these very different outcomes is the fine balance between the three rates. The higher rates of clonal growth were derived from 'age-state' (Rabotnov, 1969) counts covering a range of sites, whereas the lower rate in the shaded site was derived from the histories of individually marked plants in site B1, the undisturbed site with old coppice on limestone pavement in which the population was in marked decline (Figure 7.4).

Other very different long-term outcomes are possible in undisturbed sites, but the evidence suggests that for *N. pseudonarcissus* it is very unlikely that under conditions of constant shade (or constant lack of disturbance) the population size could increase. The best it can possibly do (Figure 7.4, B2, B3) is to maintain a stable number.

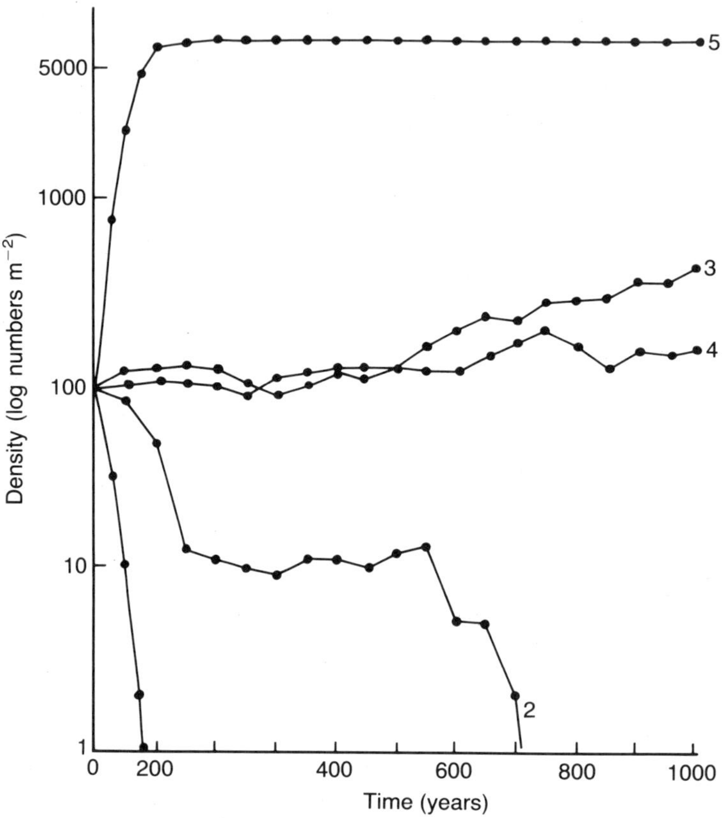

Figure 7.5 Simulated changes in population size of *N. pseudonarcissus* using probabilities calculated from field data of annual rates of adult mortality (*p*m), clonal growth (*p*v) and reproduction from seed (*p*s), all expressed as per adult per year (after Barkham and Hance, 1982).

	*p*m	*p*v	*p*s
1 shade site (a)	0.056	0.015	0.001
2 shade site (b)	0.056	0.059	0.001
3 open site (a)	0.038	0.045	0.005
4 open site (b)	0.038	0.038	0.005
5 simulated coppice cycle			
shade phase (5 years)	0.056	0.059	0.001
open phase (5 years)	0.038	0.057	0.005

Filling gaps in the ground flora

The principal results of the gap experiment (Barkham, 1992b in preparation) are given in Figure 7.6. When planted into gaps, populations of *N. pseudonarcissus* bulbs (3) change in size in a way indistinguishable from control clumps of bulbs (1). This indicates that there is nothing about the physico-chemical environment of the gap to prevent them from growing there.

Control gaps gained seedling *N. pseudonarcissus* (Figure 7.6, (10)), but normally they survived no more than one or, in a few cases, two years. This is consistent with the very low rate of adult development from seed in shaded environments found previously (*ps* in Figure 7.5; Barkham, 1980a). Seedlings and seedlings produced from seed, both controls and those planted into gaps (4, 6, 8, 9), similarly failed after a few years in every case in this experiment.

However, under different soil conditions the outcomes are strikingly different. Bulbs grown in pots (artificial gaps) sunk into the ground and filled with well-manured garden soil increased in numbers fourfold in the first three years of the experiment (Figure 7.6, (2)). Furthermore, some plants derived from seedlings (5) and from seed (7) survived through to maturity. This indicates that the characteristics of the rooting environment are also critical to what happens in a gap, whether it is through root competition, growth rates of bulbs in relation to moisture and nutrients, or in relation to herbivores and pathogens.

7.3.3 Other long-term studies

A summary of results from five other long-term studies is given in Table 7.6. This gives an indication of trends in various kinds of temperate forest that are subject to differing degrees of disturbance. Overall, if the trend is towards denser overhead canopy and deeper shade, the more abundant 'spring ephemerals' survive at the expense of the rarer species; whereas if, for whatever reason, there is increased light penetration the more abundant species increase in cover – again at the expense of rarer species. Disturbance of the forest floor including moderate grazing (for example of bramble by deer) seems to be necessary for the maintenance or increase of herbaceous species diversity. Excessive vertebrate grazing leads to impoverishment: it appears that grazing may be a factor that is frequently overlooked.

7.4 DISCUSSION

7.4.1 Differences between the environments of coppice and non-intervention stands

In order to begin to understand the varied outcomes of these two approaches to the conservation management of ancient woodland in Britain

Table 7.6 A comparison of the results of five long-term studies of change in temperate forest herbaceous vegetation

Length of study (years)	*Hough (1965)* 20	*Brewer (1980)* 50	*Metzger and Schultz (1981)* 50	*Davison and Forman (1982)* 29	*Inghe and Tamm (1985)* 38
Type of forest and location	Virgin hemlock hardwood forest, Pennsylvania	Northern hardwood forest dominated by beech and sugar maple, Michigan, USA	Northern hardwood forest dominated by sugar maple, Michigan, USA	Mixed deciduous forest dominated by oak species, New Jersey, USA	Spruce forest with pine and birch, eastern Sweden
Management	None	None	Four treatments: two clear cuts, one group selection, one individual selection	None	None mentioned
Other disturbance	None mentioned	Little: occasional wind-thrown tree	None mentioned	Regional air pollution, introduction of non-native species, hurricanes, droughts, caterpillar plagues	None mentioned
Herbaceous herbivory	White-tailed deer most important cause of ground-layer change through winter browsing	None mentioned	None mentioned	'Possible'	None mentioned

Response of herb-layer vegetation	Change from preferred palatable species to less preferred but available species. No detailed records of ground layer given	Decrease of 16 species; increase of ten, mainly spring ephemerals	Higher percentage cover of herbs in areas initially clearcut, three species increased particularly in both density and cover	Decrease from 33 to 26 species; rarer species declined or disappeared; overall increase in cover from 8% to 60% consisted of abundant species only	Overall stability of abundance and composition. None of 15 abundant species lost; of five species lost, all infrequent; and four were gained within the study period
Causes of change	Deer browsing, as above	Increased summer shade, less favourable soil-moisture conditions, competition between spring ephemerals	Initially intense disturbance caused by clearcutting favoured increased herbaceous diversity and cover	Major changes in cover due to decrease in tree canopy cover	No major changes
Conclusions	Major changes due to increased deer numbers, encouraged by wildlife/game management policies	Gain in spring ephemerals in little disturbed conditions.	Differences due to treatments minor after 50 years, mainly quantitative, rather than qualitative	Major changes in cover due to decrease in tree canopy above	Little change; gains and losses largely trivial

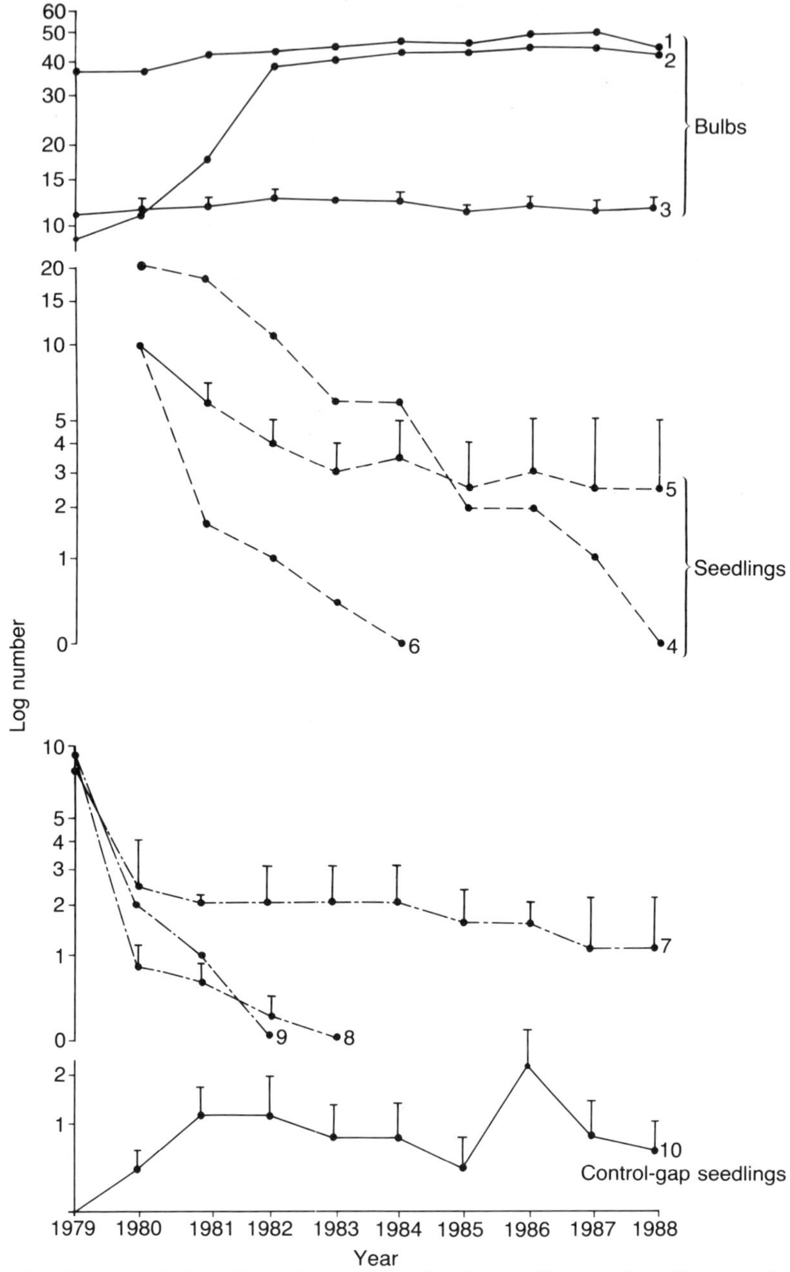

Figure 7.6 Changes 1979–1988 in the number of bulbs, seedlings and seedlings produced from seeds that were planted into spaces (gaps) between clumps of *N. pseudonarcissus* in the field. Treatments (ten bulbs, seedlings or seeds planted equidistant apart) were allocated randomly to selected gaps and replicated five times each; 25 cm-pot treatments of nine bulbs, seedlings or seeds planted equidistant apart were replicated twice. **1** Control clump of 37 bulbs; **2** nine bulbs planted in pot of garden soil (replicate pot became waterlogged and bulbs died); **3** Mean number of SE (standard error) of bulbs planted into gaps; **4** Control clump of 20 seedlings; **5** Mean number with SE of seedlings planted in pots of garden soil; **6** Mean number with SE of seedlings planted into gaps; **7** Mean number with SE of seedlings growing from seed planted in pots in garden soil; **8** Mean number with SE of seedlings growing from seed planted into gaps; **9** Seedlings growing from control group of ten seeds; **10** Mean number with SE of seedlings in control gaps.

it is important to know the differences between the environments that they create and the responses of plant species to them.

Variations over time, season and locality in the amount of light reaching the ground-layer vegetation are well known to affect species distribution (Salisbury, 1924; Blackman and Rutter, 1946) through differences between species in terms of their adaptation to light and shade (Lieth and Ashton, 1961; Packham and Willis, 1977). The cycle of light and shade which results from coppice management is different from that of an unmanaged wood. The flush of flowering of shade-tolerant perennials in the second spring following coppicing is also well known from both early and more recent studies (Salisbury, 1916, 1918, 1924; Rackham, 1975), whereas the relative effects that this has on the demography of these plants and the competition between them is little understood. The rate at which the canopy closes after cutting and clearing is probably crucial to the outcome of competition between ground flora plants. In a well-managed coppice with a high stool density this is achieved in 3–4 years. This is likely to be in marked contrast to the closing of a large natural canopy gap. Here, the open-ground flora may persist for many years and bramble (*R. fruticosus*) or bracken (*Pteridium aquilinum*) become dominant, to the detriment of the woodland shade herbs.

Important differences that are likely to be seen between coppice and unmanaged woods in terms of both the soil environment and animal activity have been itemized in Table 7.1. They are not well known and have been little quantified; therefore at this stage they are speculative. Together, these two groups of factors may go a long way to explaining the often apparently unaccountable differences in ground flora between woods. The results of observations of and experiments with soil moisture (Craib, 1929; Salisbury, 1923; Korstian and Coile, 1938; Ellison and Houston, 1958; Martin, 1968; Chapter 4) and nutrients (Pigott and Taylor, 1964; Pigott, 1982; Slade and Hutchings, 1987) and their effects on key ground flora species, and the effects of particular tree species on soil properties (Beatty, 1984), together show how important these factors are in determining the distribution, abundance and the outcome of competition between ground flora species. Results of the 'gap' experiments presented here hint at the way the survival of offspring can be determined by the soil-nutrient status of a shaded environment.

The effects of mammals grazing on the ground flora of woodlands have been grossly underestimated in Britain although they have been pointed out by some authors (Rackham, 1975, 1980; Pigott, 1982; Shirreffs, 1985) and noted as important in the temperate forests of other countries (Ellison and Houston, 1958; Hough, 1965; Tamm, 1972; Bratton, 1975; Davison and Forman, 1982; Falinski, 1986). Excluding mammals from ancient English woodlands can result in changes in relative abundance (Pigott,

1982). For example, in Monks Wood National Nature Reserve a mixed ground flora in plots fenced against deer and rabbits and dominated by bluebell and dog's mercury (*H. non-scripta* and *M. perennis*), on a variety of soils, has changed within 12 years to one dominated by bramble (*R. fruticosus*) more than 1 m high. The characteristic ancient woodland ground flora dominated by low-growing vernal perennials with showy flowers is the product of intensive selective grazing over hundreds, and probably thousands, of years.

Coppice woods dominated by *R. fruticosus* (e.g. Mason and Long, 1987) may be those in which deer-grazing is absent and/or rabbit populations have not recovered since the introduction of myxomatosis, and/or canopy re-establishment is slow. The latter may result from either low coppice growth-rates in comparison with herbaceous flora, particularly on acid soils (Salisbury, 1924), or from low stool density. Other herbaceous dominants are not necessarily unpalatable. *A. nemorosa* is known to be grazed by roe deer and to form an important part of the diet of bank voles (*Clethrionomys glareolus*) (Shirreffs, 1985) while in the Bialowieza Forest herbs are grazed by a variety of mammals in proportion to their abundance (Falinski, 1986). How these grazing effects differ between coppice and non-intervention systems is a matter for speculation (Table 7.1). Similarly, the differential effects of moles, whose activities in favourable soil cause disturbance to roots and rhizomes more or less continuously in time and space, are so far unknown. These are likely to be fruitful areas for research.

7.4.2 Changes in species composition

Understanding the present distribution of ground flora species in ancient woodland is fraught with hazard (Mitchell and Kirby, 1989), particularly because of the slow response rate to past disturbance of these plants compared with those of open habitat. Each individual plant is not only responding to a changing canopy overhead, neighbouring plants, continuous mole activity underground maybe, and the physical and chemical nature of the soil, but also to past disturbance. The latter may have been severe, long ago, and entirely unidentifiable at the present time. Indeed, only the pattern of plant distribution now remains as an historical echo – the ghost of past activities at the site. The possible range of outcomes in terms of ground flora species diversity is considerable (see for instance Metzger and Schultz, 1981; Shafi and Yarranton, 1973; Brewer, 1980; Davison and Forman, 1982).

Thus the 18–20-year period of the present study is too short a time in which to determine the outcome of particular management regimes, or indeed the differences between them. Nevertheless, we may get hints of what is to come. Figure 7.2 is suggestive of two hypotheses about the consequences of non-intervention management.

Firstly, the 'centrifugal' quality of changes in positions of sites in relation to one another suggests that, in the absence of disturbance, the vegetation is given more opportunity to express basic differences of site. Ground floras under different soil and canopy conditions become increasingly different. This result suggests that a wide variety of outcomes is possible under non-intervention management, depending on site conditions. However, a reasonable working hypothesis from the present study is that under any particular set of conditions, in the absence of significant disturbance, there will be an impoverishment of the ground flora. This has been found elsewhere, for a set of woods in which the least disturbed was floristically poorest (Middleton and Merriam, 1985), and in a single stand that had been undisturbed for about 150 years (Brewer, 1980). Where non-management disturbance occurs (air pollution, grazing, wind-throw, drought, caterpillar plagues, etc.) the trend may be opposite (Davison and Forman, 1982). The variation in ground flora associated with different coppice types and different soils adds to the complications and difficulty of predicting management outcomes.

Secondly, plots within a particular type of site (plot-group) tend to move closer together (Figure 7.2), indicating that, on a smaller scale, the vegetation of a particular site-type may become more uniform with time. It is interesting that the coppice plots also show this tendency – the ground flora being more uniform 15 years after coppicing (1987), than it was three years before (1969), which was probably only about nine years after the previous coppicing, see Figure 7.1 (*c.* 1960).

7.4.3 Changes in species status

Few species have been lost from either coppice or non-intervention sites. Those losses that have occurred are mainly trivial, such as tree and shrub seedlings of which there is a constant turnover (Figure 7.3).

Herbaceous species lost from individual plots consisted of isolated individuals in 1969. Presumably they are vulnerable to the sort of random accident to which the many individuals of a large clone make a genet less prone; Inghe and Tamm (1985) indicate similar processes operating in a Swedish forest. Included amongst losses from the coppice plots were herb paris (*Paris quadrifolia*) and twayblade orchid (*Listera ovata*) – two species that are particularly valued by conservationists and are indicators of primary woodland (Peterken, 1974). This event is probably trivial in the context of Brigsteer Wood because both species are widespread and frequent in other parts of the coppice site and elsewhere in the wood. However, it is a pointer to the probability that it is the 'rare' species that are most vulnerable to particular management choices. These may be rare in a statistical sense, as here, or rare in the more commonly understood way. Either way, these species may be the most sensitive indicators of directional changes in the environment and of the potential consequences of management choices. In North America, rare ground-layer species were found to decline or disappear over a 29-year period in undisturbed forest, while the cover of abundant species increased sevenfold (Davison and Forman, 1982).

Changes in the overall cover of ground-layer plants (Table 7.4) depend upon what is happening to the canopy overhead in the context of the particular ground conditions of the site. This means constant adjustment in the growth of herbs. The pavement sites are relatively dry and the woody canopy slow growing. There may have been a decrease in the amount of light reaching the ground as a result of a slowly increasing canopy density above. Under the yew canopy, the limestone scree appeared to become more stable over the two decades of the study, and this may account for greater ground cover of both vascular and bryophytic plants. Under oak, a very large increase in the percentage cover of vascular plants suggests that major adjustments of the ground flora are still continuing following the last major disturbance. The decline in bryophyte cover is consistent with this and with the increasing amounts of tree leaf litter at the surface. Salisbury (1924) suggested that the vegetation under coppice 'always tends to be more abundant' than under high forest. Whether or not this is true depends on the soil and woodland type. Something is known of the effect of changes in the accumulation of leaf litter (Sydes and Grime, 1981a,b) and dead wood (Falinski, 1978) on woodland plants, but not nearly enough. The varying quantity and quality of leaf litter falling from a developing overhead canopy affects the small-scale vegetation patterns of the woodland floor and the relative abundance of the species present. One of the biggest differences between coppice and non-intervention sites is likely to be that of the quantity and rate of decomposition of leaf litter. Moreover, once robust species such as *R. fruticosus* become well-established (possibly as the result of release from grazing, an increase in light, or a combination of both), litter gets trapped (Sydes and Grime, 1981a) and shading out of lower-growing species is more likely.

7.4.4 Demography and dominance in ground flora species

The demographic attributes of *N. pseudonarcissus* summarized in this chapter and reported in more detail elsewhere (Barkham, 1980a,b; 1991b in preparation) indicate the possible differing consequences for this species of coppice and non-intervention management (Table 7.7). The net result of these is that this plant can expand its population under a regime like coppicing in which there are frequent alternating periods of light and shade whereas under a non-intervention system of continuous shade it can at best hold its own in favourable sites, and, at worst it will move towards extinction (Figures 7.4 and 7.5).

Table 7.7 A summary of the likely relative consequences of coppicing and non-intervention management for the demography of *N. pseudonarcissus*

	Coppicing	*Non intervention*
Flowering	more	less
Seed production	more	less
Germination		no difference
Seedling survival	more	less
Clonal growth	more	less
Colonization of gaps	more	less
Death rate		
density-dependent	greater	less
density-independent	greater	less
Survivorship	less	greater
Population turnover rate	greater	less
Rate of gap creation	greater	less

N. pseudonarcissus at Brigsteer Wood is able to pre-empt the use of a considerable amount of ground flora space. It is a very abundant plant at that site. Yet, characteristically, it is distributed in clumps of rarely more than 50 individuals (Barkham and Hance, 1982), separated by unoccupied space or by areas occupied by other species of the ground flora guild. A few other members of the guild occupy more space but most much less. The implication of the results presented in Figure 7.5 is that the potential a species has for fully occupying available space is determined by its demography in relation to site characteristics (Angevine and Handel, 1986). Most, if not all, species of the ground flora guild flower and produce more seed in high light compared with low light conditions. Those species that depend to a greater extent on seed production to fill space with new individuals will be less effective in shaded periods than those that spread rhizomatously. Similar results have been found for woodland *Viola* species (Schellner *et al.*, 1982). There appear to be two sorts of strategy for persistence which enable certain species to cover the ground completely in a shaded environment over considerable areas given sufficient time since last disturbance:

1. Production of a superabundance of seed, even under shaded conditions, combined with limited clonal growth; exemplified by *H. non-scripta* and *Allium ursinum*.
2. Rhizomatous clonal growth from which a complete ground cover of shoots arise; exemplified by *A. nemorosa* and *M. perennis*.

These kinds of species will be the principal beneficiaries of non-intervention and their development leads to increased uniformity of the ground layer. This conclusion is supported by Struik's (1957) data and by Falinski's (1986) description of the development of the ground layer composition in a primeval forest, but it is opposite to that reached by Whitford (1949) whose work underlines the importance of the type of clonal growth of species at a particular site.

However, most perennial species characteristic of woodland shade do not have either of the two capacities listed above. Some are extremely rare in a conventional sense. Others may be common, but occupy only a minute fraction of the space available and are therefore 'rare' in the context of the average ground vegetation. In these cases, it may be reasonable to suppose that the rate of density-independent mortality is high in relation either to seed reproduction or to clonal growth. These are the species which will benefit most from coppicing in terms of maintenance of abundance. This is due to the creation of more space for colonization by the destruction of the more abundant species through the disturbance caused by coppicing, and to an enhanced sexual reproductive success in frequent periods of high illumination.

In lowland deciduous woodlands in Britain, there are visually striking differences from place to place in percentage ground occupancy of herbaceous species (or the proportion of bare ground), and also in the extent of mixing of species. In one place there may be intimate mixtures of several abundant species, in others dense pure stands of single species abutting one another. The results presented here for *N. pseudonarcissus* suggest a broad hypothesis to explain this kind of ground vegetation pattern. Subject to basic limitations of soil type, which determine the availability and tolerance of particular subsets of species, pattern is determined by four variables:

1. frequency, scale and type of disturbance;
2. length of time taken for canopy closure following large-scale disturbance;
3. demography of the species;
4. grazing.

Evidence presented here (Table 7.5) suggests that coppicing favours species like *N. pseudonarcissus* which depend on high light conditions for the successful establishment of new individuals from seed at a rate great enough to counteract mortality; and that non-intervention favours species that rarely occupy new ground as a result of seeding, but extend into it through clonal growth and an effective rhizome system. A similar explanation may account for the differences in abundance, related to management treatment, of ground-layer species found by Metzger and Schultz (1981).

7.4.5 Relevance to woodland management for conservation

It is suggested above that under prolonged conditions without disturbance the shade-tolerant dominants will eliminate other species by pre-empting space created by density-independent agents of mortality. This would seem to be the logical consequence of the abandonment over the last 100 years of the coppice system of management and may explain the large areas of abandoned coppice woods covered by nearly pure stands of *A. nemorosa*, *H. non-scripta*, and *M. perennis*. Intimate mixtures of these and other species may be found in intermediate situations, either where lack of disturbance has been of long enough duration for gaps to have been colonized but not long enough for competitive exclusion to have occurred, or where disturbance to ground vegetation resulting from woodland management has been minimized. The latter depends upon the timing of the operation and the means by which it is carried out.

The results and argument presented above suggest a rationale for reinstating a traditional system of coppice management in ancient British woodlands in order to maintain valuable and attractive mixtures of herbs. The effect is:

1. to reduce by trampling and other disturbance the competitive dominance of a few species, e.g. *M. perennis* (Rackham, 1975) thereby
2. reducing interspecific competition, and creating bare and temporarily open ground to maximize the probability of successful reproduction of rarer species that largely depend upon seed production during the open-ground phase for establishing themselves in new sites;
3. temporarily increasing species richness and competition from fast-growing herbs and grasses of open ground which subsequently die as shade increases, creating gaps for re-exploitation by shade-tolerant or shade-evading species.

An attempt is made in Table 7.8 to categorize environmental factors and processes affecting woodland ground flora into those favouring homogeneity and those favouring heterogeneity and so, in Table 7.9, to group important woodland herb species characteristic of shaded sites in British ancient woodlands into those that benefit from coppicing and those that benefit from non-intervention. These are imperfect lists, but they may allow managers to evaluate more critically the consequences of their decisions about whether or not to coppice. To be confident of these lists we need to know much more about the responses of species to management on a range of soil types, under a range of stool and tree densities, and about what happens to them under different intensities of mammal grazing and in its absence. Finally, we need to know how species react in competition with each other under different woodland environmental conditions.

Table 7.8 Factors favouring homogeneity and heterogeneity of the woodland ground flora. Homogeneity means relatively low species density; heterogeneity means relative high species density

Homogeneity	*Heterogeneity*
Recent establishment	Ancient
Soil type	Soil type
uniform	heterogeneous
nutrient-rich	low nutrient status
very acid	moderately acid to calcareous
Uniformity of overhead canopy of trees and shrubs	Heterogeneous overhead canopy of trees and shrubs
Disturbance	Disturbance
low intensity silvicultural management, or none	moderate intensity of silvicultural management (e.g. coppicing, selection thinning)
Grazing	Grazing
lack of native grazing mammals	grazing by native mammals
high intensity domestic mammal grazing	domestic mammal grazing intensity low, occasional or nil
Mole activity little or absent	Mole activity intense
Stable climatic and physico-chemical environment with little natural treefall	Some natural tree-fall and canopy gaps
Very dry or very wet sites	Freely-drained sites with high avail, able water capacity, or spatially very variable (e.g. carr woodland)

Table 7.9 Woodland perennial herbs that are likely to benefit from different kinds of conservation management. Apart from *Primula vulgaris*, species included are known to be largely confined to ancient woodlands and include most of those particularly valued by conservationists. All except *H. non-scripta*, *M. perennis* and *Viola riviniana* are listed in Table 1 of Peterken (1974)

Coppicing	*Non-intervention*	*No preference*
Adoxa moschatellina	*Allium ursinum*	*Anemone nemorosa*
Calamagrostis canescens	*Convallaria majallis*	*Hyacinthoides non-scripta*
Campanula latifolia	*Lamiastrum galeobdolon*	
Carex pendula	*Luzula sylvatica*	
Conopodium majus	*Mercurialis perennis*	
Narcissus pseudonarcissus	*Sanicula europaea*	
Orchis mascula		
Paris quadrifolia		
Primula vulgaris		
Ranunculus auricomus		
Veronica montana		
Viola reichenbachiana		
V. riviniana		

REFERENCES

Angevine, M.W. and Handel, S.N. (1986) Invasion of forest floor space, clonal architecture, and population growth in the perennial herb *Clintonia borealis*. *Journal of Ecology*, **74**, 547–60.

Ash, J.E. and Barkham, J.P. (1976) Changes and variability in the field layer of a coppice woodland, Norfolk, England. *Journal of Ecology*, **64**, 697–712.

Barkham, J.P. (1980a) Population dynamics of the wild daffodil (*Narcissus pseudo-narcissus*). I. Clonal growth, seed reproduction, mortality and the effects of density. *Journal of Ecology*, **68**, 607–33.

Barkham, J.P. (1980b) Population dynamics of the wild daffodil (*Narcissus pseudo-narcissus*). II. Changes in number of shoots and flowers, and the effect of bulb depth on growth and reproduction. *Journal of Ecology*, **68**, 635–64.

Barkham, J.P. (1991a) The effects of management on the ground flora of ancient woodland, Brigsteer Park Wood, Cumbria, England. *Biological Conservation* (in press).

Barkham, J.P. (1992b) Population dynamics of the wild daffodil (*Narcissus pseudo-narcissus*) IV. Clumps and gaps. (In preparation.)

Barkham, J.P. and Hance, C.E. (1982) Population dynamics of the wild daffodil (*Narcissus pseudonarcissus*) III. Implications of a computer model of 1000 years of population change. *Journal of Ecology*, **70**, 323–44.

Beatty, S.W. (1984) Influences of microtopography and canopy species on spatial patterns of forest understory plants. *Ecology*, **65**, 1406–19.

Blackman, G.E. and Rutter, A.J. (1946) Physiological and ecological studies in the analysis of plant environment. I. The light factor and the distribution of bluebell (*Scilla non-scripta*) in woodland communities. *Annals of Botany*, **10**, 361–90.

Brewer, R. (1980) A half-century of changes in the herb layer of a climax deciduous forest in Michigan. *Journal of Ecology*, **68**, 823–32.

Bratton, S.P. (1975) The effect of the European wild boar, *Sus scrofa*, on gray beech forest in the Great Smoky Mountains. *Ecology*, **56**, 1356–66.

Bratton, S.P. (1976) Resource division in an understory herb community: responses to temporal and microtopographic gradients. *American Naturalist*, **110**, 679–93.

Brown, A.H.F. and Oosterhuis, L. (1981) The role of buried seed in coppice woods. *Biological Conservation*, **21**, 19–38.

Collins, B.S., Dunne, K.P. and Pickett, S.T.A. (1985) Responses of forest herbs to canopy gaps, in *The Ecology of Natural Disturbance and Patch Dynamics* (eds S.T.A. Pickett and P.S. White), Academic Press, Orlando, pp. 217–34.

Connell, J.H. (1978) Diversity in tropical rainforests and coral reefs. *Science*, **199**, 1302–10.

Cook, R.E. and Lyons, E.E. (1983) The biology of *Viola fimbriatula* in a natural disturbance. *Ecology*, **64**, 654–60.

Craib, I.J. (1929) *Some aspects of soil moisture in the forest*, Yale University School of Forestry Bulletin, 25, Yale University.

Cummings, I.P.F. (1982) *Bradfield Woods: bonfires and their effects on the flora*, MSc thesis, University of Salford.

Davison, S.E. and Forman, R.T.T. (1982) Herb and shrub dynamics in a mature oak forest: a thirty-year study. *Bulletin of the Torrey Botanical Club*, **109**, 64–73.

Ehrenfeld, J.G. (1980) Understory response to canopy gaps of varying sizes in a mature oak forest. *Bulletin of the Torrey Botanical Club*, **127**, 29–41.

Ellison, L. and Houston, W.R. (1958) Production of herbaceous vegetation in openings and under canopies of western aspen. *Ecology*, **39**, 337–45.

Falinski, J.B. (1978) Uprooted trees, their distribution and influence in the primeval forest biotope. *Vegetation*, **38**, 175–83.

Falinski, J.B. (1986) *Vegetation Dynamics in Temperate Lowland Primeval Forests: Ecological Studies in Bialowieza Forest*. W. Junk, Dordrecht.

Geiger, R. (1965) *The Climate Near the Ground*, Harvard University Press, Cambridge, Massachusetts.

Handel, S.N. (1978) The competitive relationship of three woodland sedges and its bearing on the evolution of ant-dispersal of *Carex pedunculata*. *Evolution*, **32**, 151–63.

Harrison, A.F. (1970) *Soil organic phosphorus – its mineralisation and its importance in the nutrition of woodland plants. A literature review*. Merlewood Research and Development Paper 23, Nature Conservancy, Grange-over-Sands.

Hearney, A.W. (1983) *The feeding ecology of red and roe deer in Thetford Forest*. M. Phil thesis, University of East Anglia, Norwich.

Hill, M.O. (1979) *DECORANA: a Fortran Program for Detrended Correspondence Analysis and Reciprocal Averaging*. Cornell University, Ithaca (NY).

Hough, A.F. (1965) A twenty-year record of understory vegetational change in a virgin Pennsylvania forest. *Ecology*, **46**, 370–3.

Inghe, O. and Tamm, C.O. (1985) Survival and flowering of perennial herbs. IV: The behaviour of *Hepatica nobilis* and *Sanicula europaea* on permanent plots during 1943–1981. *Oikos*, **45**, 400–20.

Korstian, C.F. and Coile, T.S. (1938) *Plant competition in forest stands*. Duke University School of Forestry Bulletin 3.

Lamb, J. (1989) *The ecological effects of the Great Storm of October 1987*, Nature Conservancy Council, CSD Report No. 937, Peterborough.

Levin, S.A. and Paine, R.T. (1974) Disturbance, patch formation, and community structure. *Proceedings of the National Academy of Science, USA*, **71**, 2744–7.

Lieth, H. and Ashton, D.H. (1961) The light compensation points of some herbaceous plants inside and outside deciduous woods in Germany. *Canadian Journal of Botany*, **39**, 1255–9.

Ludolf, C., Payne, S. and Robertson, P. (1988) The effect of pheasant release pens and strawed rides on ancient woodland ground flora. *The Game Conservancy Review of 1988*, 42–6.

Martin, M.H. (1968) Conditions affecting the distribution of *Mercurialis perennis*, L in certain Cambridgeshire woodlands. *Journal of Ecology*, **56**, 777–93.

Mason, C. and Long, S. (1987) Management of lowland broadleaved woodland, Bovingdon Hall, Essex, in *Conservation Monitoring and Management* (Ed. R. Matthews), Countryside Commission, Cheltenham, pp. 37–42.

Mellanby, K. (1971) *The Mole*, Collins, London.

Metzger, F. and Schultz, J. (1981) Spring ground layer vegetation 50 years after harvesting in northern hardwood forests. *American Midland Naturalist*, **105**, 44–50.

Middleton, J. and Merriam, G. (1985) The rationale for conservation: problems from a virgin forest. *Biological Conservation*, **33**, 133–45.

Minckler, L.S. and Woerheide, J.D. (1965) Reproduction of hardwoods 10 years after cutting as affected by site and opening size. *Journal of Forestry*, **63**, 103–7.

Mitchell, P.L. and Kirby, K.J. (1989) *Ecological Effects of Forestry Practices in Long-established Woodland and their Implications for Nature Conservation*, Oxford Forestry Institute Occasional Paper, 39, Department of Plant Sciences, University of Oxford, Oxford.

Moore, M.C. and Vankat, J.L. (1986) Responses of the herb layer to the gap dynamics of a mature beech maple forest. *American Midland Naturalist*, **115**, 336–47.

Packham, J.R. and Willis, A.J. (1977) The effects of shading on *Oxalis acetosella*. *Journal of Ecology*, **65**, 619–42.

Peterken, G.F. (1974) A method for assessing woodland flora for conservation using indicator species. *Biological Conservation*, **6**, 239–45.

Peterken, G.F. (1981) *Woodland Conservation and Management*, Chapman and Hall, London.

Peterken, G.F. and Backmeroff, C. (1988) *Long term monitoring in unmanaged woodland nature reserves*, Research and Survey in Nature Conservation, 9, Nature Conservancy Council, Peterborough.

Pigott, C.D. (1982) The experimental study of vegetation. *New Phytologist*, **90**, 389–404.

Pigott, C.D. and Taylor, K. (1964) The distribution of some woodland herbs in relation to the supply of nitrogen and phosphorus in the soil. *Journal of Ecology*, **52** (supplement), 175–85.

Rabotnov, T.A. (1969) On coenopopulations of perennial herbaceous plants in natural coenoses. *Vegetatio*, **19**, 87–95.

Rackham, O. (1975) *Hayley Wood: Its History and Ecology*, Cambridgeshire and Isle of Ely Naturalists' Trust, Cambridge.

Rackham, O. (1976) *Trees and Woodland in the British Landscape*, Dent, London.

Rackham, O. (1980) *Ancient Woodland*, Edward Arnold, London.

Runkle, J.R. (1985) Disturbance regimes in temperate forests. In *The Ecology of Natural Disturbance and Patch Dynamics* (eds S.T.A. Pickett and P.S. White), Academic Press, London, pp. 17–33.

Salisbury, E.J. (1916) The oak–hornbeam woods of Hertfordshire, parts 1 and 2. *Journal of Ecology*, **4**, 83–117.

Salisbury, E.J. (1918) The oak–hornbeam woods of Hertfordshire, parts 3 and 4. *Journal of Ecology*, **6**, 14–52.

Salisbury, E.J. (1924) The effects of coppicing as illustrated by the woods of Hertfordshire. *Transactions of the Hertfordshire Natural History Society*, **18**, 1–21.

Schellner, R.A., Newell, S.J. and Solbrig, O.T. (1982) Studies on the population biology of the genus *Viola*. IV. Spatial pattern of ramets and seedlings in three stoloniferous species. *Journal of Ecology*, **70**, 273–290.

Shafi, M.I. and Yarranton, G.A. (1973) Diversity, floristic richness, and species evenness during a secondary (post-fire) succession. *Ecology*, **54**, 897–902.

Shirreffs, D.A. (1985) Biological flora of the British Isles. *Anemone nemorosa* L. *Journal of Ecology*, **73**, 1005–20.

Shure, D.J. and Phillips, D.L. (1987) Litter fall patterns within different-sized disturbance patches in a southern Appalachian mountain forest. *American Midland Naturalist*, **118**, 348–57.

Slade, A.J. and Hutchings, M.J. (1987) The effects of nutrient availability on foraging in the clonal herb *Glechoma hederacea*. *Journal of Ecology*, **75**, 95–112.

Struik, G.J. (1957) *The distribution and reproduction of some herbaceous plants in a maple forest.* MS thesis, University of Wisconsin.

Sydes, C. and Grime, J.P. (1981a) Effects of tree leaf litter on herbaceous vegetation in deciduous woodland. I. Field investigations. *Journal of Ecology*, **69**, 237–48.

Sydes, C. and Grime, J.P. (1981b) Effects of tree leaf litter on herbaceous vegetation in deciduous woodland. II. An experimental investigation. *Journal of Ecology*, **69**, 249–62.

Tamm, C.O. (1972) Survival and flowering of perennial herbs III. The behaviour of *Primula veris* on permanent plots. *Oikos*, **23**, 159–66.

Thompson, J.N. (1980) Treefalls and colonisation patterns of temperate forest herbs. *American Midland Naturalist*, **104**, 176–84.

Thompson, J.J. and Willson, M.F. (1978) Disturbance and dispersal of fleshy fruits. *Science*, **200**, 1161–3.

Veblen, T.T. (1985) Stand dynamics in Chilean *Nothofagus* forests, in *The Ecology of Natural Disturbance and Patch Dynamics* (eds S.T.A. Pickett and P.S. White), Academic Press, London, pp. 35–51.

Vitousek, P.M. (1985) Community turnover and ecosystem nutrient dynamics, in *The Ecology of Natural Disturbance and Patch Dynamics* (eds S.T.A. Pickett and P.S. White), Academic Press, London, pp. 325–33.

Whitford, P.B. (1949) Distribution of woodland plants in relation to succession and clonal growth. *Ecology*, **30**, 199–208.

Whitbread, A. (1989) Nature's stormy cycle. *Natural World*, **26**, 29–30.

8

The effects of changing management on seed banks in ancient coppices

A.H.F. BROWN and SUSAN J. WARR

8.1 INTRODUCTION

The presence of dormant seeds in soils (that are able to germinate given the right conditions) is a phenomenon obvious to any farmer or gardener; and in common with the soils of most other vegetation types, those of woodlands also contain viable seeds (Roberts, 1981; Leck *et al.*, 1989). The density and species composition of soil seed banks depend on the rate at which seeds enter and are lost from the soil. In frequently disturbed habitats, seed banks are large and the species composition of the seed bank and the vegetation is usually similar, e.g. in arable fields (Jensen, 1969; Wilson *et al.*, 1985). In undisturbed habitats seed banks are smaller and there is generally less correspondence between the species present in the seed bank and the vegetation, e.g. in old undisturbed forests (Kellman, 1974; Johnson, 1975; Whipple, 1978). Seed banks are often dominated by early successional species, e.g. the weeds of agricultural soils. In forest soils it has been shown that the seeds of some species can survive entire forest rotations (Granstrom, 1988) but others only survive for a few decades (Hill and Stevens, 1981). Late successional species, e.g. forest herbs and most trees, generally have short-lived seeds which do not survive for long periods in the soil (Strickler and Edgerton, 1976; Petrov, 1977; Brown and Oosterhuis, 1981; Staaf *et al.*, 1987).

The long history of coppicing in many British woodlands – responsible for a good deal of their floristic diversity and conservation interest – mostly came to an end during this century, especially since World War II. Apart from those woodlands grubbed for agriculture, many were abandoned or converted to conifers. In these woods the diverse flora becomes impoverished through increased shade and probably also because of the resulting soil changes. To what extent can the seed banks of these former coppices continue to provide a potential for their floristic recovery (given restoration of appropriate conditions) under these changes in management? Following a summary of the role of seed banks in coppices as a basis for comparison, this chapter describes a number of studies in which the seed banks of neglected and coniferized coppice woods (in south-east and south-west England and in the west Midlands) have been examined.

The presence of at least some of the seeds in soils can be relatively easily tested by placing samples of the soils in suitable containers in a glasshouse and providing appropriate conditions of light, moisture and warmth, thus stimulating the breaking of dormancy and the germination of the seeds. The extent to which such germination tests enable the presence of all seeds to be determined may be questionable, although it has been claimed (Moore and Wein, 1977; Gross, 1990), given appropriate conditions for germination, that the test provides a good estimate of the species composition of the seed bank. It is commonly used in seed bank studies (Roberts, 1981) and has been used in all studies referred to below.

8.2 SITES AND METHODS

The data presented or referred to here were obtained from the sites shown in Figure 8.1. All are ancient woodlands, formerly or, in a few cases, recently coppiced. Further details of the sites are given in Table 8.1 and in the subsequent text. Field sampling and germination test procedures were identical with or based on those of Brown and Oosterhuis (1981), except that in all of these subsequent studies the litter layer was also retained and tested for its content of viable seeds.

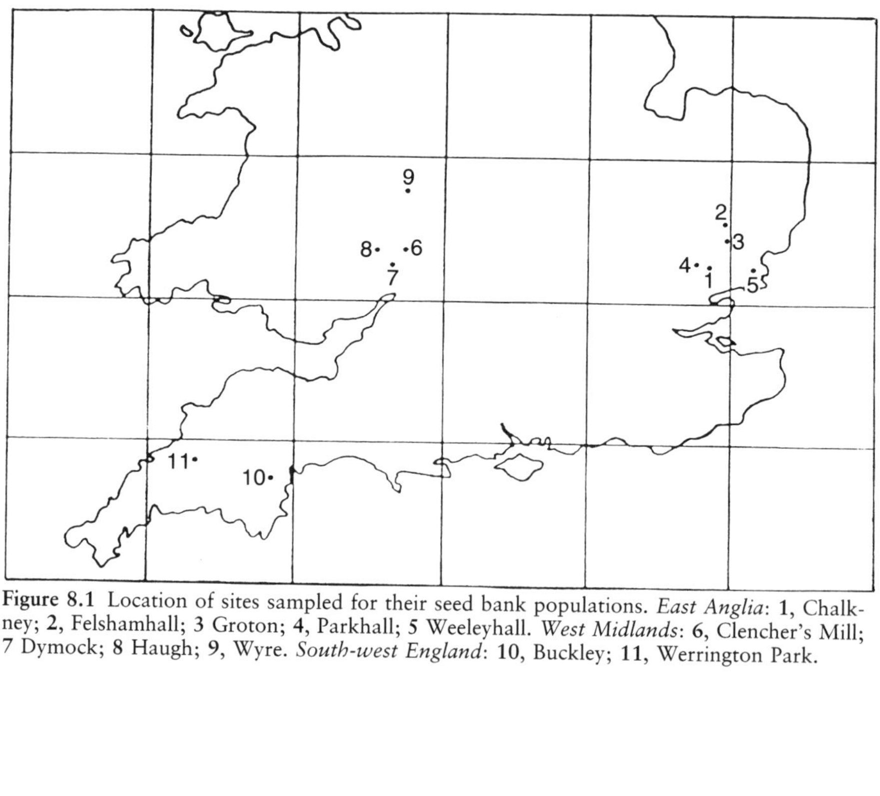

Figure 8.1 Location of sites sampled for their seed bank populations. *East Anglia*: **1**, Chalkney; **2**, Felshamhall; **3** Groton; **4**, Parkhall; **5** Weeleyhall. *West Midlands*: **6**, Clencher's Mill; **7** Dymock; **8** Haugh; **9**, Wyre. *South-west England*: **10**, Buckley; **11**, Werrington Park.

Table 8.1 Details of sites sampled and/or surveyed for their seed bank populations

Name *County* *National Grid Reference*	*Ownership*	*Brief site description*
Chalkney Wood Essex TL 8727	Forestry Commission and Essex County Council	Forestry Commission part converted to conifers *c.* 1960; remainder coppice mainly of lime, hornbeam and sweet chestnut. Soils derived from acid glacial materials with some loess. pH of sampled areas 4.2–5.0
Felshamhall Wood Suffolk TL 9357	Royal Society for Nature Conservation (managed by Suffolk Wildlife Trust)	Coppice with standards, the underwood and field layer very diverse. Soils derived from variable depths of sand and loess overlying chalky boulder clay. pH in sampled areas mainly 5.2–5.7
Groton Wood Suffolk TL 9743	Suffolk Wildlife Trust	Coppice. Soils mainly derived from loess, pH mainly 4.3–4.8
Parkhall Wood Essex TL 7528	Private	Formerly neglected coppice being brought back into management; some converted to conifers *c.* 1950. Soil pH 3.9–4.1 (up to 5.0)
Weeleyhall Wood Essex TM 1520	Private	Coppice with standards, some converted to conifers in the early 1970s. Soil pH 3.5–4.1
Clencher's Mill Wood Herefordshire/ Worcestershire SO 7235	Private	Mixed coppice with standards, part converted to conifers and chestnut coppice. Brown-earth soils over limestone, pH mainly 5.4–6.1 (up to 6.7); very diverse ground flora
Dymock Forest Gloucestershire SO 6828	Forestry Commission	Largely converted since the 19th century to conifers or oak high forest. Soils brown-earths, pH mainly 4.2–4.8 (3.9 under old Norway spruce)
Haugh Wood Herefordshire/ Worcestershire SO 5936	Forestry Commission	Largely converted over a long period to oak or beech high forest, or to conifers. Soils poorly drained silty clays, pH 4.5–5.2. Ground flora particularly diverse
Wyre Forest Herefordshire/ Worcestershire and Shropshire SO 7575	Forestry Commission and private	Much of the Forestry Commission part has long been converted to conifers. Otherwise coppice mainly of oak on brown podzolic soils; some brown-earth soils and gleys. pH mainly 4.1–4.7
Buckley Wood Devon SX 8774	Private	Part hazel coppice under oak standards, part conifers planted *c.* 1970. Shallow, but base-rich soils over limestone
Werrington Park Wood Cornwall SX 3187	Private	Former oak coppice converted to conifers from 1935 onwards giving a range of ages. Soils mainly acid brown podzols

8.3 SEED BANKS IN ANCIENT COPPICES

Germination tests were used to assess the seed banks in soils of ancient coppices in East Anglia. Areas of coppice, uncut for 30 to 40 years (referred to here as 'neglected'), in Chalkney, Felshamhall, Groton, Parkhall and Weeleyhall Woods in north Essex and south Suffolk were sampled as described in Brown and Oosterhuis (1981). The mineral soils were sampled at two depths from the most densely shaded parts of these woods. In this initial study, the surface L, F and H layers were discarded on the basis that these would mainly contain seeds recently shed on to, or recently dispersed into, the area, rather than the long-persisting buried seeds which were of predominant interest. The soil depths sampled were 0–5 cm and 5–15 cm and the germination test continued for two years. The species appearing in the combined results for the five woods and two depths are shown in Table 8.2.

On grouping these species according to their tolerance or otherwise to shade (Table 8.2), it may be seen that by far the largest group is that of the light-demanding species and much the smallest group comprises deep-shade species, with species of medium shade occurring in intermediate numbers. Furthermore, few of the ground flora plants present in the same densely shaded areas of neglected coppice from which the samples were taken appeared in the germination test (Table 8.3). In fact, less than 40%, on average, of the seedling species for any given wood currently occurred as vegetation anywhere in the old coppice areas of that wood. Evidently the population of seed bank species is, to a large extent, a different one from that of the remaining shade-tolerant vegetation. Therefore, what population does it represent? Comparison of the seed bank germinants with the lists of species known to appear after coppicing (but which subsequently gradually disappear again as they become shaded out by the redeveloping coppice growth (Salisbury, 1918; Brown and Oosterhuis, 1981)) indicates a close correspondence (Table 8.4). From this and other evidence (Brown and Oosterhuis, 1981), it may be inferred that the seed banks of these woods are the main source of what Salisbury (1924) termed the 'marginal' species which reappear afresh each time the coppice is cut – providing light and greater fluctuations of temperature that help break seed dormancy and encourage their germination.

The lack of correspondence between the seed bank population and the existing closed canopy vegetation commonly occurs in a variety of vegetation types (grasslands: Chippindale and Milton, 1934; Major and Pyott, 1966; Hayashi and Numata, 1971; wetlands: Van der Valk and Davis, 1976; Smith and Kadlec, 1985, Wilson *et al.*, in press; woodlands: Oosting and Humphries, 1940; Petrov, 1977; Pratt *et al.*, 1984). In relation to coppice, how should this non-conformity be interpreted? Thompson and

Table 8.2 Species germinated from soil samples (grouped according to shade tolerance) showing total number of occurrences. Sample area: neglected coppice, East Anglia; 27 sampling positions, with data combined from 0–5 cm and 5–15 cm depths (within an area of 900 cm^2)

	Number of seedlings		*Number of seedlings*
Light-demanding species			
Agrostis canina	113	*J. effusus*	4205
A. stolonifera	327	*Lotus corniculatus*	15
Anthoxanthum odoratum	105	*Plantago major*	1
Betula spp.	1335	*Polygonum aviculare*	1
Carex pallescens	71	*P. convolvulus*	23
C. pilulifera	50	*P. lapathifolium*	1
Centaurium erythraea	11	*P. nodosum*	2
Chenopodium album	39	*P. persicaria*	1
Cirsium palustre	28	*Ranunculus flammula*	6
Cytisus scoparius	2	*Rumex obtusifolius*	26
Digitalis purpurea	276	*Sagina procumbens*	11
Epilobium adenocaulon	42	*Salix* spp.	4
Festuca spp.	34	*Solanum nigrum*	34
Gnaphalium uliginosum	1	*Stellaria media*	1
Holcus lanatus	26	*Trifolium* spp.	1
Hypericum hirsutum	3	*Tripleurospermum maritimum inodorum*	1
H. humifusum	34	*Veronica chamaedrys*	6
H. pulchrum	20	*V. officinalis*	60
H. tetrapterum	54	*V. serpyllifolia*	7
Isolepis setacea	68		
Juncus bufonius	139		
Species tolerant of some shade			
Carex pendula	10	*P. trivalis*	27
C. remota	14	*Potentilla sterilis*	8
C. sylvatica	1	*Ranunculus repens*	8
Corydalis claviculata	13	*Rosa* spp.	1
Deschampsia cespitosa	2	*Rubus idaeus*	9
Euphorbia amygdaloides	1	*Sambucus nigra*	11
Fragaria vesca	3	*Scrophularia nodosa*	4
Luzula pilosa	19	*Silene* spp. (*dioica*?)	118
Lysimachia nemorum	57	*Solidago virgaurea*	1
Moehringia trinervia	33	*Stachys sylvatica*	1
Poa nemoralis	5	*Veronica montana*	45
Species tolerant of heavy shade			
Ajuga reptans	32	*Rubus fruticosus*	446
Primula vulgaris	3	*Viola* spp.	16
Rubus caesius	1		

Table 8.3 Most frequent species in ground flora of neglected coppice areas. Combined data for five woods in north Essex/south Suffolk as percentage occurrences in sample plots of 200 m^2 (n=57)

Species	%	*Presence in germination test*
Rubus fruticosus	100	√
Anemone nemorosa	70	
Hyacinthoides non-scripta	60	
Lonicera periclymenum	56	
Mercurialis perennis	51	
Viola spp.	49	√
Poa nemoralis/trivialis	43	√
Dryopteris filix-mas	32	
Circaea lutetiana	29	
Dryopteris dilatata	29	
Geum urbanum	29	
Ranunculus ficaria	29	

Table 8.4 Relation of species found in the mineral soil seed bank of neglected coppices (north Essex/south Suffolk) to the more common species in the ground flora of newly cut coppices. (From Brown and Oosterhuis, 1981)

	Ground flora of newly coppiced areas		*Soil seed bank*
Species	*Frequency* for combined data from three woods in north Essex/ south Suffolk (1975)*	*Salisbury's assessment of frequency† for sessile oak–hornbeam woods in Hertfordshire (1918)*	*Presence of seedlings in two-year germination test*
Agrostis capillaris	8	(r)	
A. stolonifera	7	f-c	+
Ajuga reptans	5	f-c	+
Anemone nemorosa	7	–	
Anthoxanthum odoratum	–	f-vc	+
Arctium spp.	13	(o-f)	
Betula spp. seedlings	15	common	+
Carex sylvatica	8	–	+
Centaurium erythraea	(1)	f-vc	+
Chamaenerion angustifolium	13	(o)	
Cirsium arvense	17	–	
C. palustre	18	f-ab	+
C. vulgare	9	–	
Deschampsia cespitosa	8	–	+
Digitalis purpurea	(+)	l-c	+
Epilobium hirsutum	10	–	
E. montanum	21	+-c	

Table 8.4 *Continued*

Species	*Ground flora of newly coppiced areas*		*Soil seed bank*
	Frequency for combined data from three woods in north Essex/south Suffolk (1975)*	*Salisbury's assessment of frequency† for sessile oak–hornbeam woods in Hertfordshire (1918)*	*Presence of seedlings in two-year germination test*
Euphorbia amygdaloides	10	(r-lc)	+
Fragaria vesca	(3)	f-c	+
Geum urbanum	12	(r-f)	
Gnaphalium sylvaticum	6	(r-lf)	+
Holcus lanatus	9	f-vc	+
H. mollis	(2)	f c	
Hypericum humifusum	(4)	f-c	+
Juncus bufonius	6	(o)	+
J. effusus/conglomeratus	7	lf-lc	+
Lamiastrum galeobdolon	–	o-lab	
Lonicera periclymenum	15	c-ab	
Luzula pilosa	9	o-fc	+
Lysimachia nemorum	5	c-ab	+
Mercurialis perennis	19	(lc)	
Moehringia trinervia	7	f-c	+
Plantago major	6	(r)	+
Poa annua	8	(o)	
P. nemoralis *P. trivialis*	17	–	+
Primula vulgaris	13	(o)	+
Potentilla sterilis	8	(o-f)	+
Prunella vulgaris	13	(r-f)	
Ranunculus repens	5	(o-vr)	+
Rubus fruticosus	21	c-ab	+
Rumex sp.	10	(o-f)	+
Sambucus nigra	9	(r-lc)	+
Sonchus asper	10	–	
Stachys sylvatica	9	(o)	+
Stellaria media	11	(r)	+
Urtica dioica	11	(r)	
Veronica officinalis	(4)	f-fc	
V. montana	7	f-fc	+
V. serpyllifolia	7	(r-rr)	+
Viola sp. (*riviniana*?)	16	(f)	+

*Number of sample plots (maximum 32) in which species occurred. Five or more occurrences used as criterion of 'more common'.

†ab, abundant; vc, very common; fc, fairly common; c, common; f, frequent; o, occasional; rr, rather rare; vr, very rare; l, locally. Assessments of frequency > f used as criterion of 'more common'. (In either column, any lesser frequencies are also included in brackets.)

Grime (1979) have shown that seed-bearing plants can be divided into those in which at least a proportion of the seeds are, or become, dormant and form a persistent seed bank, through subsequently becoming buried; and those in which dormancy, if present, is transient and quickly broken, resulting in germination within the first year. The persistent seed banks are regarded as an adaptation by light-demanding plants to situations in which sufficiently open conditions are only intermittent, brought about by occasional disturbance of the vegetation cover and/or the soil. In general, this class of seeds also has a requirement for light if dormancy is to be broken. Transient seed banks, in contrast, tend to be associated with more permanently closed vegetation of various sorts, including forest (Thompson and Grime, 1979).

Clearly, the presence of ancient woods which have been regularly coppiced over long periods, with alternating conditions of closed forest canopy and the disturbance of both canopy and soil at coppicing, has enabled floras employing both of these strategies to develop. Because the coppice regrowth is so rapid, the open periods immediately after coppicing are sufficiently short that the forest shade flora is not appreciably inhibited by competition, excessive insolation or otherwise, but they are long enough for the complete cycle of germination, flowering and seeding of the light-demanding species to occur. In the coppices, therefore, there appear to be two complementary floras:

1. An intermittent marginal flora of light-demanding species and those of partial shade adapted to scrub and forest edges but which are also well-suited to a relatively frequent coppicing cycle. In seral terms, they can be regarded as early and mid-successional species (Buckley and Knight, 1989) both groups of which 'are adapted to accumulate persistent reservoirs of buried seeds' (Donelan and Thompson, 1980) as a strategy for surviving the dark periods.
2. Late-successional or forest species, adapted to persist in the vegetative phase, relatively few of which produce a seed bank. In common with some other species with only transient seed banks (Pickett and McDonnell, 1989) they use a strategy of shade tolerance or shade avoidance via the vernal habit, and in many cases also employ storage organs such as bulbs and corms to help withstand periods of low light levels. Table 8.5 gives examples of shade species, lacking a persistent seed bank, which commonly occur in coppices.

What are the capabilities of these complementary floras, with their different strategies, to cope with changing coppice management such as neglect and conversion to conifers?

Table 8.5 Examples of shade species, commonly occurring in coppices, which lack a persistent seed bank, i.e. their seeds do not persist beyond the start of the next growing season. (From Grime *et al.*, 1988)

Allium ursinum	*Geum urbanum*
Anemone nemorosa	*Hyacinthoides non-scripta*
Arum maculatum	*Lonicera periclymenum*
Circaea lutetiana	*Mercurialis perennis*
Conopodium majus	*Ranunculus ficaria*
Geum rivale	*Sanicula europaea*

8.4 EFFECTS OF NEGLECT

By surveys of existing vegetation, those species which tend to disappear from the ground flora in the absence of coppicing can be determined. Initially, such surveys were carried out in East Anglia within the same group of woods as those from which the five seed bank study areas were drawn, comparing the effects of neglect (and 'coniferization') with remaining areas of worked coppice (Brown, 1979). Table 8.6 lists those species which had either disappeared or were on the brink of disappearing under neglect. As indicated in the table, virtually all were present in the seed bank – having occurred in the germination tests; potentially, therefore, these 'missing' species were replaceable from this reservoir of propagules. That such potential for recovery can be realized in practice is indicated by data from another study area in south-west England (Buckley Wood, Devon). Table 8.7 compares the greatly increased flora in an area of neglected coppice-with-standards, a few years after felling, with the vegetation in an adjacent uncut area. In the latter part of the wood, only the shade flora has persisted – mostly of species lacking a seed bank. Conversely, many of the additional species appearing in the felled area are those to be found in the seed bank of one part or another of this wood (see Table 8.10).

Table 8.6 Shade species in which neglect causes a marked reduction or loss (East Anglian coppices)

**Carex sylvatica*	**Lysimachia nemorum*
Circaea lutetiana	**Primula vulgaris*
**Epilobium montanum*	**Scrophularia nodosa*
**Euphorbia amygdaloides*	**Urtica dioica*
**Fragaria vesca*	**Veronica montana*
Hedera helix	**Veronica officinalis*
**Hypericum hirsutum*	*Vicia* sp.
**Luzula pilosa*	

*Occurring in germination tests and classified by Grime *et al.* (1988) as having, or probably having, a persistent seedbank (except for *E. amygdaloides* which they do not list).

Table 8.7 Comparison of floras in cut and uncut portions of neglected coppice with standards, Buckley Wood, Devon

Species	*Uncut*	*Cut*
Ajuga reptans		√
Anemone nemorosa	√	√
Arctium minus		√
Arum maculatum	√	√
Cirsium arvense		√
C. vulgare		√
Conopodium majus	√	√
Epilobium spp.		√
Euphorbia amygdaloides		√
Fragaria/Potentilla		√
Galium aparine		√
G. odoratum		√
Glechoma hederacea		√
Hyacinthoides non-scripta	√	√
Hypericum hirsutum		√
H. montanum		√
Iris foetidissima	√	√
Lamiastrum galeobdolon		√
Mercurialis perennis	√	√
Myosotis sylvatica		√
Poa trivialis		√
Primula vulgaris	√	√
Ranunculus ficaria		√
R. repens		√
Rubus fruticosus		√
Senecio jacobaea		√
Solanum dulcamara		√
Stachys sylvatica		√
Taraxacum officinale		√
Veronica chamaedrys		√
V. montana		√
Viola riviniana	√	√

Thus, the ground flora and seed banks of ancient coppices appear to be adapted not only to a regime of regular cutting on a relatively short cycle, but also to one of longer cycles, or neglect, for three or four decades at least. In the long history of these ancient coppices it would be remarkable indeed if there had not been great variation in the frequency of cutting with inevitable periods of neglect. The vegetation has, it may be assumed, developed in relation to these vicissitudes and the seed bank provides the potential to allow recovery of those species lost from the vegetation through neglect provided the length of time before removal of the canopy is not excessive. Very long periods of neglect, at least on the acidic soils of the Tavistock Estate woodlands (bordering Devon and Cornwall), have led to losses of coppice ground flora species from both the vegetation and the seed bank. Many species had disappeared from the seed bank after *c.* 70 years of neglect, such that the 'potential for ground flora recovery no longer exists in the seed bank of this oak coppice' (Harris and Kent, 1987b). Darby (1986) also found reductions in the seed banks of these woods after *c.* 70 years, after 100 years only two species, *Digitalis purpurea* and *Hypericum pulchrum*, remained, and after 200 years only *H. pulchrum* was left. These findings are entirely consistent with the conclusions of Hill and Stevens (1981) that in temperate forests many seeds persist for *c.* 50 years followed by a period of relatively rapid decline, such that after 100 years, other things being equal, most of the original seed population would have disappeared. However, over such a time-scale occasional disturbances, the presence of refugia enabling persisting plants to flower and seed, or gradual immigration of seeds from elsewhere might be expected to permit some replenishment of the seed bank.

8.5 EFFECTS OF CONIFERS

8.5.1 Evidence from vegetation surveys

Vegetation surveys of former ancient coppices converted to conifers more than 20 years previously were also carried out within the sample group of East Anglian woodlands (Brown, 1979). Numbers of stands of any given conifer species were too few to study their separate effects: the results therefore refer to the effect of conifers in general, combined from Douglas fir (*Pseudostuga menziesii*), Norway spruce (*Picea abies*) and western red cedar (*Thuja plicata*). Not surprisingly, in addition to the species found to disappear under neglect, a further portion of the ground flora was also shown to be no longer present in the vegetation (Table 8.8). In contrast to the species missing under neglect, very few of these additional lost species had occurred in the germination tests and virtually all are members of the list given in Table 8.5, i.e. they are known not to form a persistent seed bank (Grime *et al.*, 1988). Furthermore most are species that are of particular conservation interest which tend to occur, with varying degrees of strictness, only or mainly in ancient woodland sites in eastern England (Peterken, 1974).

As these conifer stands age and the canopy is thinned or eventually felled, will these lost species be able to recolonize? They are mostly absent from the seed bank and their recolonizing ability is almost certainly very low: shade species tend to have heavy seeds with dispersal limited to a few metres or even only a few centimetres (e.g. *Oxalis acetosella*: up to 1 m (Packham, 1978); *Hyacinthoides non-scripta*: average 40 cm (Knight, 1964)), which provides one possible explanation for the scarcity of at least

Table 8.8 Shade species unaffected by neglect, but markedly reduced or lost in conifer-planted coppices in East Anglia

†*Anemone nemorosa*
Arum maculatum
†*Conopodium majus*
Corydalis claviculata
†*Geum rivale*
†*Mercurialis perennis*
**Moehringia trinervia*
†*Oxalis acetosella*
†*Ranunculus auricomus*
†*Sanicula europaea*

*Occurring in germination tests
†Species associated with ancient woodland sites in eastern England

some of the forest shade flora in secondary woods. It seems likely, therefore, that these species will not reappear once lost – even if suitable conditions are otherwise restored. Evidence in support of this comes from further vegetation surveys from 'coniferized' former ancient coppices in a third study area in the west Midlands comprising four sites: Wyre Forest, Dymock Forest, Haugh Wood and Clencher's Mill Wood (Brown *et al.*, 1979). Here, older conifer stands and those which had completed a full rotation and had been felled were also available for comparison. Although the mean numbers of marginal species per sample plot (Table 8.9) showed some recovery with opening of the canopy in the older stands, and after felling, this was not the case for the shade species, nor for those species associated with ancient sites; if anything, both categories continued to decline even following clear-felling, possibly through long-term soil changes brought about by the conifers. Hence, the presence of even-aged conifers may lead to further losses of species from the ground flora which are

Table 8.9 Number of species per sample plot (of 200 m^2) found in the vegetation of coppices or former coppices converted to conifer plantations. Means of four woods in the west Midlands, 1977–1978

	Coppice 'control' (n=40)	*Conifers (15–30 years) (n=52)*	*Conifers (30–60 years) (n=68)*	*Felled conifers (n=32)*
Shade spp.	14.8	11.9	11.0	10.4
Marginal spp.	6.0	4.6	5.4	8.5
Total spp.	26.7	20.4	21.3	24.5
Ancient woodland spp.*	5.6	4.2	3.4	2.5

*i.e. all groups, combined, of those given by Peterken (1974)

unlikely to reappear by natural immigration and which are not potentially replaceable from the seed bank – even supposing the seed bank itself remains intact under a conifer regime.

8.5.2 Seed banks under conifers

Comparisons were made of seed banks in areas of closed-canopy conifers with remaining coppice areas, and between different ages of conifers. They indicated that in the short term, the presence of conifers *per se* of some 20 years or so probably has little, if any, effect on the seed bank, if the results from both Devon and Essex given in Table 8.10 are typical. Data from Buckley Wood provide a comparison between coniferized areas and areas of coppice with standards. Part of the wood had been felled and replanted 20 years ago with hybrid larch (*Larix eurolepis*) (the present ground vegetation currently being restricted to shade species) and Douglas fir (now with no ground vegetation at all except for a few straggling *Hedera helix*). Seed bank data for both conifer areas have been combined, as have the comparative results for unfelled and very recently felled parts of the coppice with standards. There is no evidence of any depletion of the seed bank under these conifers: nor does there appear to be any particular floristic distinction between the 0–5 cm soil seed banks found under the two types of management. Similar evidence is provided by the Essex data. These show the combined seed bank (0–15 cm) for former coppice areas of Parkhall and Chalkney Woods which had been converted to conifers about 18 years before sampling, and the comparable seed bank for the remaining coppice areas of these two woods.

For older stands, however, evidence from both the west Midlands and south-west England does indicate some decline in the species composition of seed banks under coniferous stands older than about 50 years, relative to younger ones (Table 8.11). These reductions in seed bank species numbers in the older conifer stands again conform with the conclusions (Hill and Stevens, 1981) referred to in connection with neglect, i.e. that *c.* 50 years represents the period up to which the seed population remains large, but after which decline accelerates.

Therefore, even-aged conifers can not only eliminate some of those shade species from the vegetation which have no seed bank to act as a source of potential replacements, but also in old stands (as in coppices after sufficiently long periods of neglect) their presence appears to lead to losses from the seed bank itself: hence the potential for recovery of the light-demanding and/or intermediate shade species is also reduced. It has yet to be determined whether, under older conifers, these losses from the seed bank merely reflect the length of time (relative to the normal longevity of the bulk of the seeds) before felling and accompanying light allows renewal

Table 8.10 Comparisons of seed banks in conifer-planted (*c.* 20 year) former coppices and remaining coppice areas: total numbers of seedlings in germination

Chalkney/Parkhall (combined date) soil sample depth: 0–15 cm (n=60*)			*Buckley Wood soil sample depth: 0–5 cm* (n=50*)		
Species	*Neglected coppice*	*Conifers*	*Species*	*Coppice with standards*	*Conifers*
Agrostis canina	2	5	*Agrostis* spp.	13	1
A. capillaris	–	18	*Ajuga reptans*	2	
A. stolonifera	17	67	*Anagallis arvensis*		2
Ajuga reptans	1	–	*Arctium* spp.		6
Betula sp.	691	189	*Betula* spp.	150	59
Carex sp. (? *pallescens*)	26	1	*Buddleja davidii*	2	6
Carex sp. (? *pendula*)	8	1	*Cardamine hirsuta*	1	
C. sylvatica	–	6	*Carex* spp.		5
Carpinus betulus	–	1	*Centaurium erythraea*	3	3
Centaurium erythraea	11	1	*Cerastium fontanum*	4	
Cirsium palustre	15	20	*Chamaenerion augustifolium*	6	3
C. vulgare	–	1	*Chenopodium album*	2	
Deschampsia cespitosa	–	1	*Cirsium arvense*	12	24
Digitalis purpurea	1	–	*C. vulgare*	25	14
Epilobium sp. (? *adenocaulon*)	42	53	*Clematis vitalba*	3	
E. montanum	8	23	*Eupatorium cannabinum*		6
Festuca sp.	4	–	*Euphorbia amygdaloides*	3	12
Fragaria vesca	3	1	*Geum urbanum*		2
Gnaphalium uliginosum	1	–	*Holcus* sp. (? *lanatus*)		2
Holcus lanatus	–	13	*Hypericum* spp. (mainly *H. perforatum*, some *H. montanum*)	14	232
Hypericum hirsutum	3	5	*Hypochoeris* spp.	1	
H. humifusum	29	6	*Iris foetidissima*	2	2
H. pulchrum	13	20	*Moehringia trinervia*	1	5
H. tetrapterum	54 (all from 1 sample)	1	*Poa trivialis*	18	27
			Potentilla sterilis	1	2
Isolepis setacea	2	–	*Primula vulgaris*	3	2
Juncus bufonius	20	36	*Ranunculus repens*	3	1
J. effusus	1988	1039	*Rubus fruticosus*	38	93
Lotus corniculatus	6	–	*Rumex* spp.	6	48
L. pedunculatus	–	1	*Scrophularia nodosa*		2
Luzula sp. (?*pilosa*)	6	2	*Senecio jacobaea*	14	

Table 8.10 *Continued*

Chalkney/Parkhall (combined date) soil sample depth: 0–15 cm (n=60*)		
Species	*Neglected coppice*	*Conifers*
Lysimachia nemorum	11	35
Moehringia trinervia	31	–
Plantago major		2
Poa nem./triv.	24	1
P. pratensis	–	1
Polygonum convolvulus	22	6
P. lapathifolium	1	–
P. nodosum	2	–
P. persicaria	1	–
Potentilla sterilis	7	2
Primula vulgaris	3	2
Prunella vulgaris	–	20
	(all from 1 sample)	
Ranunculus repens	5	–
Ribes sp.	–	1
Rosa sp.	1	–
Rubus fruticosus	148	432
R. idaeus	7	35
Rumex sp. (? *conglom./sanguin.*)	25	2
Sagina procumbens	11	40
Salix sp.	2	1
Sambucus nigra	1	1
Scrophularia nodosa	3	1
Silene dioica	114	1
	(all from 1 sample)	
Solanum nigrum	18	4
Sonchus asper	–	31
Stachys sylvatica	1	–
Stellaria media	1	–
Taraxacum officinale	–	3
Tripleurospermum inodorum	1	–
Veronica chamaedrys	6	1
V. montana	34	1
V. officinalis	107	159
V. serpyllifolia	3	2
Viola sp.	–	2
	50 spp.	49 spp.

Buckley Wood soil sample depth: 0–5 cm (n=50*)		
Species	*Coppice with standards*	*Conifers*
S. sylvaticus	3	
Sonchus asper	15	3
Silene spp.	1	4
Stellaria media		12
Taraxacum officinale	2	4
Urtica dioica	36	52
Veronica chamaedrys	3	5
V. montana		6
V. officinalis		3
V. persica	1	
	31 spp	32 spp

*Samples bulked in groups of five for germination tests

Table 8.11 Numbers of species per stand in the seed banks under different ages of conifers planted on former ancient coppice sites

	Tree species and stand age	*Number of species in vegetation*		*Number of species found in seed banks, at specific soil depths*		
			Litter	*0–5 cm*	*5–15 cm*	*0–15 cm*
Haugh Wood*	Douglas fir (*Pseudostuga menziesii*) 22 years	few, if any	33	57	52	68
	Douglas fir (*Pseudostuga menziesii*) 52 years	17	29	41	27	48
Werrington Park†	Sitka spruce (*Picea sitchensis*) 28 years	0	11	20		
	Sitka spruce (*Picea sitchensis*) 29 years	4 (seedlings)	12	19		
	Sitka spruce (*Picea sitchensis*) 42 years	8	9	18		
	Japanese larch (*Larix kaempferi*) 42 years	8	12	15		
	European larch (*Larix decidua*) 53 years	8	11	11		

*35 field samples per stand and depth bulked in groups of five for germination tests.
†50 field samples per stand bulked in groups of five for germination tests.

tests of the seed bank species, or whether there is some other effect of conifers on the soil conditions that in turn further influences the longevity of buried seeds.

8.5.3 Seed longevity and soil acidity

Because the presence of conifers can lead to acidification of the surface soil, the question of whether seed longevity is influenced by soil pH is of some relevance.

Milton (1936, 1939 and 1943) concluded that seed survival was best under conditions of low pH; Champness and Morris (1948) also found high viable seed numbers in peaty and acid soils. However, as Harper

(1957) has pointed out this may be because species characteristic of acid habitats tend to produce a high density of seeds (e.g. *Juncus* spp., *Agrostis* spp. and *Calluna vulgaris*) and not necessarily because seed survival is better under acid conditions. Brown and Oosterhuis (1981) found a positive correlation between both species numbers and seed numbers in the seed bank and pH; and Hill and Stevens (1981) noted greater numbers of viable seeds in mineral soils than in peat. Staaf *et al.* (1987) studied seven different mature beech forests in Sweden with a range of soil pHs. Although their data also showed far fewer species and seeds in the seed banks of the acid sites, these differences could equally well be attributed to differences in floras between the sites. The question of whether soil acidity enhances or diminishes longevity of buried seeds has therefore to remain open; it is not impossible that there is an optimum pH in this respect, longevity being reduced by both higher and lower soil pHs.

8.5.4 Uneven-aged conifers

Uneven-aged conifers, exemplified by the Bradford-Hutt system in the Tavistock Estate Woodlands, have been shown to retain the diversity of the ground flora via the presence of all successional stages permanently and regularly throughout the wood, analogous to a regular coppicing system (Harris and Kent, 1987a). Such a silvicultural system may well also enable the seed bank diversity to be better retained than in even-aged conifers. At Tavistock, because the time which has elapsed since initiation of the system in 1959 is appreciably less than the suggested 50-year criterion for seed longevity, this likely benefit (which is dependent on the dispersal of seeds over the short distances between the sub-units of the Bradford-Hutt system) has yet to be demonstrated (Harris and Kent, 1987b).

8.6 CONCLUSIONS

Seeds of many of the typical plants of coppices, especially those light-demanding species which reappear, albeit temporarily, after coppicing are present in the soil seed bank of coppice woodlands. Whilst such a seed bank persists, it provides a mechanism for the floristic recovery of a vegetation that has become reduced in diversity through shading, once the source of the shading (the canopy) is removed. Although seeds of a few species may not remain viable more than *c.* 30 years, many can apparently persist for up to 50 years. Beyond this period, decline in numbers of species is probably fairly rapid such that after a century, seeds of only a very few species are likely to remain alive. Any form of management – whether absence of coppicing or conversion to conifers – that involves the long-term presence of a canopy will prevent renewal of the seed bank via the cycle of germination, flowering and seeding. If this period is less than *c.* 50 years, e.g. neglect of around 40 years, the seed bank, as has been shown, will remain largely intact, enabling floristic recovery to take place. Where it exceeds this approximate period, there are likely to be losses from the seed bank, and its capacity to provide a full floristic recovery of the vegetation following the eventual removal of the canopy will be impaired. It is not clear whether conifers, by the additional changes that they make to soil conditions (pH and other chemical and physical properties, organic matter content, etc.) have any further effect on seed longevity in the soil, beyond that of the shading effect.

An important group of the shade flora of coppices, in contrast, produces seed which germinates (or disappears through other causes) soon after it is shed, and does not enter the seed bank. These species are adapted to survive vegetatively under a deciduous canopy via a degree of shade tolerance or shade avoidance. Whilst normally withstanding neglect, many are unable to survive the more extreme and continuous shading of an evergreen coniferous canopy. Because seed dispersal in most of such species is very restricted, and because they are not present in the seed bank, there is no obvious mechanism to enable them to re-colonize a site and the evidence suggests that they do not.

ACKNOWLEDGEMENTS

Some of the work described here formed part of a PhD research project sponsored by Polytechnic South West. We should like to thank Ken Thompson and Martin Kent who supervised that project, and who suggested improvements to earlier versions of this chapter.

REFERENCES

Brown, A.H.F. (1979) *Changes in the vegetation of ancient coppicewoods in East Anglia resulting from altered management*, Nature Conservancy Council, Peterborough.

Brown, A.H.F. and Oosterhuis, L. (1981) The role of buried seed in coppicewoods. *Biological Conservation*, **21**, 19–38.

Brown, A.H.F., Pearce, N.J. and Robertson, S.M.C. (1979) *Management effects in lowland coppicewoods: vegetation changes resulting from altered management in ancient coppicewoods in the West Midlands*, Nature Conservancy Council, Peterborough.

Buckley, G.P. and Knight, D.G. (1989) The feasibility of woodland reconstruction, in *Biological Habitat Reconstruction* (Ed. G.P. Buckley), Belhaven Press, London, 171–88.

Champness, S.S. and Morris, K. (1948) The population of buried viable seeds in relation to contrasting pasture and soil types. *Journal of Ecology*, **36**, 149–73.

Chippindale, H.G. and Milton, W.E.J. (1934) On the viable seeds present in the soil beneath pastures. *Journal of Ecology*, **22**, 508–31.

Darby, C.D. (1986) *The dynamics of buried seed banks beneath woodlands with particular reference to* Hypericum pulchrum. Unpublished PhD thesis, Plymouth Polytechnic.

Donelan, M. and Thompson, K. (1980) Distribution of buried viable seeds along a successional series. *Biological Conservation*, **17**, 297–311.

Granstrom, A. (1988) Seed banks at six open and afforested heathland sites in southern Sweden. *Journal of Applied Ecology*, **25**, 297–306.

Grime, J.P., Hodgson, J.G. and Hunt, R. (1988) *Comparative Plant Ecology*, Unwin Hyman, London.

Gross, K.L. (1990) A comparison of methods for estimating seed numbers in the soil. *Journal of Ecology*, **78**, 1079–93.

Harper, J.L. (1957) The ecological significance of dormancy and its importance in weed control. *Proceedings of the international congress on crop protection, 4th Hamburg*, **1**, 415–20.

Harris, M.J. and Kent, M. (1987a) Ecological benefits of the Bradford-Hutt system of commercial forestry. I: Ground flora and the light climate. *Quarterly Journal of Forestry*, **81**, 145–57.

Harris, M.J. and Kent, M. (1987b) Ecological benefits of the Bradford-Hutt system of commercial forestry. II: The seed bank and the ground flora species phenology. *Quarterly Journal of Forestry*, **81**, 213–24.

Hayashi, I. and Numata, M. (1971) Viable buried seed populations in the *Miscanthus* and *Zoysia* type grasslands in Japan – ecological studies on the buried seed population in the soil related to plant succession. VI. *Japanese Journal of Ecology*, **20**, 243–52.

Hill, M.O. and Stevens, P.A. (1981) The density of viable seed in soils of forest plantations in upland Britain. *Journal of Ecology*, **69**, 693–709.

Jensen, H.A. (1969) Content of buried seed in arable soil in Denmark and its relation to the weed population. *Dansk Botanisk Arkiv.*, **27**, 1–55.

Johnson, E.A. (1975) Buried seed populations in the subarctic forest east of Great Slave Lake, Northwest Territories. *Canadian Journal of Botany*, **53**, 2933–41.

Kellman, M.C. (1974) Preliminary seed budgets for two plant communities in coastal British Columbia. *Journal of Biogeography*, **1**, 123–33.

Knight, G.H. (1964) Some factors affecting the distribution of *Endymion nonscriptus* (L.) Garcke in Warwickshire woods. *Journal of Ecology*, **52**, 405–21.

Leck, M.A., Parker, V.T. and Simpson, R.L. (eds) (1989) *The Ecology of Soil Seed Banks*, Academic Press, London.

Major, J. and Pyott, W.T. (1966) Buried, viable seeds in two California bunchgrass sites and their bearing on the definition of a flora. *Vegetatio*, **13**, 253–82.

Milton, W.E.J. (1936) The buried viable seeds of enclosed and unenclosed hill land. *Welsh Plant Breeding Bulletin, Series H.*, No. 14.

Milton, W.E.J. (1939) The occurrence of buried viable seeds in soils at different elevations and on a salt marsh. *Journal of Ecology*, **27**, 149–59.

Milton, W.E.J. (1943) The buried viable seed content of a midland calcareous clay soil. *Empire Journal of Experimental Agriculture*, **20**, 155.

Moore, J.M. and Wein, R.W. (1977) Viable seed populations by soil depth and potential site recolonization after disturbance. *Canadian Journal of Botany*, **55**, 2408–12.

Oosting, H.J. and Humphreys, M.E. (1940) Buried viable seeds in a successional series of old field and forest soils. *Bulletin of the Torrey Botanical Club*, **67**, 253–73.

Packham, J.R. (1978) Biological flora of the British Isles: *Oxalis acetosella* L. *Journal of Ecology*, **66**, 669–93.

Peterken, G.F. (1974) A method for assessing woodland flora for conservation using indicator species. *Biological Conservation*, **6**, 239–45.

Petrov, V.V. (1977) Reserve of viable plant seeds in the uppermost soil layer beneath the canopies of coniferous and small-leaved forests. *Vestnik Moskovskogo Universiteta. Biologiya*, **32**, 43–55.

Pickett, S.T.A. and McDonnell, M.J. (1989) Seed bank dynamics in temperate deciduous forest, in *Ecology of Soil Seed Banks* (eds. M.A. Leck, V.T. Parker and R.L. Simpson), Academic Press.

Pratt, D.W., Black, R.A. and Zamora, B.A. (1984) Buried viable seed in a ponderosa pine community. *Canadian Journal of Botany*, **62**, 44–52.

Roberts, H.A. (1981) Seed banks in soils. *Advances in Applied Biology*, **6**, 1–55.

Salisbury, E.J. (1918) The oak–hornbeam woods of Hertfordshire, Parts III and IV. *Journal of Ecology*, **6**, 14–52.

Salisbury, E.J. (1924) The effects of coppicing as illustrated by the woods of Hertfordshire. *Transactions of the Hertfordshire Natural History Society*, **18**, 1–21.

Smith, L.M. and Kadlec, J.A. (1985) The effects of disturbance on marsh seed banks. *Canadian Journal of Botany*, **63**, 2133–7.

Staaf, H., Jonsson, M. and Olsen, L. (1987) Buried germinative seeds in mature beech forests with different herbaceous vegetation and soil types. *Holarctic Ecology*, **10**, 268–77.

Strickler, G.S. and Edgerton, P.J. (1976) Emergent seedlings from coniferous litter and soil in eastern Oregon. *Ecology*, **57**, 801–7.

Thompson, K. and Grime, J.P. (1979) Seasonal variation in the seed banks of herbaceous species in ten contrasting habits. *Journal of Ecology*, **67**, 893–921.

Van der Valk, A.G. and Davis, C.B. (1976) The seed banks of prairie glacial marshes. *Canadian Journal of Botany*, **54**, 1832–8.

Whipple, S.A. (1978) The relationship of buried, germinating seeds to vegetation in an old-growth Colorado subalpine forest. *Canadian Journal of Botany*, **56**, 1505–9.

Wilson, R.G., Kerr, E.D. and Nelson, L.A. (1985) Potential for using weed seed content in the soil to predict future weed problems. *Weed Science*, **33**, 171–5.

Wilson, S.D., Moore, D.R.J. and Keddy, P.A. (in press) The relationship between seed banks and vegetation in a freshwater marsh in Ontario, Canada. *Journal of Applied Ecology*.

Part Five

EFFECTS OF COPPICING ON BIRDS

9

Effects of coppice management on woodland breeding birds

R.J. FULLER

9.1 INTRODUCTION

A frequent justification for the use of coppicing in woodland conservation is that it benefits breeding birds, particularly the nightingale (*Luscinia megarhynchos*). This view stems partly from a study at Ham Street Woods National Nature Reserve, Kent, which concluded that: 'a return to the traditional mixed coppice-with-standards management on nature reserves, and other woods where financial return is deemed to be less urgent than conservation, would be desirable in the long-term interests of the nightingale in Britain' (Stuttard and Williamson, 1971). Surprisingly, this remained the sole study of the effects of coppice management on bird populations until the late 1980s when the British Trust for Ornithology (BTO) collated data from a series of coppice woods. This chapter reviews these findings with the purpose of evaluating coppice management as a conservation tool in managing woodland songbird populations.

The specific aims are: (a) to review trends in the composition and species richness of breeding bird communities in relation to age of coppice, (b) to describe aspects of coppice management that influence breeding bird communities, particularly with respect to breeding long-distance migrants (tree pipit (*Anthus trivialis*), warblers and nightingale), and (c) to compare bird communities in actively coppiced woods with those in derelict coppice and broadleaved high forest. Derelict coppice is defined here as coppice that has been abandoned and allowed to grow well beyond the age at which it was commercially mature.

9.2 STUDY AREAS, MATERIALS AND METHODS

Responses of birds to coppicing were assessed from the distribution of birds during the breeding season in relation to factors such as the age, habitat structure and plant species composition of coppice. No attempt was made to assess breeding success.

In the convention used here, coppice is aged according to the number of full summers of growth, so that coppice in its first spring or summer after cutting is termed year 0 but in the following spring it becomes year 1. Coppice compartments within which the regrowth of underwood is of a uniform age are termed 'panels', and typically range in size from 0.2 ha to 2.5 ha.

Information was available from the BTO's Common Birds Census (Marchant *et al.*, 1990), giving the distribution of birds within more than 20 woodland census plots, each containing active coppice. Several of these plots were unsuitable for the present analysis – either because the area of coppice was very small or the available information on the vegetation and history of coppicing was inadequate. The Common Birds Census plots were supplemented by two special studies, one at Bradfield Woods, Suffolk, and one at Shrawley Wood, Worcestershire. The information here is taken from a total of 17 plots distributed as follows: Essex three, Gloucestershire three, Kent five, Northamptonshire one, Oxfordshire two, Suffolk one, Sussex one, Worcestershire one. For nine of these there was sufficient information to summarize occurrence of species over the first nine years of growth (Appendices 9.1 and 9.2). Four of the plots with large areas of coppice and comprehensive information on management were treated as case studies, each being published as a separate account of the responses of birds to the development of coppice. Details of these four case study sites are given in Table 9.1. In these woods, complete canopy closure of the coppice had occurred seven to nine years after cutting.

In all cases the distribution of birds within the woods was estimated using a territory mapping method (Marchant, 1983; International Bird Census Committee, 1969) in which the locations and activities of birds were mapped on 1:2500 maps. Each observation of a bird was termed a 'registration'. Territories which spanned more than one panel were divided between panels according to the proportion of registrations in each. Results were combined from different years for each coppice year-class by dividing the total number of territories in that year-class by the total area to give an index of density for each species expressed as territories per ten hectares. In the case of Ham Street Woods, densities were calculated as registrations per hectare because many of the panels at this site were extremely small, making estimation of territory densities of doubtful validity (Fuller *et al.*, 1989b). Further details of methods are given in the references listed in Table 9.1. Caution is needed when comparing the densities of species between the woods because the censuses were conducted by different observers in different years. It is also possible that factors such as variation in the surroundings of the woods influenced the populations of birds found within them. Therefore, only the broadest comparisons of bird communities can be drawn here.

Most of the analyses focus on songbirds – which are defined as all passerines, excluding Corvidae and starling (*Sturnus vulgaris*). The reason

Table 9.1 Details of four case study coppice woodlands

Name *County* *National Grid Reference*	*Type of coppice*	*Study site area (ha)*	*Years censused*	*Reference*
Longbeech Wood* Kent TQ95	Sweet chestnut (*Castanea sativa*) No standards	22	1975–1984	Fuller and Moreton (1987)
Brasenose Wood Oxfordshire SP50	Hazel (*Corylus avellana*), hawthorn (*Crataegus monogyna*), blackthorn (*Prunus spinosa*), aspen (*Populus tremula*) Many large standards	18	1980–1987	Fuller and Steel (1990)
Bradfield Woods Suffolk TL95	Alder (*Alnus glutinosa*), ash (*Fraxinus excelsior*), birch (*Betula* spp.), hazel (*Corylus avellana*) Few large standards	62	1987	Fuller and Henderson, in press
Ham Street Woods Kent TR03	Hornbeam (*Carpinus betulus*), chestnut (*Castanea sativa*), hazel (*Corylus avellana*) Moderate numbers of large standards	24	1969–1973	Fuller *et al.* (1989)

* The densities of birds in Longbeech Wood used in all analyses in this paper differ from those presented in Fuller and Moreton (1987). Here, the densities have been recalculated using the procedure described in the text.

for this is that songbirds tend to have relatively small, discrete territories which can be related to the scale on which coppice management is usually carried out.

9.3 SPECIES TRENDS IN RELATION TO AGE OF COPPICE

Within any wood there are, of course, differences between species of birds in their distribution across the coppice cycle. Those occurring in the early (0–2) years and middle years (3–8) are listed in Appendices 9.1 and 9.2. Amongst species that can be associated strongly with the youngest growth are tree pipit, dunnock (*Prunella modularis*), whitethroat (*Sylvia communis*) and yellowhammer (*Emberiza citrinella*). The middle growth can hold increasing numbers of wrens (*Troglodytes troglodytes*), robins (*Erithacus rubecula*), nightingales, blackbirds (*Turdus merula*), several warblers and bullfinch (*Pyrrhula pyrrhula*). Bird communities in older growth stands (nine years and over) are typically dominated by robin and tits (*Paridae*).

The distributions of eight species in four coppiced woods are summarized in Figures 9.1 and 9.2. These species illustrate the extent to which habitat distribution can differ between woods. Note that Ham Street Woods are only comparable with the other three woods in the pattern, not the magnitude, of the responses shown by the eight species. An important point to bear in mind is that the observed distribution of a species within coppice may be a consequence of population level. Scarce species may be confined to their preferred habitats, whilst abundant species may occupy not only their preferred habitats but also be forced into suboptimal ones. Similarly, a species such as wren, which is subject to much variation in overwinter mortality, may occupy a different range of year-classes at different population densities. Such phenomena are difficult to detect and control.

Wren and robin avoided the years immediately following cutting in all woods. Robin densities were consistently highest during, or after, canopy closure but wren showed rather more variation between woods, being very scarce in Longbeech. Blackbird trends also varied between woods: e.g. in Bradfield it was mainly associated with the middle-aged growth but elsewhere it was relatively abundant in the later stages. In Brasenose and Ham Street, great tit (*Parus major*) numbers increased, albeit slightly, with the age of coppice. There was rather more concordance amongst the migrant species (Figure 9.2), with all species showing peak numbers in young and middle growth. The strength of the responses varied considerably, with garden warbler (*Sylvia borin*) and chiffchaff (*Phylloscopus collybita*) weak in Longbeech and willow warbler (*P. trochilus*) weak in Brasenose.

Further evidence given in Appendices 9.1 and 9.2 indicates that there are substantial differences between woods both in the timing and magnitude of the responses shown by many species to coppice management. The likely causes of this variation are discussed below.

9.4 COMMUNITY TRENDS IN RELATION TO AGE OF COPPICE

9.4.1 Community composition

Three broad phases can be discerned in the bird communities of coppice: the first covers the years immediately following cutting when the coppice is sparse; the second embraces the canopy closure period and the third, post canopy closure. The three phases can be termed 'establishment', 'canopy closure' and 'maturation', respectively (Fuller and Moreton, 1987). The exact timing and the species composition of the phases differ from wood to wood, but species most strongly associated with the phases include:

establishment:	tree pipit, whitethroat;
canopy closure:	nightingale, garden warbler, blackcap (*Sylvia atricapilla*), chiffchaff, willow warbler;
maturation:	robin.

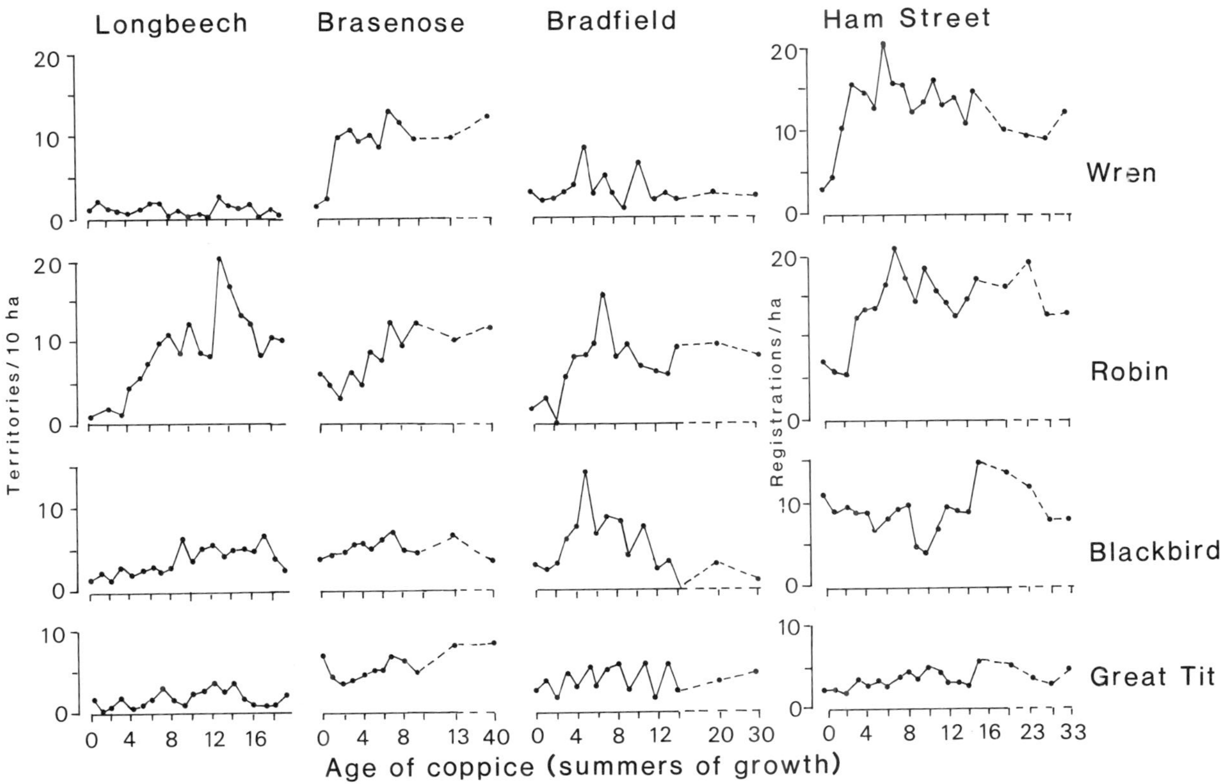

Figure 9.1 Abundance of four resident bird species in relation to age of the coppice in four woods. Note that for Ham Street Woods abundance is measured as registrations per hectare, but for the other woods it is measured as territories per 10 hectares. For details of the woods and the years they were censused see Table 9.1. Results from the different years have been averaged (see text).

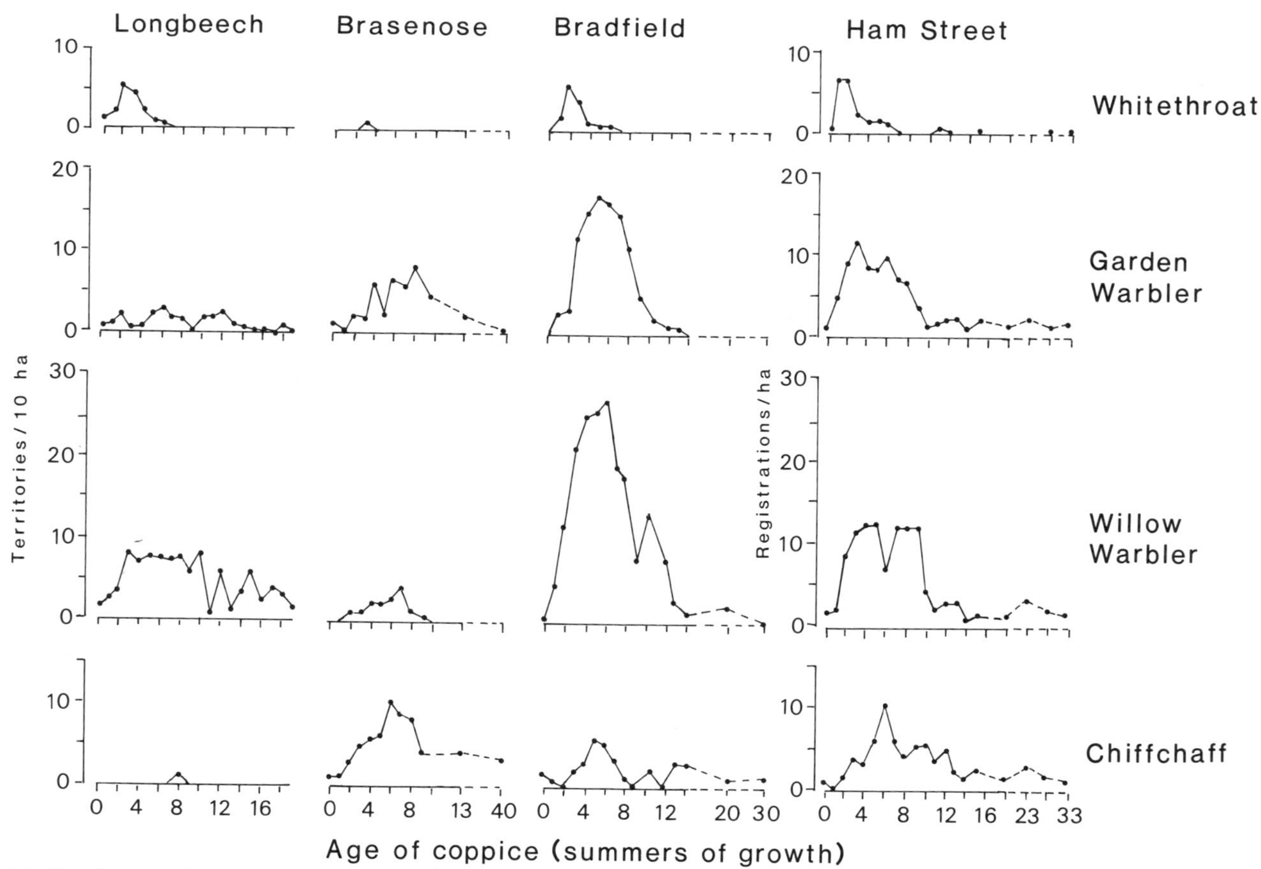

Figure 9.2 Abundance of four migrant bird species in relation to age of the coppice in four woods. Note that for Ham Street Woods abundance is measured as registrations/ha, but for the other woods it is measured as territories/10 ha. For details of the woods and the years they were censused see Table 9.1. Results from the different years have been averaged (see text).

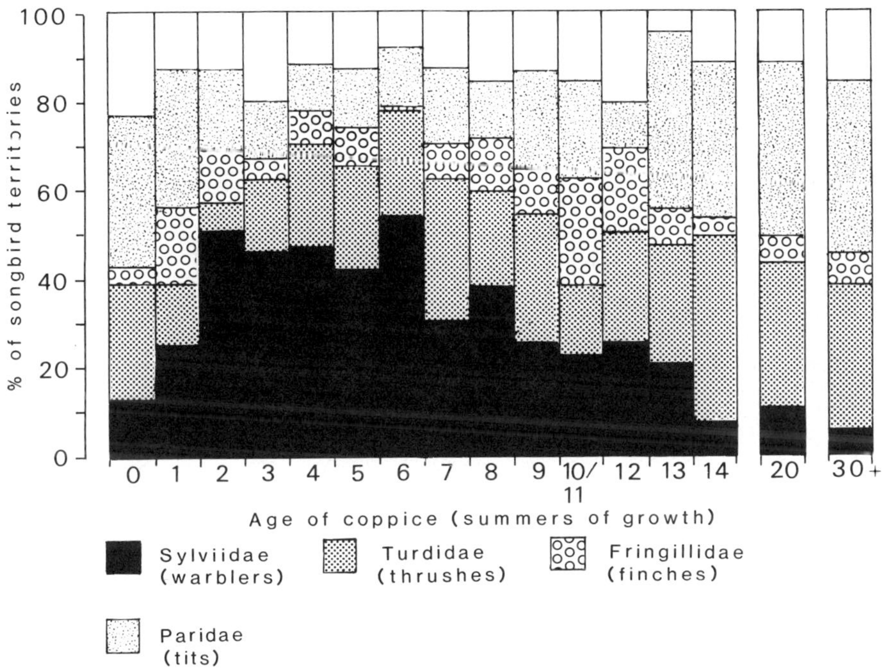

Figure 9.3 Percentage composition of the songbird communities in relation to age of the coppice in Bradfield Woods.

The broad composition of bird communities in Bradfield Wood is shown in Figure 9.3. The most striking feature was that the proportion of warblers was very low in the youngest and oldest classes of coppice. The greatest contribution of the warblers was between years 2 and 6 (Figure 9.4). The proportions of tits and thrushes also increased following canopy closure with these two families consistently contributing over 60% of the breeding songbirds in coppice of 13 years and over. These patterns of community composition were broadly representative of those in the other case study sites. The primary difference was that in the sweet chestnut (*Castanea sativa*) coppice of Longbeech Wood, warblers, thrushes, tits and finches contributed less than 40% of the birds in the first two years compared with more than 75% in Bradfield. This was partly because tree pipits were abundant in the young chestnut.

Coppice woodland can support exceptionally high densities of breeding migrant birds. These include six species which winter south of the Sahara (turtle dove (*Streptopelia turtur*), tree pipit, nightingale, whitethroat, garden warbler, willow warbler) and two which winter around the Mediterranean (blackcap and chiffchaff). With the exception of turtle dove, which can make considerable use of post canopy closure growth (Fuller and

Figure 9.4 Bradfield Woods, Suffolk in 1987, showing six-year coppice to the left, and 12-year coppice to the right of the ride. The former hold high densities of warblers and a nightingale territory; the latter hold few warblers and the breeding birds were dominated by robins and tits. (Photograph: R.J. Fuller.)

Moreton, 1987), all these migrant species are strongly associated with young or mid-growth (Appendices, Figures 9.2 and 9.3). This pattern of habitat selection is illustrated clearly by the ratio of migrant to resident songbirds which in all woods declines steeply soon after canopy closure (Figure 9.5). Where tree pipits are abundant in the earliest years, as at Longbeech Wood, the ratio shows an almost linear decline through the coppice cycle (Figure 9.5) but elsewhere the ratio is low in the first two years.

For each of the four case study woods similarities between the different years of coppice growth were assessed in terms of the composition of their songbird communities using species densities in two-way indicator species analysis (TWINSPAN) (Gauch, 1982). The year-classes were classified into more or less discrete temporal groups although the exact classification differed substantially between woods (Figure 9.6). In all cases the primary division isolated the youngest growth and/or the middle-aged growth. Old growth (post canopy closure) was always discrete from the youngest growth with the exceptions of year 12 in Bradfield and year 0 in Ham Street. The

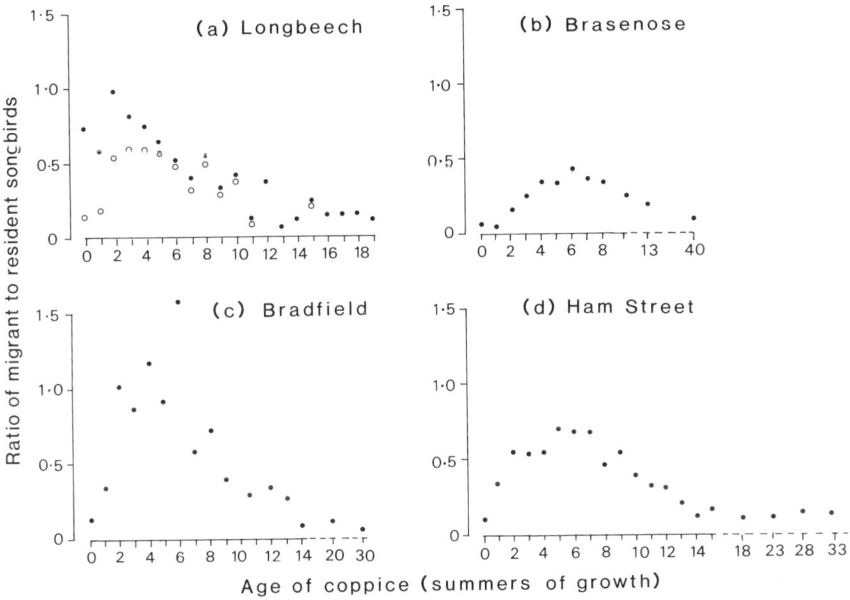

Figure 9.5 Ratios of migrant to resident songbird territories in four woods in relation to age of the coppice. Open circles for Longbeech Wood exclude tree pipit records.

establishment, canopy closure and maturation phases are identified in Figure 9.5.

9.4.2 Total density and species richness

Trends in total density of songbird territories with age of coppice varied substantially among the woods (Figure 9.7). In the sweet chestnut of Longbeech Wood, density was more or less even throughout the cycle, although it was lowest in the oldest growth. In the mixed coppices of Brasenose and Bradfield, however, densities were highest in the middle stages (years 3–7), although this pattern was by far the most marked in Bradfield. Higher densities were evident in the two mixed coppices than in the chestnut coppice.

Trends in species richness have been described in two woods: Longbeech (Fuller and Moreton, 1987) and Bradfield (Fuller and Henderson, in press). In both cases, richness in each year-class of coppice was corrected by rarefaction for variations in the areas of coppice available and in the numbers of individual birds. At Longbeech, richness slightly but significantly decreased with coppice age. At Bradfield, richness was highest in the mid-stage of growth (3–8 years) and was a consequence of the very high

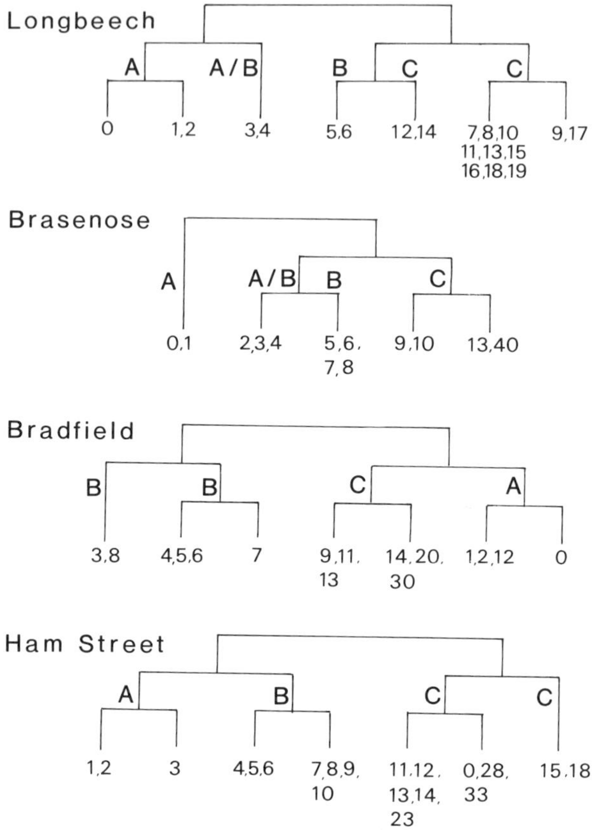

Figure 9.6 Similarity between coppice year-classes in the composition of their breeding song-bird communities in four different woods as shown by TWINSPAN. A, establishment phase; B, canopy closure phase; C, maturation phase.

densities of birds supported by that stage. When the numbers of individuals were standardized by rarefaction, no variation in species richness was detected with coppice age.

9.5 FACTORS INFLUENCING BIRD COMMUNITY COMPOSITION

9.5.1 Vegetation structure

The temporal changes in coppice bird communities broadly relate to a sequence in vegetation structure involving a gradient in canopy cover and a cycle of development and decline in the field and shrub layers. This sequence of events matches the temporal changes in bird communities closely. The high density of low foliage (<2 m) only occurs for a very small

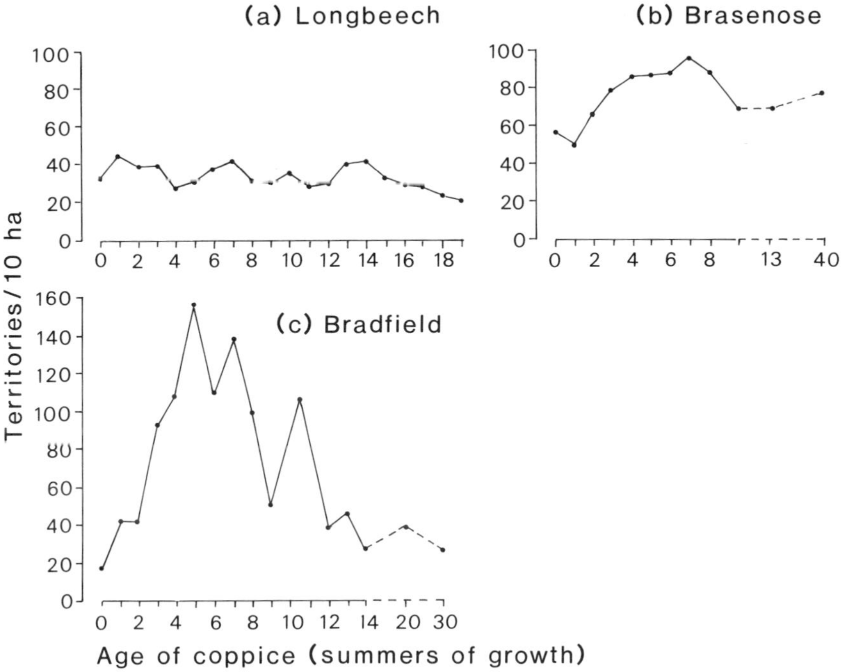

Figure 9.7 Total densities of songbirds in three woods in relation to age of the coppice.

number of years (Fuller *et al.*, 1989; Fuller and Henderson, in press). Unless there is a very meagre density of stools, this low vegetation rapidly thickens following cutting, usually reaching maximum density between years 3 and 5. This vegetation is composed partly of the low coppice bushes themselves and partly of other shrubs that established themselves between the stools during the open phase. As the canopy closes and the coppice gains height, this low vegetation rapidly thins until by year 8 or 9 the low foliage is extremely sparse (Figure 9.8). This cycle of vegetation structure drives the changes in bird communities described above. Some species have a requirement for very open habitats (e.g. tree pipit), others for dense low vegetation (nightingale and many warblers).

The probable importance of the shrub layer as an influence on bird distribution within coppice woods was further emphasized by work in Bradfield Woods (Fuller and Henderson, in press). The distribution of birds was examined in relation to a range of variables describing structure and composition of the vegetation. The analysis was based on 40 panels spread through the coppice cycle from freshly cut to more than 30 years old. Spatial variation in the density of the shrub layer was the single factor that most frequently explained distribution patterns of birds. Out of eight species examined with stepwise regression, the distributions of six were best

Figure 9.8 The interior of *c.* 14-year coppice at Bradfield Woods, Suffolk, showing a very old ash stool. Coppice of this age lacks the low, vigorous and dense vegetation required by many summer visitors. (Photograph: R.J. Fuller.)

explained by the shrub layer, each being more abundant where the shrub layer was well developed. These species (with the percentage variance in their densities explained by the shrub layer given in parentheses) were dunnock (55%), nightingale (22%), blackbird (26%), garden warbler (53%), blackcap (30%) and willow warbler (67%). Wren and robin were the other species examined, for which density of standard trees and canopy cover, respectively, were the single best variables. Variations in the shrub layer also gave the best explanations of the species richness (30%), density of migrants (64%) and overall density of songbirds (46%) in the panels.

It follows that any management activity that reduces the vigour of the shrub layer is likely to have a negative impact on the densities of many bird species within the coppice. Heavy browsing pressure from deer would have the same effect (see Chapter 12).

9.5.2 Tree species composition

Although gross changes in the woodland structure are of overwhelming importance in community dynamics within the coppice cycle, tree species composition certainly introduces variation into the bird communities. This is clearest when comparing chestnut with other types of coppice. Wren, chiffchaff and garden warbler were very scarce in the chestnut of Longbeech Wood compared with the three mixed coppice woods (Figures 9.1 and 9.2). Based on other woods documented in the Common Birds Census, chestnut coppice would seem to carry relatively small populations of warblers, with the exception of willow warbler and perhaps whitethroat. Chestnut coppice is also characterized by a distinctive assemblage of species in the establishment phase (Appendix 9.1). Species characteristic of chestnut but not of mixed coppice are yellowhammer, tree pipit and, perhaps, linnet (*Carduelis cannabina*). It is possible that the occurrence of these species is an effect of site rather than tree species. Chestnut may be grown more frequently on well-drained soils than mixed coppice and these species may have a preference for such sites.

Three factors seem relevant in explaining the low abundance of summer visitors in chestnut. First, chestnut may be a poor habitat for insectivorous birds because the biomass of invertebrates tends to be low in chestnut compared with other coppice species (Hill *et al.*, 1990). Second, the structure of chestnut coppice may be simpler than that of much mixed coppice. Chestnut casts a heavy shade and the density of stools is generally higher than in much mixed coppice, causing the rapid shading out of any vegetation beneath. Third, chestnut is generally grown on acid soils which may be less fertile than soils at many mixed coppice sites. A link has been established between soil productivity and the number of small birds in woodlands by Newton *et al.* (1986).

Birch (*Betula* spp.) saplings frequently invade young chestnut coppice but the amount of birch varies considerably depending on soil type and, probably, stool density. Most of these birch trees are shaded out as the chestnut canopy closes. It would be interesting to know whether those panels with relatively large amounts of birch hold the most songbirds.

Even within mixed coppice woodlands it is likely that tree species composition will influence bird distribution. Tree species differ in their attractiveness to feeding birds (Holmes and Robinson, 1981; Peck, 1989) presumably because of variations in the amount and type of food they provide, but also partly because different species are adapted to foraging in different types of tree. One might expect, therefore, that those panels richest in tree species (assuming that vegetation structure has been controlled) would be used by more species of birds. Indeed, there is some evidence that tree species composition does influence distribution of birds within Bradfield

Woods (Fuller and Henderson, in press). At the scale of the entire wood, bird distribution was strongly related to vegetation structure (see above). However, when the analyses were repeated separately for both young (3–8 years) and old (10–30 years) coppice, there was evidence of tree species effects, with birch having a negative effect on several measures of bird abundance. Studies of foraging, rather than of territory locations, are necessary to gain a better understanding of the role of tree species composition.

9.6 NIGHTINGALES IN COPPICE

The nightingale, probably more than any other bird, is popularly associated with coppice woodlands. Yet, in a national survey of the species in 1976 only 14% of the territories were recorded as in coppice (Hudson, 1979). Furthermore, the response of nightingales to coppice woodland is extremely variable. This is illustrated in Figure 9.9. Out of 17 coppiced sites with potentially suitable habitat – dense coppice three to eight years old (see also Bayes and Henderson, 1988) – only two held substantial nightingale populations. None of the four chestnut sites regularly held breeding nightingales. No sites with less than 3 ha of habitat available per year supported the species regularly, although some held a pair occasionally. Three of the woods in Figure 9.9 which had no nightingales in their coppice, did hold substantial numbers in non-coppiced habitats. As many of the woods included in Figure 9.9 were recently reinstated coppice regimes, it is possible that second or third rotation coppices may provide better quality habitats

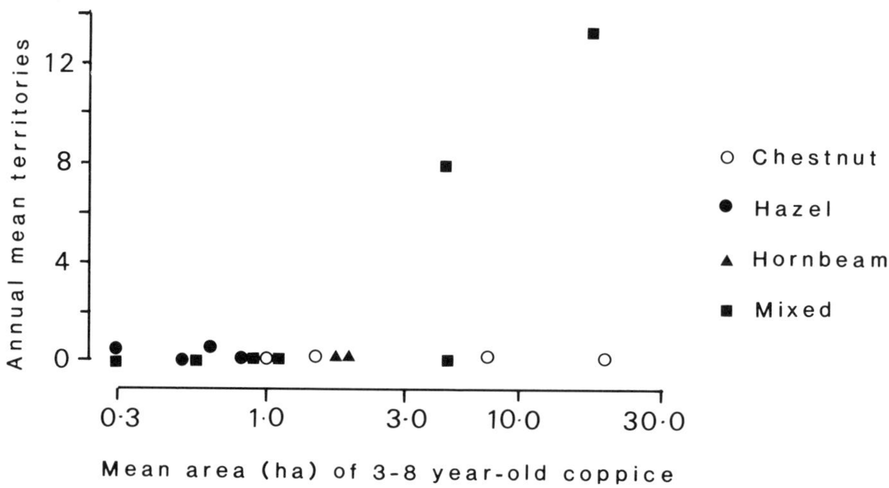

Figure 9.9 Breeding populations of nightingales in 17 coppiced woods in relation to the amount of potentially suitable habitat available (coppice of 3–8 years growth). ○, chestnut; ●, hazel; ▲, hornbeam; ■, mixed.

for nightingales than do first rotation coppices. Where nightingales are present in coppice, then continuation of coppicing may be essential for retaining the birds at that site. Whether reinstatement of coppicing can encourage nightingales to return to a site once they have left is less certain. It is possible that individuals show strong fidelity to sites (cf. Wiens *et al.*, 1986) and that birds do not colonize new sites readily, particularly ones distant from their natal areas.

9.7 COPPICE MANAGEMENT AND BIRDS

Managing coppice involves manipulating five variables: rotation length, density of standards, density of stools, size and location of panels, and species composition. Here I outline the likely effects, particularly on migrant songbirds, of varying these parameters. Management guidelines are available in Fuller and Warren (1990). I focus on the migrants because in general these are the species for which active coppice is particularly important. The effects of variations in coppice management on warblers were summarized by Fuller and Warren (1991). Tree species composition was discussed in section 9.5.2 and is not repeated here.

9.7.1 Rotation length

The rotation length determines the amount of establishment and canopy closure-phase habitat available. In many coppice woods species with narrow habitat preferences (i.e. those confined to one particular stage of coppice growth) are confined to the establishment or canopy closure phases (Fuller and Moreton, 1987; Fuller *et al.*, 1989). No bird species are confined to the maturation phase (i.e. the post canopy closure period). Long rotations do not benefit any particular species, nor do they enhance populations of breeding migrants, nor the overall diversity of songbirds. The effect of manipulating rotation length can be modelled where bird densities are available for different year-classes of coppice. Ideally this should take into account annual variations and any site-related effects.

An example is given in Figure 9.10 for Bradfield Woods where the effects on the overall warbler population of three different hypothetical rotations are predicted, based on warbler densities in 1987. With the entire wood under a very short rotation of 12 years or less, the predicted warbler population exceeds 200 pairs. Such short rotations are, however, an unrealistic proposition for the entire wood. Rotations of 20 to 25 years would still create habitats for some 120 to 140 pairs of warblers but with rotations exceeding 30 years the likely population would be less than 100 pairs. By cutting part of the wood on a short rotation (say 12 years) with a longer rotation on the remainder, the warbler population could be increased sub-

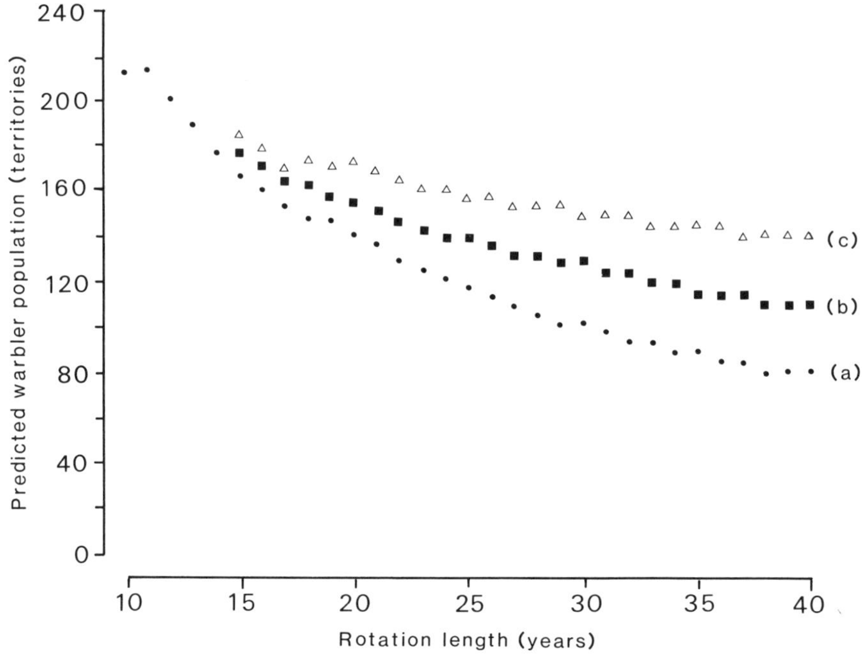

Figure 9.10 Predicted breeding populations of warblers (Sylviidae) in Bradfield Woods under three types of coppice rotation. In each case rotation lengths of up to 40 years were simulated: (a) a single rotation with the entire wood cut on a 10, 11, 12 . . . 40-year rotation; (b) a 'split' rotation with 25% of the wood cut on a 12-year rotation and the remainder on a 15–40-year rotation; (c) a split rotation with 50% of the wood cut on a 12-year rotation.

stantially for any given length of cycle in the remainder of the wood. The present objective of maintaining such a 'split' rotation in Bradfield Woods is clearly a good strategy for warblers. From an ornithological point of view it is desirable to devote as large a proportion of the wood as possible to the short rotation and to keep the long rotation to a minimum time limit.

9.7.2 Density of standard trees

Effects of standards on coppice bird communities are complex and poorly understood. There are at least four possible effects:

1. the provision of nest sites for hole-nesting birds and canopy nesters such as Corvidae;
2. the provision of additional foraging sites for insectivorous birds such as chaffinches (*Fringilla coelebs*) and tits;

3. the provision of song posts (e.g. for tree pipits);
4. the suppression of the shrub layer where the number of standards is large.

There has been no work on the first three aspects so they are not dealt with in detail here. It is likely, however, that where there are many standards, canopy feeders such as blue tits (*Parus caeruleus*) and chaffinches will be numerous throughout the coppice cycle; in contrast they will be scarce or absent in young coppice which has no standards.

There is empirical evidence to show that standards can affect the quality of coppice as a habitat for breeding migrants. In Ham Street Woods, Fuller *et al.* (1989) found that most migrant species were more abundant in young coppice with few standards (<25/ha) than in that with many standards (>50/ha). This was particularly marked for whitethroat. Resident species showed no such differences. Whether this effect was due to the standards inhibiting coppice growth (i.e. the shrub layer upon which the migrants depend) or due to the birds simply preferring a more open habitat is not clear, although the latter seems quite likely in the case of the whitethroat. There was clear evidence from Bradfield Woods, however, that the height and cover of coppice was considerably suppressed beneath large standards (Table 9.2).

Table 9.2 Coppice growth of hazel (*Corylus avellana*) and ash (*Fraxinus excelsior*) in relation to presence of standard trees in Bradfield Woods, Suffolk. Values given are paired sample means; sample sizes were equal for both categories*

	Close to standard	*Away from standard*	P (*Wilcoxon test*)
Hazel stool height (1 year, n=58)	2.2 m	2.3 m	0.09
Hazel stool height (2 year, n=27)	2.0 m	2.2 m	0.29
Hazel stool height (4 year, n=30)	2.4 m	2.8 m	0.03
Ash stool height (2 year, n=18)	3.2 m	4.1 m	<0.01
Coppice cover (0–4 years, n=30)	36%	53%	<0.01

*Based on data collected in 1987 by R.J. Fuller and C.M. Ray. Two types of measurements were made. First, three areas with clusters of five to nine oak standards were selected; these were in panels of 1, 2 and 4 years growth. For ash and hazel stools within 5 m of the standards, heights were measured to the nearest 0.5 m. The heights of a comparable number of stools were then measured that were not under the canopy of a standard. Second, all oak standards (>40 cm diameter at breast height) growing in 0, 1, 2 and 4-year coppice were visited. Paired samples were used as follows: percent coppice cover within 5 m of the tree was estimated by eye and compared with coppice cover within a 5 m radius of a nearby point, at the same distance from the panel edge, but not under another standard.

The effects of standards on the underwood are probably complex, e.g. a moderate density of standards may have the effect of slowing down coppice growth and hence prolonging the period over which the shrub layer is suitable for migrants; too high a density of standards would result in very patchy, thin coppice regrowth. Fuller *et al.* (1989) found that migrants, most strikingly nightingales, were more abundant in coppice of 3 to 7 years growing beneath a high density of standards. They suggested this might be a consequence of bramble (*Rubus fruticosus*) and low shrubs persisting for longer in the vicinity of standards, perhaps because there is a phase when light is stronger beneath the standards than beneath the coppice canopy itself. Stuttard and Williamson (1971) went so far as to suggest that nightingales required coppice with a fairly high density of standards, although they did not control for the age of the coppice. At Bradfield Woods nightingales were abundant in coppice of 4–7 years' growth with very low densities of large standards.

9.7.3 Density of stools

A high density of stools may lead to rapid canopy closure, giving little opportunity for other plants to establish themselves between the stools. The consequences of this for birds are uncertain. On the other hand, a very low density of stools is definitely disadvantageous. This is a common feature of derelict coppices due to the deterioration of the hazel (*Corylus avellana*). Without restocking, the underwood, if cut, will be extremely patchy and may never close canopy. This type of coppice structure is a poor habitat for breeding migrants (Fuller and Steel, 1990).

9.7.4 Size and location of panels

The area requirements of most small woodland birds have not been well documented but most of the summer visitors hold breeding territories of at least 0.5 ha. It is desirable, therefore, not to cut panels less than this size. Especially where only small areas can be cut each year, it is sensible practice to cut panels in a sequence so that consecutive year-classes are adjacent. This will increase the available patch size for many species. No positive effects of panel size on the density of any species in closed-canopy coppice could be detected in Bradfield Woods in 1987, but an edge effect in vegetation structure was detected where young coppice adjoined old growth (R.J. Fuller, unpublished data). The shrub layer was suppressed at the edge of young coppice presumably as a result of shading or deer browsing. Within the old coppice, however, the shrub layer appeared to be slightly denser adjacent to the open edge – perhaps as a consequence of light penetration. Such edge effects have implications for the design of coppice woods where the creation of large areas of shrubby growth to benefit populations of migrant birds is a management objective. Any reduction of the shrub layer in young panels due to such edge effects would be increasingly severe in small panels, especially those less than 0.5 ha (Fuller *et al.*, 1989a).

9.8 MANAGED COPPICE COMPARED WITH OTHER TYPES OF BROADLEAVED WOODLAND

No direct comparisons have been made of the bird communities associated with actively managed coppice, derelict coppice and high forest. Some general observations are possible, although it must be remembered that considerable variation in bird communities occurs within coppice and high forest. Derelict coppice carries very similar bird communities to those in managed coppice that has grown several years beyond canopy closure. These stands have low species richness and are strongly dominated by robin and tits. They lack the warblers that characterize the young and middle-aged coppice but they also hold lower densities of hole-nesting birds than do many high forest stands (Fuller, 1988). Those warblers that do use old coppice are mainly confined to those edges adjoining younger growth (Fuller *et al.*, 1989).

Taking a long-term view of derelict coppice, it seems likely that its bird communities would potentially be enriched by either reversion to active coppice or by promotion to high forest. Support for this comes from counts of birds in coppiced stands of lime (*Tilia cordata*) at Shrawley Wood where derelict coppice held lower numbers of birds than recently thinned (singled) coppice and, especially, recently cut coppice (Fuller, 1990a).

High forest systems, because they are managed on much longer rotations, do not offer continuity of such large areas of potentially suitable habitat for breeding migrants as do coppice woods. Also, there may be differences of habitat structure in the young growth of coppice and high forest that influence the responses of warblers and other migrants (Fuller and Warren, 1991). Within mature lowland high forest, densities of warblers often tend to be rather low with the external edges holding the largest populations (Fuller, 1988). Conversely, hole-nesters such as nuthatches (*Sitta europaea*) and woodpeckers are rarely abundant within coppice – they generally reach higher densities in high forest and wood pasture. No species of bird is confined to coppice habitats in Britain and no particularly rare species are dependent on it. However, all species of woodland birds in southern England will breed in coppice woods – with the possible exceptions of redstart (*Phoenicurus phoenicurus*) and wood warbler (*Phylloscopus sibilatrix*).

9.9 CONCLUSIONS

This chapter has emphasized three aspects of bird communities in coppice woodlands: (a) the enormous variation that exists between coppice woods, (b) the profound influence of coppice management, and (c) the fundamental differences between coppice and high forest.

The great variation in coppice bird communities makes it difficult to predict the response of birds to coppicing where it is being started afresh. In general, the response of breeding migrants to coppice rotations is strongest where the management is conducted in such a way that a dense shrub layer is created around the canopy closure period. However, nightingales often do not colonize fresh coppice coupes, even where the habitat appears suitable, although the response of warblers seems rather more predictable.

Coppicing can create a spectrum of woodland habitats ranging from open scrubby vegetation, with no large trees at all, to woods heavily stocked with standards where the canopy above the coppice is almost closed. This habitat gradient is probably matched by a spectrum of bird communities, ranging from those containing species typically associated with scrub to those more associated with high forest. Somewhere along this gradient may lie communities which are rich both in young-growth species and old-growth species. Research is needed to assess where this balance between standards and underwood might lie. A multidisciplinary research project involving experimental manipulation of densities of standard trees growing over coppice could be extremely rewarding.

The results presented here are based on counts of breeding birds in different ages of coppice growth. Whilst this approach is adequate for determining broad patterns of habitat selection it provides little understanding of the ways in which birds really use such patchy environments. How do individuals respond to the considerable variety of vegetation structures offered by even small areas of coppice woodland? Do some species select one age of coppice for nesting and another for feeding? Are panel edges selected or avoided? The foregoing discussion may help to formulate hypotheses which need to be followed up by detailed studies.

Neither active coppice nor high forest bird assemblages are intrinsically more interesting than the other, and a balanced strategy for woodland conservation should seek a place for both. An example of this approach is the long-term policy developed for ten woodland reserves owned by the Berkshire, Buckinghamshire and Oxfordshire Naturalists' Trust (Fuller, 1990b). These woods cover 358 ha, of which it is intended that some 17% should be managed as coppice, 67% as high forest and 15% as non-intervention areas. The treatment for each wood has been selected to suit its particular wildlife interest, while a practical balance of woodland types has been sought for the whole estate. This policy should produce a complementary range of bird and other animal communities within the Trust's woodland estate.

In the wider countryside coppicing may be set to make something of a modest comeback, stimulated by the interest in farm woods and energy crops, and by the announcement in 1990 of planting and management grants for coppice. What might be the implications for birds? The fast growing 'biomass' crops, including non-native alders, hybrid willows and hybrid poplars, are unlikely to support bird populations of conservation interest, although research on this is desirable. The trees are grown at extremely high density and they close canopy far more rapidly than 'traditional' coppice. Insect populations, and hence food availability for many birds, are probably low. Where existing woods are being brought into a coppice rotation the information presented here and in Fuller and Warren (1990) may help in creating woods that are both commercially viable and rich in birds and other wildlife. In summary, the desirable features are an underwood not dominated by sweet chestnut, a moderate density of standards, the operation of short or 'split' rotations, panels no smaller than 0.5 ha, and a continuity of young and middle-aged growth.

ACKNOWLEDGEMENTS

My thanks go to the following people whose Common Birds Census plots have contributed towards this paper: K.G. Adsett, L.H. Bricknell, Mr and Mrs B. Flack, C.J. Glover, Bernard Moreton, N.W. Owens, the late M.J. Penistan, R.S. Pitcher, Jeff Stedman, David Steel, Peter Stuttard, Mrs C. Taylor, Vyvyan Veal, R.L.C. Williamson. I also thank Harry Green, Andrew Henderson and Carolyne Ray for help with fieldwork. Dr David Hill kindly commented on a draft of this chapter and Liz Murray drew the illustrations. The Common Birds Census and the BTO's programme of research on woodland birds are funded by the Joint Nature Conservation Council and by the Department of the Environment for Northern Ireland.

APPENDIX 9.1 Occurrence of bird species in the establishment phase (0–2 years of growth) of coppice cycles in nine woods.

Key: +, strong association with the establishment phase; –, presence; 0, absence.

Species	*Chestnut*			*Hornbeam*	*Mixed coppice*				
	1	2	3	4	5	6	7	8	9
Turtle dove (*Streptopelia turtur*)	–	0	–	–	0	0	0	–	–
Nightjar (*Caprimulgus europaeus*)	0	–	0	0	0	0	0	0	0

Tree pipit (*Anthus trivialis*)	+	+	−	0	0	−	0	0	0
Wren (*Troglodytes troglodytes*)	−	−	−	−	+	−	−	−	−
Dunnock (*Prunella modularis*)	−	+	+	−	0	−	−	+	−
Robin (*Erithacus rubecula*)	−	−	−	−	−	−	−	−	−
Nightingale (*Luscinia megarhynchos*)	0	0	0	0	0	0	0	−	0
Blackbird (*Turdus merula*)	−	−	+	−	+	−	−	+	−
Song thrush (*Turdus philomelos*)	−	−	−	+	0	−	−	+	−
Whitethroat (*Sylvia communis*)	+	−	0	0	0	+	0	+	−
Garden warbler (*Sylvia borin*)	−	−	−	0	0	−	−	−	−
Blackcap (*Sylvia atricapilla*)	0	−	0	−	−	−	−	−	−
Chiffchaff (*Phylloscopus collybita*)	0	−	0	−	0	0	−	−	−
Willow warbler (*Phylloscopus trochilus*)	−	−	−	−	0	−	−	−	−
Marsh tit (*Parus palustris*)	0	0	−	−	−	+	−	−	−
Willow tit (*Parus montanus*)	0	0	0	−	0	0	−	0	0
Blue tit (*Parus caeruleus*)	−	−	−	−	−	−	−	−	−
Great tit (*Parus major*)	0	−	−	−	0	+	−	−	−
Chaffinch (*Fringilla coelebs*)	−	−	+	−	−	+	+	+	−
Greenfinch (*Carduelis chloris*)	0	0	0	+	0	−	0	+	−
Linnet (*Carduelis cannabina*)	+	0	0	0	0	0	0	−	0
Bullfinch (*Pyrrhula pyrrhula*)	−	−	−	−	0	−	−	−	−
Yellowhammer (*Emberiza citrinella*)	+	+	+	0	0	0	0	+	0

Notes

1. Strong associations were defined as species recorded in all three year-classes and where at least one of the two years with the highest density (for year-classes 0–8) was in the establishment phase.
2. Birds were counted in all woods using territory mapping and in most cases abundance was estimated from the density of registrations (Fuller *et al.*, 1989b).
3. The nine woods were: **1** Sweet chestnut coppice, Kent, 22 ha, censused 1975–84; **2** Sweet chestnut coppice, Kent, 40 ha, 1970–75; **3** Sweet chestnut coppice, Kent, 5 ha, 1979–85; **4** Hornbeam coppice, Essex, 4 ha, 1978–85; **5** Mixed coppice, Gloucester, 3 ha, 1976–84; **6** Mixed coppice, West Sussex, 5 ha, 1980–86; **7** Mixed coppice, Oxfordshire, 18 ha, 1980–87; **8** Mixed coppice, Kent, 24 ha, 1969–73; **9** Mixed coppice, Suffolk, 62 ha, 1987.

APPENDIX 9.2 Occurrence of bird species in the canopy closure phase (3–8 years of growth) of coppice cycles in nine woods

Key: +, strong association with the canopy closure phase; –, presence; 0, absence. For scientific names of birds see Appendix 9.1.

Species	*Chestnut*			*Hornbeam*	*Mixed coppice*				
	1	2	3	4	5	6	7	8	9
Turtle dove	–	+	–	–	0	–	0	+	+
Nightjar	–	0	0	0	0	0	0	0	0
Tree pipit	–	–	–	0	0	–	0	0	0
Wren	–	+	+	+	–	+	+	+	+
Dunnock	–	+	+	+	–	+	+	+	+
Robin	+	+	+	+	+	+	+	+	+
Nightingale	–	0	0	0	0	–	0	+	+
Blackbird	+	+	+	+	+	+	+	+	+
Song thrush	+	–	–	+	–	+	+	+	+
Whitethroat	–	–	0	0	–	–	–	–	–
Garden warbler	–	–	–	–	0	+	+	+	+
Blackcap	–	–	–	+	–	+	+	+	+
Chiffchaff	–	+	–	+	0	–	+	+	+
Willow warbler	+	+	+	+	0	+	+	+	+
Marsh tit	0	–	+	+	–	–	–	–	+
Willow tit	–	0	0	–	0	0	–	–	–
Blue tit	–	+	+	+	+	+	+	+	+
Great tit	–	+	+	+	0	+	+	+	+
Chaffinch	–	–	–	+	–	+	–	–	+
Greenfinch	–	–	–	–	0	–	0	–	+
Linnet	–	–	0	0	0	–	0	–	0
Bullfinch	+	–	–	+	–	–	+	+	+
Yellowhammer	+	+	+	0	0	0	0	–	–

Notes
Strong associations were defined as species recorded in all six year-classes and where one of the two years' growth with the highest density (for years 0–8) was in the canopy-closure phase.
See also notes 2 and 3 in Appendix 9.1.

REFERENCES

Bayes, K. and Henderson, A. (1988) Nightingales and coppiced woodland. *RSPB Conservation Review*, **2**, 47–9.

Fuller, R.J. (1988) A comparison of breeding bird assemblages in two Buckinghamshire clay vale woods with different histories of management, in *Woodland Conservation and Research in the Clay Vale of Oxfordshire and Buckinghamshire*, Research and Survey in Nature Conservation, 15 (eds. K.J. Kirby and F.J. Wright, Nature Conservancy Council, Peterborough, pp. 53–65.

Fuller, R.J. (1990a) Responses of birds to lowland woodland management in Britain: opportunities for integrating conservation with forestry. *Sitta*, **4**, 39–50

Fuller, R.J. (1990b) *An integrated policy for the long-term management of*

BBONT's woodland nature reserves. Berkshire, Buckinghamshire and Oxfordshire Naturalists' Trust, Oxford.

Fuller, R.J. and Moreton, B.D. (1987) Breeding bird populations of Kentish sweet chestnut (*Castanea sativa*) coppice in relation to age and structure of the coppice. *Journal of Applied Ecology*, **24**, 13–27.

Fuller, R.J. and Steel, D. (1990) Coppicing in Brasenose Wood, Oxfordshire: the response of breeding birds. *Fritillary*, **1**, 5–15.

Fuller, R.J., Stuttard, P. and Ray, C.M. (1989b) The distribution of breeding songbirds within mixed coppiced woodland in Kent, England, in relation to vegetation age and structure. *Annales Zoologici Fennici*, **26**, 265–75.

Fuller, R.J. and Warren, M.S. (1990) *Coppiced Woodlands – their Management for Wildlife*, Nature Conservancy Council, Peterborough.

Fuller, R.J. and Warren, M.S. (1991) Conservation management in ancient and modern woodlands: responses of fauna to edges and rotations, in *The Scientific Management of Temperate Communities for Conservation*, 31st Symposium, British Ecological Society, (eds. I.F. Spellerberg, F.B. Goldsmith and M.G. Morris), Blackwell Scientific Publications, Oxford, 445–71.

Gauch, H.G. (1982) *Multivariate Analysis in Community Ecology*, Cambridge University Press, Cambridge.

Hill, D., Roberts, P. and Stork, N. (1990) Densities and biomass of invertebrates in stands of rotationally managed coppice woodland. *Biological Conservation*, **51**, 167–76.

Holmes, R.T. and Robinson, S.K. (1981) Tree species preferences of foraging insectivorous birds in a northern hardwoods forest. *Oecologia*, **48**, 31–5.

Hudson, R. (1979) Nightingales in Britain in 1976. *Bird Study*, **26**, 204–12.

International Bird Census Committee (1969) Recommendations for an international standard for a mapping method in bird census work. *Bird Study*, **16**, 249–55.

Marchant, J.H. (1983) *BTO Common Birds Census Instructions*, British Trust for Ornithology, Tring.

Marchant, J.H., Hudson, R., Carter, S.P. and Whittington, P. (1990) *Population Trends in British Breeding Birds*, British Trust for Ornithology, Tring.

Newton, I., Wyllie, I. and Mearns, R. (1986) Spacing of sparrowhawks in relation to food supply. *Journal of Animal Ecology*, **55**, 361–70.

Peck, K. (1989) Tree species preference shown by foraging birds in forest plantations in northern England. *Biological Conservation*, **48**, 41–57.

Stuttard, P. and Williamson, K. (1971) Habitat requirements of the nightingale. *Bird Study*, **18**, 9–14.

Wiens, J.A., Rotenberry, J.T. and Van Horne, B. (1986) A lesson in the limitations of field experiments: shrubsteppe birds and habitat alteration. *Ecology*, **67**, 365–76.

10

Coppice management for pheasants

C.E. BEALEY and P.A. ROBERTSON

10.1 INTRODUCTION

The pheasant (*Phasianus colchicus*) is the most numerous gamebird in Britain. It comprises over 80% of all quarry species shot and is the mainstay of lowland game shooting with at least eight million birds shot per year.

Pheasants exhibit seasonal variation in habitat selection: woodlands are used extensively during the winter but the birds move out into growing crops during the spring and summer (Ridley, 1983; Hill and Robertson, 1988). During the winter pheasants select certain types of woodland in preference to others. Densities, as shown by trapping success, are significantly higher in areas of scrub and woodland with a well-developed shrub layer than in bare or grazed woodland, hedgerows or open fields during the winter (Robertson, 1985).

Apart from providing the main overwintering sites for pheasants, woodlands are also an integral part of male pheasant breeding territories and the birds select woodland edges with well-developed shrub layers (Lachlan and Bray, 1976; Ridley, 1983; Hill and Robertson, 1988). Lachlan and Bray (1976) examined the tree and shrub species and structural composition of woodland within male breeding territories and concluded that species composition was relatively unimportant provided the structure of the woodland was suitable.

Pheasant shooting is a widespread winter activity in many areas of lowland Britain. It is popular with many landowners and can be an important influence on their attitudes to woodland planting, design and management. Piddington (1980) estimated that pheasant shooting took place on 58% of agricultural properties with this rising to 88% of those of more than 400 ha. She also found that 33% of owners had planted or retained coverts, belts or spinneys for game.

Similarly, a questionnaire survey of Country Landowner Association members (Cobham Resource Consultants, 1983) found that 67% of respondents claimed that game interests were a reason for retaining existing woods, less than 10 ha in size, while 56% indicated game was also a reason for planting new woodland. Game was second only to beauty in the landscape as an incentive for woodland management, being given more frequently than either timber production or wildlife conservation. More recently, Ludolf *et al.* (1989) sent a questionnaire to 400 contributors to The National Game Census, a self-selected sample of estates with a keen interest in shooting, requesting details of woodland management carried out to encourage pheasants. From 150 replies received, 81% carried out some form of management for pheasants with over 30% using new woodland planting, felling, restocking and coppicing as game management tools.

Advocates of game management have long claimed that shooting can lead to benefits for conservation (Simpson, 1907; Gray, 1986; Page, 1987). We suggest that pheasant shooting provides a financial and personal incentive for landowners to carry out many forms of management that would otherwise be uneconomic but which have side benefits for other species. Coppicing is a technique of known value for the conservation of certain rare or endangered species, but its use has seriously declined in recent years (Warren, 1976; Peterken, 1981). Moreover, although coppicing is used by a significant proportion of shoot managers, in many cases they do not reinstate a full coppice cycle and may only manage relatively small areas of a wood (e.g. 25 m × 25 m). Coppicing may occur sporadically in one or two sites, but the continuity of a full rotation with a mixture of age classes is often absent.

In the first part of this chapter we describe recent studies to quantify the woodland habitat requirements of the pheasant and to predict the role of coppicing in providing suitable conditions for the species. Secondly, we present preliminary findings on the effect of re-introducing small-scale coppicing on the ground flora of two woodlands. This chapter draws on a recent report which contains full details of the methods and analyses used (Robertson *et al.*, 1989).

10.2 WOODLAND HABITAT SELECTION BY PHEASANTS DURING THE WINTER

To investigate the use of woodlands by pheasants during the winter ten estates were visited on shoot days and, from a position behind the guns, the number of pheasants flushed from cover in each drive was recorded. Data were collected from 155 different areas of woodland. On a subsequent visit to each area the cover provided by the vegetation in each drive was measured at various height levels, both within the wood and around its edge. This involved recording the presence or absence of vegetation at 0–1, 1–10, 11–30, 31–100, 101–200, 201–500 and >500 cm above the ground within a 30 cm × 30 cm column at 60 points within the woodland selected on a stratified random basis. This presence or absence data (percentage of transect points where cover was present in that height category) was then combined to give a figure for the percentage cover at each height category for the woodland (a) edge and (b) interior. All percentages were transformed to arcsines for later analysis. Other recorded variables included the area of each wood, the length of woodland edge, the extent of artificial feeding, whether the wood was predominantly broadleaved, coniferous or mixed and the number of birds released. This last variable was an average figure for each estate as it proved impossible to obtain reliable figures for individual woods. Of the ten sites visited releasing occurred on six, the remainder relying purely on wild birds.

The measures of vertical structure proved to be strongly intercorrelated. To avoid problems of interpretation resulting from stepwise multiple regressions using intercorrelated variables the measures of vertical structure were considered as a group. If one was selected in the stepwise regression, the remainder were then forced into the model. Forcing intercorrelated variables into an equation may be statistically questionable, but in this case the measures of vertical structure form a biologically meaningful group. The inclusion of a number of them in isolation would oversimplify the situation. The aim of this analysis was not to maximize the explained variance but rather to produce a robust model for predictive purposes.

In a forward stepwise multiple regression, constrained through the origin with $P<0.05$ to enter or remove variables, pheasant density was negatively related to woodland area ($P<0.001$) and to the measure of cover, 100–200 cm in height ($P<0.001$). Subsequent forcing of all seven measures of internal vertical structure into the model led to a significant improvement in the relationship ($F_{7,145} = 36.72$, $P<0.001$).

The number of pheasants released in each area was not a significant variable. Although pheasant shooting is largely supported by the release of hand-reared birds, the numbers that remain in an area until the shooting season appear to be determined by the suitability of the habitat and the extent of supplementary feeding. The data was collected from well-managed pheasant estates. Unmanaged areas of farmland almost certainly contain lower densities.

These findings are supported by other studies which have highlighted the importance of vertical structure within woodland (Ridley, 1983; Robertson, 1985). In addition, the preference of pheasants for small woodlands can be related to the high degree of selection for woodland edges described by studies of radiomarked pheasants, small woods having a higher edge to area ratio (Robertson *et al.*, 1989).

**F* = standard statistical abbreviation

10.3 PREDICTING THE POTENTIAL HOLDING CAPACITY FOR PHEASANTS OF COPPICE WOODLAND

The data used to relate pheasant density to the range of variables described above were collected from a wide range of woodland types. No attempt was made to determine pheasant density in particular woodland stand-types or age-classes. However, the measures of vertical structure proved to be a major factor influencing density. Such measures can be used in a predictive sense by using the regression equation developed in the previous section to calculate a relative index of attractiveness in different woodlands of known management. To this end data on the vertical structure were collected from woodlands in the same way as described earlier. These were then entered into the regression equation with the area of the wood arbitrarily set at 1 ha and assuming that no supplementary feeding was being conducted. This allowed relative indices of attractiveness based on vertical structure to be compared between different stand-types. The regression equation used was as follows:

$$\text{Pheasant Density (Birds/ha}(\log_{10}+1)) = (x^{*}-0.5242) + (y^{*}0.2525)\ (a^{*}0.0030) + (b^{*}0.0021) + (c^{*}0.0006) + (d^{*}0.0042) + (e^{*}0.0112) + (f^{*}0.0006) + (g^{*}0.0097)$$

where x is the woodland area in ha ($\log_{10} + 1$); y is the presence or absence of feeding (1 or 0); and a–g are values for percentage cover of vegetation at various heights above ground (arcsin transformed): a, 0–1 cm; b, 0–10 cm; c, 11–31 cm; d, 31–100 cm; e, 101–200 cm; f, 201–500 cm; g, >500 cm.

Data on vertical structure were collected from an area of managed hazel (*Corylus avellana*) coppice at Clarendon Park near Salisbury, Wiltshire, and from managed hornbeam (*Carpinus betulus*) coppice in the Ham Street Woods near Ashford, Kent, in February and March 1990. Data were collected from coupes with 0, 1, 2, 3, 4, 5, 6 and >10 years of regrowth at Clarendon and 0, 1, 3, 5, 6, 8, 11, 15, 18 and 21 years of regrowth in Ham Street.

The indices of relative attractiveness for the age-classes of coppice in each wood are presented in Figure 10.1. In both types of coppice a similar pattern emerged. The suitability of each area rose in the first few years after cutting, reaching a peak at three years for the hazel and six years for the slower growing hornbeam, before declining to moderate levels in the older age-classes.

This analysis only considers the suitability of isolated coppice compartments for the cover they provide. In actuality the boundaries between older coppice and newly cut areas may also provide extra 'woodland edges' for pheasants, thus increasing the attractiveness of the whole wood compared

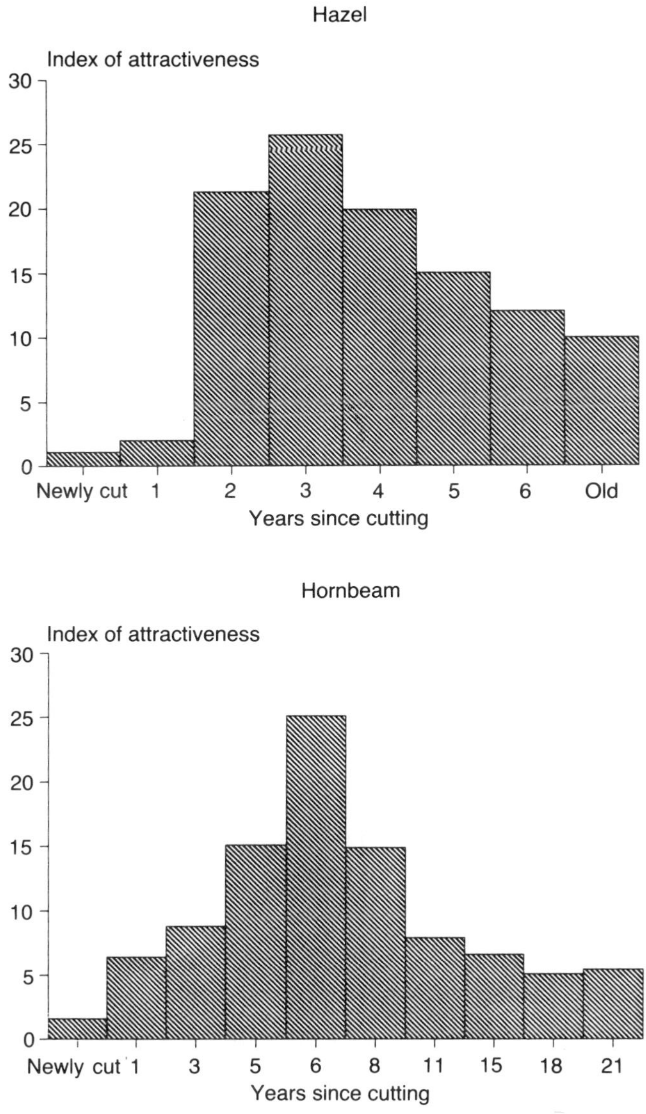

Figure 10.1 The relative attractiveness to pheasants, based on vertical structure, of different age classes of hazel and hornbeam coppice. This assumes each age class constituted of a single block of 1 ha surrounded by open ground and that no supplementary feeding took place.

with a similar unmanaged area. While predictive analyses oversimplify the factors influencing winter pheasant density, two points are clear. First, managed coppice provides a more attractive habitat to the pheasant than unmanaged coppice. Secondly, if the attractiveness is to be maximized then short rotations, in these examples consisting of three to four years in hazel and around six years in hornbeam, would maximize both the area at the most attractive phase and also the length of boundary between shrubby and newly cut areas. Soil type and stocking densities of coppice and the presence of standards will also affect the results.

10.4 THE EFFECTS OF COPPICING FOR GAME MANAGEMENT ON WILDLIFE

Following the decline in active coppice management many small woods on farmland are now relatively unmanaged or derelict, and many of the most valuable remaining sites for wildlife are now protected and managed by conservation bodies. However, this approach in itself can do little to maintain the species once associated with coppice in the wider countryside. Pheasant management can provide one incentive for landowners to maintain or re-introduce active coppicing. Although there are notable cases where complete coppice cycles are maintained by an interest in pheasant shooting, the majority of coppicing on pheasant shoots now involves relatively small areas cut on a sporadic basis. The value of coppicing for pheasants is recognized by many landowners but it is a labour-intensive technique and only relatively small areas at strategic places within woods are managed.

To determine ways in which standard pheasant management could be improved from a conservation viewpoint, a system of woodland monitoring was established in six relatively unmanaged woods in 1987. After the re-introduction of woodland management for pheasants, the effects on a range of different wildlife groups were monitored over the following years. The following sections focus on the effects of re-introducing coppice management on the ground flora of two of the sites.

10.4.1 Study sites and method

Edmondsham Wood. This 7 ha wood in Dorset lies on heavy clay soils, parts of which are waterlogged throughout the year. The tree layer is dominated by oak (*Quercus robur*) standards over degenerate hazel and ash (*Fraxinus excelsior*) coppice giving way in the wetter areas to alder (*Alnus glutinosa*) and sallows (*Salix cinerea* and *S. caprea*). The ground layer contains a high number of plant species typically restricted to ancient woodland, indicating a wood of ancient, semi-natural origin (Peterken, 1981).

Buttermead Copse. This 2.5 ha wood in Hampshire is on level ground where clay overlies chalk. Although more freely drained than the previous site it also consists of oak standards over degenerate hazel and ash coppice but with an admixture of blackthorn (*Prunus spinosa*), field maple (*Acer campestre*) and hawthorn (*Crataegus monogyna*). The ground flora also suggests a long history of woodland continuity and contains many ancient woodland indicator species.

In neither wood had there been any significant management for at least 20 years. The coppice growth, particularly the hazel, was over-mature and in some cases individual stools had begun to disintegrate through windthrow. Heavy shade cast by the understorey had led to considerable suppression of the ground flora. In general, the two sites were typical of neglected coppice-with-standards woods common in central southern England.

At Edmondsham, three 25 m × 25 m 'skylights' were cut for game by coppicing the understorey in the winters of 1987–1988 and 1988–1989 while two sections of the central ride were widened to 20 m over the same period (Figure 10.2). All cutting involved coppicing the hazel stools within the treatment areas, but no standards were removed. In Buttermead Copse a single coupe of 1 ha was cut in the winter of 1987/88 from the hazel coppice, retaining a low density of uncut ash stools. A thin strip of uncut stools was left around the edge of the coupe to provide a wind-break and visual screen (Figure 10.3). Some standards were felled during the winter of 1988–1989.

Although the treatments were unreplicated each wood had a large 'control' area for comparison where no management was carried out. The two woods were intended to represent contrasting scales of management, one typical of small-scale felling for gamebirds, the other reflecting more traditional coppice management, but on a scale also likely to improve conditions for the shoot.

The ground flora of each wood was surveyed twice yearly in May and July. Recorders walked a stratified transect route through each of the treatment and control areas, recording the presence of ground flora species in 0.25 m × 0.25 m quadrats on a seven-point modified Domin scale.

10.4.2 Results

Edmondsham Wood. A total of 76 vascular plant species (including tree seedlings) were recorded over the first three years of study. Changes in mean total species richness per quadrat did not differ significantly between the treatment and control sections of the wood in the two springs following

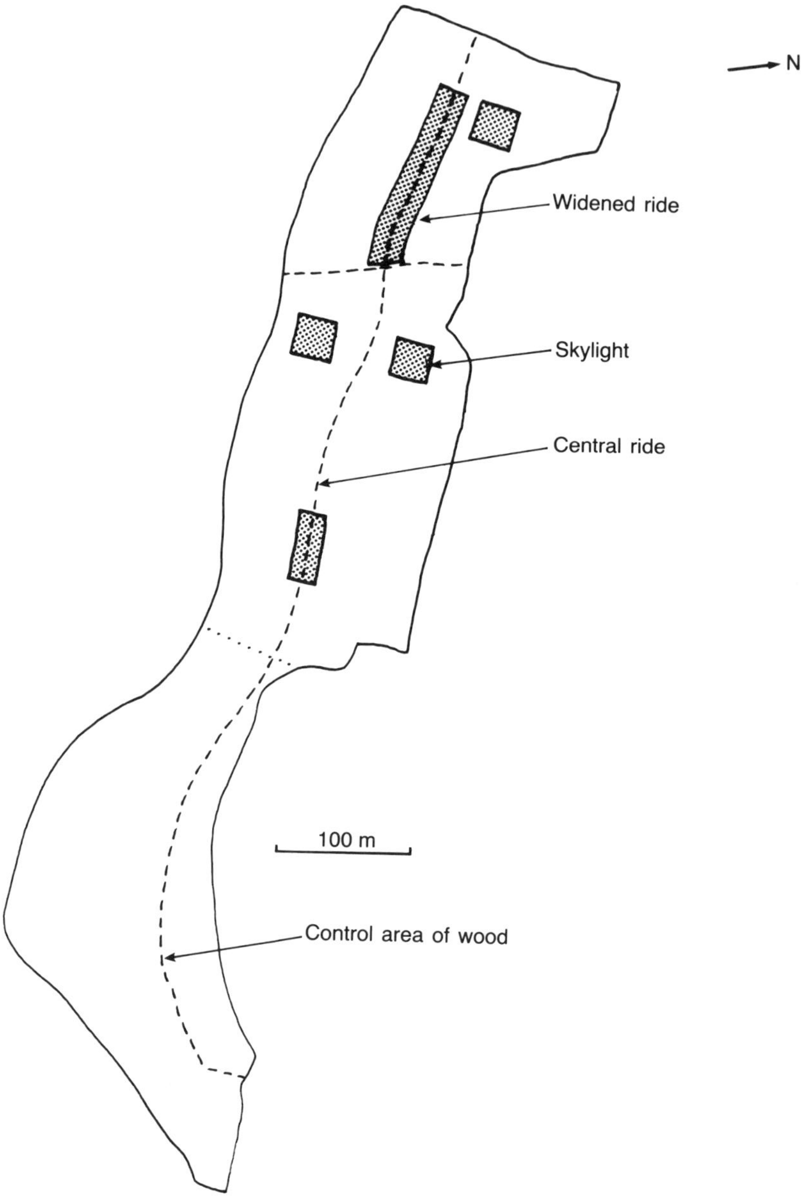

Figure 10.2 Map of Edmondsham Wood showing areas managed as widened rides and 'skylights' (shaded), and control area.

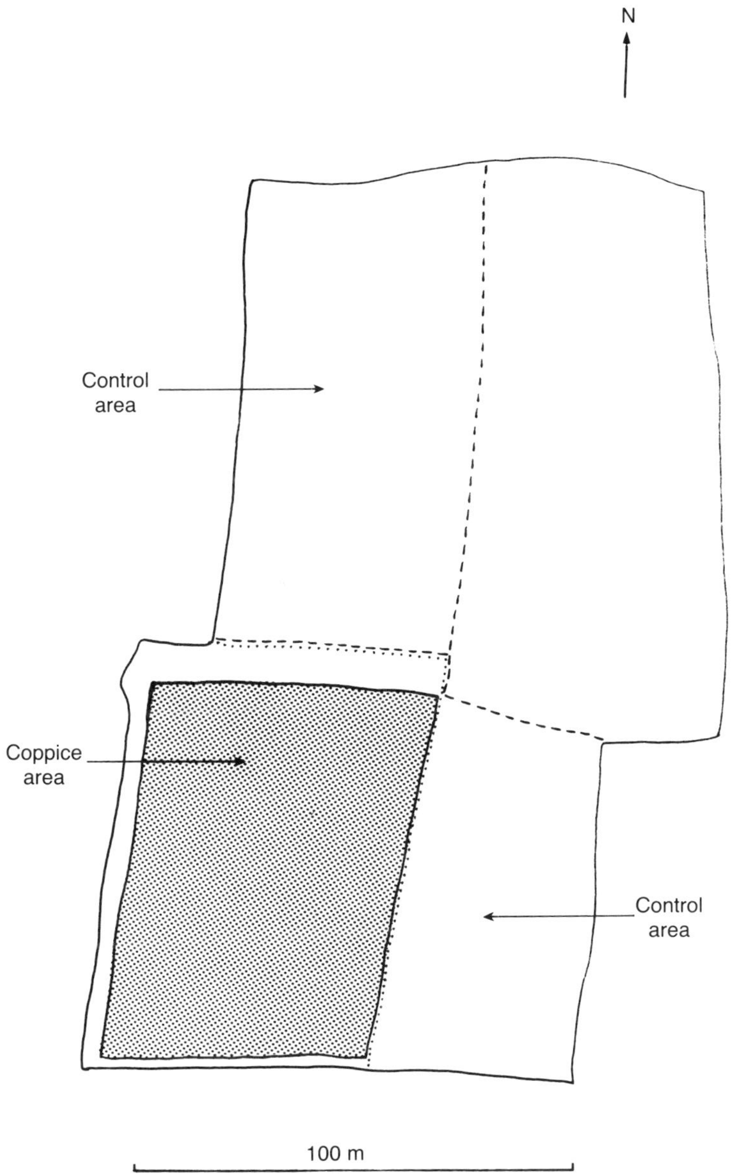

Figure 10.3 Map of Buttermead Copse showing coppice areas (shaded) and controls.

Table 10.1 List of vascular plant species present in the field layer at Buttermead Copse, Hampshire (75 species in total, 18 spring only, 15 summer only; o, ancient woodland indicator species (NCC South); s, recorded in spring; u, recorded in summer; +, shade tolerant adventives and ruderals)

Adoxa moschatellina[os]
A. capillaris[u+]
Agrostis stolonifera[u+]
Ajuga reptans[su+]
Anemone nemorosa[os]
Angelica sylvestris[su+]
Anthoxanthum odoratum[s+]
Arctium lappa[s+]
A. minus[u+]
A. agg.[s+]
Arum maculatum[s]
Betula (seedling)[u]
Brachypodium sylvaticum[u]
Cardamine hirsuta[s+]
C. pratensis[s+]
Carex remota[os]
C. sylvatica[ou]
Cerastium fontanum[s]
Cirsium arvense[su+]
C. palustre[s+]
C. vulgare[su+]
Conopodium majus[os]
Convallaria majalis[os]
Corylus avellana (seedling)[s]
Deschampsia cespitosa[su+]
Dryopteris filix-mas[u]
Epilobium montanum[s+]
Euphorbia amygdaloides[osu]
Filaginella uliginosa[u+]
Filipendula ulmaria[s+]
Fraxinus excelsior (seedling)[su]
Galeopsis tetrahit[su+]
Galium aparine[su+]
G. palustre[su+]
Geranium robertianum[su]
Geum urbanum[su]
Glechoma hederacea[su]
Holcus lanatus[su+]
H. mollis[osu+]
Hyacinthoides non-scripta[osu]
Hypericum perforatum[u]
H. tetrapterum[su]
Juncus bufonius[su+]
J. effusus[su+]
Lamiastrum galeobdolon[osu]
Lychnis flos-cuculi[su+]
Mercurialis perennis[su]
Moehringia trinervia[osu]
Myosotis sp.[s]
Oxalis acetosella[osu]
Poa annua[su]
P. nemoralis[osu]
P. trivialis[su]
Polygonatum multiflorum[osu]
Potentilla sterilis[osu]
Primula vulgaris[osu]
Polygonum persicaria[u+]
Prunus spinosa (seedling)[su]
Pteridium aquilinum[u+]
Ranunculus ficaria[s+]
Rosa sp. (seedling)[s]
R. repens[u+]
Rubus fruticosus agg.[su]
Rumex obtusifolius[u+]
R. sanguineus[su]
Sambucus migra (seedling)[su]
Scrophularia nodosa[os+]
Stachys sylvatica[su]
Stellaria holostea[u]
S. media[su+]
Urtica dioica[su]
Veronica chamaedrys[s+]
V. montana[osu]
Viola reichenbachiana[osu]
V. riviniana[su]

Table 10.2 List of vascular plant species present in the field layer at Edmondsham, Dorset (76 species in total, 15 spring only, 11 summer only; o, ancient woodland indicator species (NCC South); s, recorded in spring; u, recorded in summer; +, shade-tolerant adventives and ruderals)

Adoxa moschatellina^{os}
Agrostis stolonifera
Ajuga reptans^{su+}
Anemone nemorosa^{osu}
Angelica sylvestris^{su+}
Arum maculatum^{s}
Athyrium filix-femina^{u}
Blechnum spicant^{os}
Bromus ramosus^{os}
Carex paniculata^{su}
C. remota^{osu}
C. sylvatica^{osu}
Chrysosplenium oppositifolium^{osu}
Circaea lutetiana^{su}
Cirsium palustre^{s+}
Conopodium majus^{os}
Convallaria majailis^{os+}
Convolvulus arvensis^{s}
Deschampsia cespitosa^{su+}
Crataegus (seedling)^{u}
Digitalis purpurea^{u+}
Dryopteris dilatata^{su}
D. filix-mas^{su}
Euphorbia amygdaloides^{os}
Equisetum arvense^{u+}
Fragaria vesca^{su}
Fraxinus excelsior (seedling)^{su}
Galium aparine^{su+}
G. odoratum^{osu}
G. palustre^{su+}
G. saxatile^{s+}
Geranium robertianum^{su}
Geum urbanum^{su}
Glechoma hederacea^{su}
Glyceria fluitans^{s}
Hedera helix^{su}
Holcus mollis^{su+}
Hyacinthoides non-scripta^{osu}
Ilex aquifolium^{su}
Juncus effusus^{s+}
Lamiastrum galeobdolon^{osu}
Lamium purpureum^{u+}
Lonicera pericylmenum^{su}
Luzula pilosa^{os}
Lysimachia nemorum^{osu}
L. nummularia^{u}
Melica uniflora^{osu}
Melampyrum pratense^{ou}
Mentha aquatica^{s}
Milium effusum^{osu}
Oxalis acetosella^{osu}
Poa pratensis^{s+}
P. trivalis^{su}
Potentilla sterilis^{osu}
Primula vulgaris^{osu}
Pteridium aquilinum^{su+}
Quercus (seedling)^{s}
Ranunculus ficaria^{s+}
R. repens^{su+}
Ribes sylvestre^{osu}
Rosa arvensis^{osu}
Rubus fruticosus agg^{su}
R. idaeus^{s}
Sanicula europaea^{os}
Sorbus aucuparia (seedling)^{u}
Stachys sylvatica^{su}
Stellaria holostea^{su}
S. media^{su+}
Valeriana dioica^{u+}
Veronica chamaedrys^{s+}
V. montana^{osu}
Viburnum opulus (seedling)^{u}
Viola reichenbachiana^{osu}
V. riviniana^{su}

management. To assess more subtle effects, the occurrence of two 'functional' groups of species were examined: (a) ancient woodland indicator species (AWIs), from a list of 100 species most strongly associated with ancient woodland in southern England drawn up by the Nature Conservancy Council; and (b) shade tolerant 'weeds'. This latter group included ruderals, marginals and open field species, which contained many annuals, biennials or short-lived perennials. This list was largely drawn from Peterken (1981) and from a knowledge of the habitat and ecology of each species. The two lists of species are given in Tables 10.1 and 10.2. Mean numbers of AWIs recorded in quadrats were examined in the treatment and control sections over the three years in spring and summer respectively (Figure 10.4(a) and (b)). Some small changes occurred over time but the trends within the two areas were similar. For the weed species the relative changes were greater in summer compared with the spring but this was mainly a result of changes in the control sections (Figure 10.4(c) and (d)).

Buttermead Copse. A total of 75 species were found over the three years. In the spring both the control and treatment areas showed similar trends in species richness but there was a significant increase in treatment species richness in the second summer after treatment. For the AWIs, there was a slight but non-significant depression in species richness in the summer treatment areas (Figure 10.5(b)), but the weed species significantly increased in the treatment areas in both summers following management (Figure 10.5(d)).

10.4.3 Community trends

Few species or groups demonstrated clear responses to management – as shown in the previous section. Such complicated responses, if present, might be better detected by analysis at the community level. Data sets were therefore combined for all three years for each site and at each sampling period. Detrended Correspondence Analysis or DECORANA (Hill and Gauch, 1980) was carried out on these data sets, by pooling all treatment and control quadrats, enabling an effective between-year and between-group comparison within a single ordination. All samples can therefore be compared in 'common ordination space'. Resulting sample ordination scores for each group and year were averaged to give group means, and these were plotted on the first two DECORANA axes which accounted for most variability among the data. Axis 1 appears to reflect soil conditions, running from damp to more acidic soils. Axis 2 reflects life strategy of the species, separating AWIs from fast colonizers and weeds.

For Edmondsham, the treatment and control group scores showed a high degree of variation over the three years with no trends either for the spring or summer data. This reflected the inherent botanical variability within the

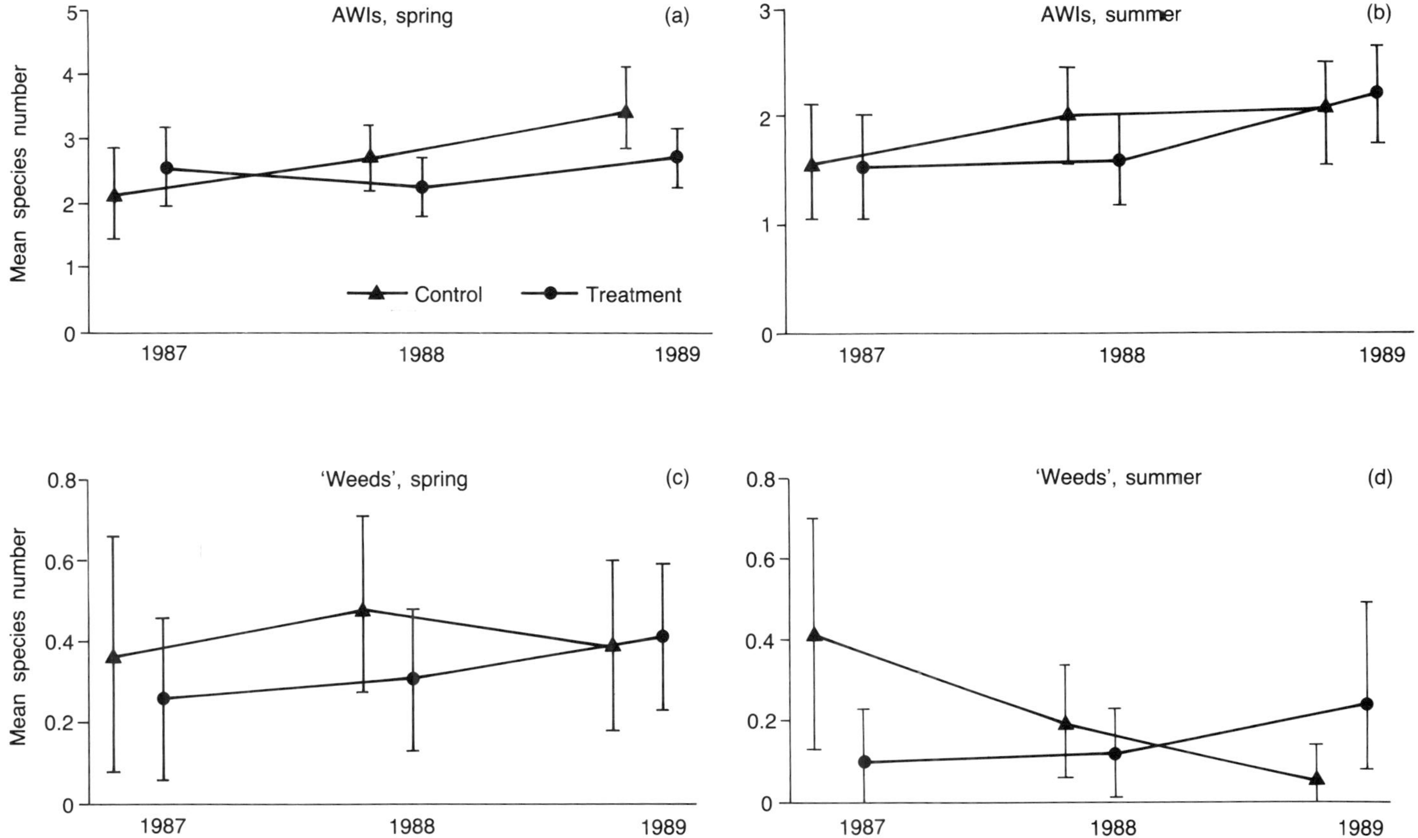

Figure 10.4 Mean number of 'weed' or ancient woodland indicator species (AWIs) recorded per quadrat during spring and summer at Edmondsham Wood, 1987–1989 (±95% confidence limits).

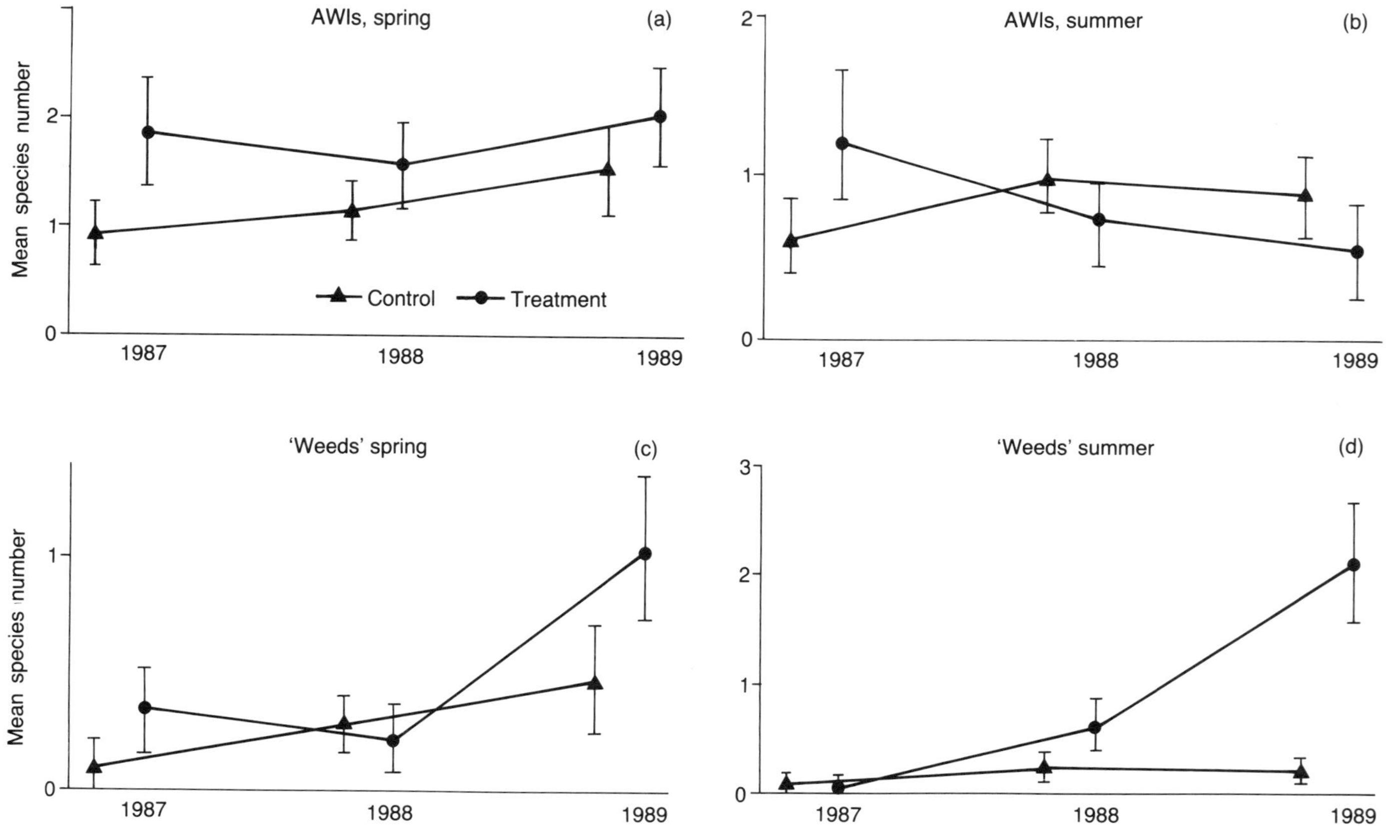

Figure 10.5 Mean number of 'weed' or ancient woodland indicator species (AWIs) recorded per quadrat during spring and summer at Buttermead Copse, 1987–1989 (±95% confidence limits).

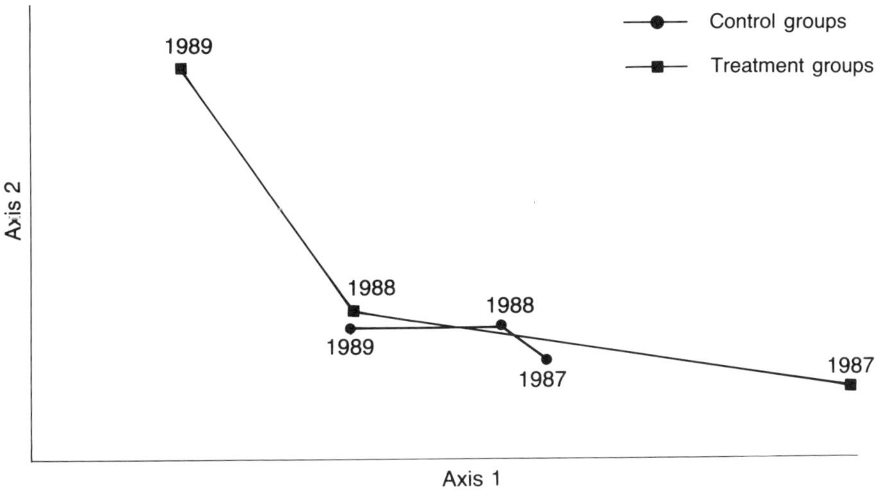

Figure 10.6 Changes between years in the mean DECORANA ordination sample scores in the control and treatment areas in Buttermead Copse (summer data only) (•, managed areas; ■, control areas).

wood, but also the ineffectiveness of removing coppice while retaining a relatively unbroken canopy of standards. For Buttermead, spring ordination scores showed much the same pattern. A plot of summer group ordination scores, however, demonstrated a clear change in sample scores for treatment groups compared with minor changes for control groups as shown in Figure 10.6.

An interpretation of these changes is afforded by the species ordination (Figure 10.7). This shows a fairly clear separation of species into three groups: AWIs, shade-bearing fast colonizers, and weeds. It is interesting to note the inclusion of some AWIs such as *Carex sylvatica*, *Holcus mollis* and *Euphorbia amygdaloides* within the weeds cluster of the ordination, which are species commonly flourishing in the open environment of newly cut coppice. This group is clearly differentiated from another cluster of AWIs along Axis 2 of the ordination, containing species such as *Oxalis acetosella*, *Hyacinthoides non-scripta* and *Polygonatum multiflorum*. These species all flowered more vigorously in the spring immediately after coppicing and before the increase in growth of the former AWI group took place (from summer 1988 to summer 1989).

10.4.4 Discussion

Clear differences between Edmondsham and Buttermead were found, particularly in relation to the number of weed species and their changes in

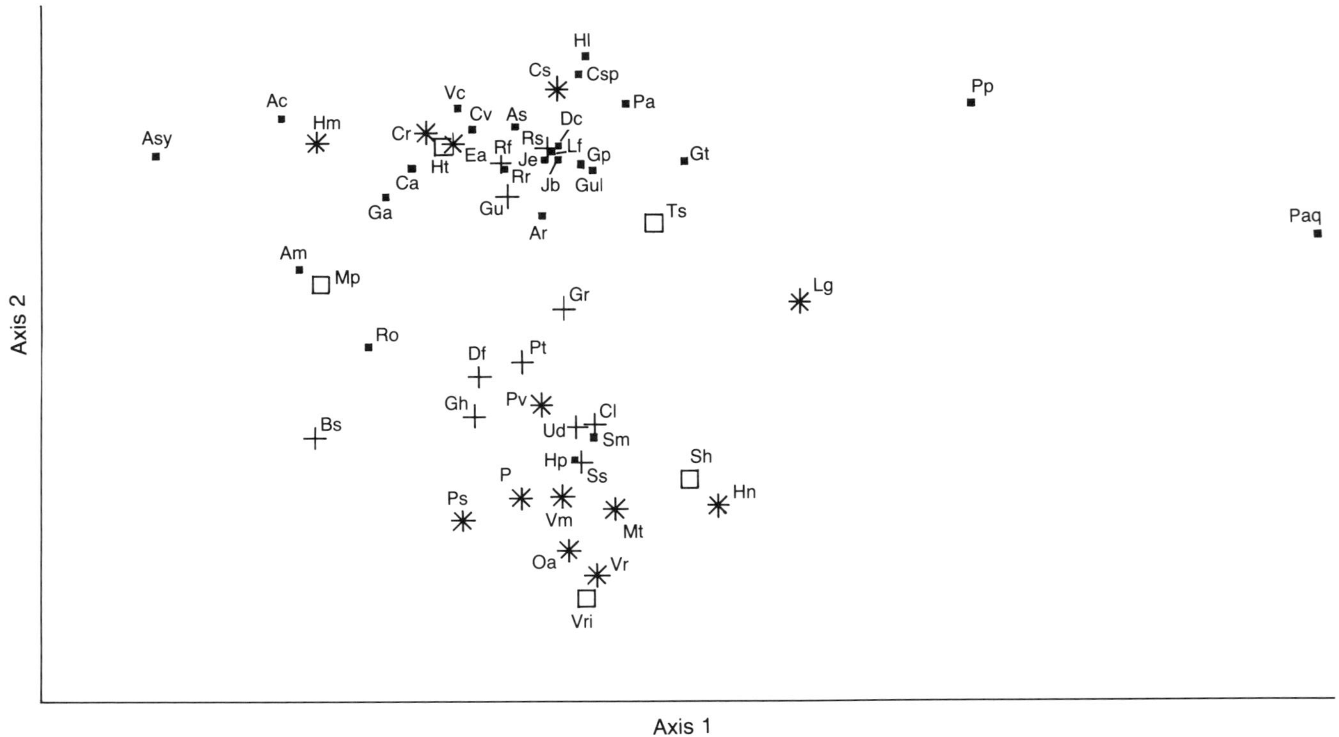

Figure 10.7 DECORANA species ordination from Buttermead Copse showing weeds; fast colonizing woodland species, ancient woodland indicator species and other non-classified species. Species initials correspond to the species names in Table 10.1 (●, weeds; +, fast colonizers; *, AWIs; ●, others).

abundance following management. At this preliminary stage, the main conclusions are that the small scale of coppice management carried out at Edmondsham has, as yet, had little noticeable effect on the flora because of the high density of remaining standards, while the much larger coupe size at Buttermead has been associated with an increase in the summer abundance of weed species, many of which were already present in the seed bank. Although thinning of the understorey by coppicing produced small increases in light level at the field layer at Edmondsham, these were insufficient to obtain a significant vegetation response. Small gap sizes, where the canopy height is equal to or greater than gap height, means that direct sunlight rarely reaches the ground, even on the north edge of the gap at mid-day (Anderson, 1989).

10.5 CONCLUSIONS

Coppicing provides a method by which the attractiveness to pheasants of currently unmanaged woodlands can be increased. This should provide an incentive for landowners to manage their woodlands in this way, with benefits for a wide range of other species. However, current coppicing for game is often carried out on a relatively small scale. The 25 m × 25 m 'skylight' cuttings used at Edmondsham are typical of many of the smaller plots used to create flushing points, gun stands or small areas of shrubby cover in later years. Although they can be effective from a shoot manager's point of view they do not, in this study at least, lead to significant increased ground flora diversity. This small scale of management is largely dictated by economics, but if somewhat larger coppicing coupes can be demonstrated to lead to benefits for conservation they may be more frequently used.

As discussed at the beginning of this chapter pheasant shooting has the potential to encourage landowners to use coppicing. However, more information is required on the minimum coupe sizes that should be cut to produce benefits not just for pheasants but also for conservation. Although the areas managed may be relatively small in any particular wood, it is one method through which coppicing may be maintained in the wider countryside where it is all too often uneconomic. If we are to address the consequences of the decline in coppice management and the effects that this has had on a wide range of woodland species we must not think purely in terms of nature reserves or 'fire brigade' actions for particular species. What is also needed is a programme to provide sound advice to the landowners who control most of the neglected coppice. This must be framed in terms that are economically acceptable and that provide them with direct benefits – for instance through improving their pheasant shooting.

REFERENCES

Anderson, M.A. (1989) Opportunities or habitat enhancement is commercial forestry practice, in *Biological Habitat Reconstruction* (Ed. G.P. Buckley), Behlhaven Press, London, pp. 129–46.

Cobham Resource Consultants (1983) *Countryside Sports: Their Economic Significance*, The Standing Conference on Countryside Sports, Reading.

Gray, N. (1986) *Woodland Management for Pheasants and Wildlife*, David and Charles, London.

Hill, D.A. and Robertson, P.A. (1988) *The Pheasant; Ecology, Management and Conservation*, Blackwell, Oxford.

Hill, M.O. and Gauch, H.G. (1980) Detrended correspondence analysis: an improved ordination technique. *Vegetatio*, **42**, 47–58.

Lachlan, C. and Bray, R.P. (1976) Habitat selection by cock pheasants in spring. *Journal of Applied Ecology*, **13**, 691–704.

Ludolf, I.C., Robertson, P.A. and Payne, S. (1989) *The effect of pheasant release pens and strawed feed rides on the ground flora of ancient woodland.* Report to the Nature Conservancy Council, The Game Conservancy, Fordingbridge, Hampshire.

Page, R. (1987) *The Fox and the Orchid: Country Sports and the Countryside*, Quiller Press, London.

Peterken, G.F. (1981) *Woodland Conservation and Management*, Chapman and Hall, London.

Piddington, H.R. (1980) *Shooting and Fishing in Land Use: A Study of Economic, Conservation and Recreation Aspects*, Department of Land Economy, University of Cambridge, Cambridge.

Ridley, M.W. (1983) *The mating system of the pheasant* Phasianus colchicus, D. Phil. thesis, University of Oxford.

Robertson, P.A. (1985) Habitat selection and the relative distributions of wild and released pheasants (*Phasianus colchicus*) as determined by winter trapping. *Journal of World Pheasant Association*, **10**, 87–99.

Robertson, P.A., Woodburn, M.I.A., Bealey, C.E. Ludolf, I.C., and Hill, D.A. (1989) *Pheasants and Woodlands: Habitat Selection, Management and Conservation.* Report to the Forestry Commission, The Game Conservancy, Fordingbridge, Hampshire.

Simpson, J. (1907) *Game and Game Coverts*, Pawson and Brailsford, London.

Warren, M.S. (1976) *The Dorset Woodlands: Their History and Conservation*, MSc thesis, University College of London.

Part Six

EFFECTS OF COPPICING ON MAMMAL POPULATIONS

11

The effects of coppice management on small mammal populations

JOHN GURNELL, MARTIN HICKS and STEVE WHITBREAD

11.1 INTRODUCTION

Although the population dynamics of woodland small mammals have been intensively studied (Flowerdew *et al.*, 1985), little work has been carried out with respect to the effects of forest management on small mammal species richness and abundance. It is widely known that the structure and species composition of vegetation in the different horizontal layers within a woodland have a considerable influence on the small mammal community (Hansson, 1977; Gurnell, 1985; Mazurkiewicz, 1986). It follows that an actively managed coppice woodland, with its mosaic of woodland stands differing markedly in plant structure and, but to a lesser extent, composition, provides an excellent model for studying the impact of such variation on small mammal communities. In the main, small mammals are vagile animals with good dispersal abilities. Therefore, with reference to these animals each coppice stand can be considered to be a large or macro-patch of habitat.

In association with the recent revival of interest in coppicing as a form of woodland management, especially for nature conservation, ecologists have started to look at the effects of such management practices on small mammal communities. So far this work has largely been descriptive and has tended to concentrate on small mammal populations within coppice stands of different ages. Studies are only just starting on the dynamic interplay of small mammals between coppice stands, and on the effects on small mammal populations of boundaries or ecotones between neighbouring coppice stands, rides and ride edges.

In this account we review the current understanding of small mammals within working coppice and compare this with our knowledge of small mammals in less intensively managed high forest. In particular, we emphasize the important features of coppice vegetation in relation to the biology of small mammals, and their communities defined by the common method of sampling using Longworth live-traps (Gurnell and Flowerdew, 1990). Other species of small mammal, concentrating on common dormice (*Muscardinus avellanarius*) and squirrels (*Sciurus* spp.) are considered separately.

Compared with those in Europe and North America, British small mammal communities are reasonably simple and consist of only a few species (Gurnell, 1985). Moreover, there is a regional variation in the possible species composition of mammals in woodland communities as a result of the different geographical distributions of mammals within the British Isles (Corbet and Harris, 1990). For example, common dormice, harvest mice (*Micromys minutus*) and yellow-necked mice (*Apodemus flavicollis*) are found in south-east England, but not Scotland.

11.2 COPPICE VEGETATION AND ITS GENERAL SIGNIFICANCE FOR SMALL MAMMAL POPULATIONS

Coppicing clearly has dramatic effects on vegetation structure and composition (see Chapters 3 and 7) which in turn is influenced by the coppice species managed, soil, climate and other site factors. The vegetation associated with different stand types may vary widely and this, too, may influence patterns of community organization in small mammal populations. To provide a framework for discussion we consider a typical mixed coppice-with-standards woodland. The presence of standard trees will affect the amount of food available each year to small mammals, particularly tree flowers, tree seeds and invertebrates.

Perhaps the most obvious consequence of cutting all the coppice within a defined area or coupe is the complete removal of nearly all the canopy cover, and the disturbance to the ground during felling and extraction. With the loss of canopy there is also a loss of associated sources of food important for certain species of small mammal (e.g. seeds, flowers, phytophagous insects). The barren landscape which results, however, disappears very quickly with the rapid spread of grasses, rushes and herbs. By the end of the second year the woodland floor is covered by a field layer up to a height of one or two metres and coppice regrowth is well under way (Figure 11.1). This provides a habitat with abundant ground cover, an altered microclimate, high densities of grasses, herbs and their associated invertebrates, all of which improve conditions for many ground-dwelling small mammals.

As the coppice continues to grow to form a canopy, there is a succession of vegetation types related to the amount of light reaching the ground and this continues until the time of the next cutting. Not only are there marked changes in growth stage (see Chapter 3) according to the age and density of coppice, there are also striking seasonal changes that affect the vegetation (Figure 11.1). Therefore, for small mammals, the three-dimensional complexity of the habitat changes through time, which in turn influences the amount of arboreal activity possible. It is worth making a special mention of bramble (*Rubus fruticosus*), which can be significantly affected by both

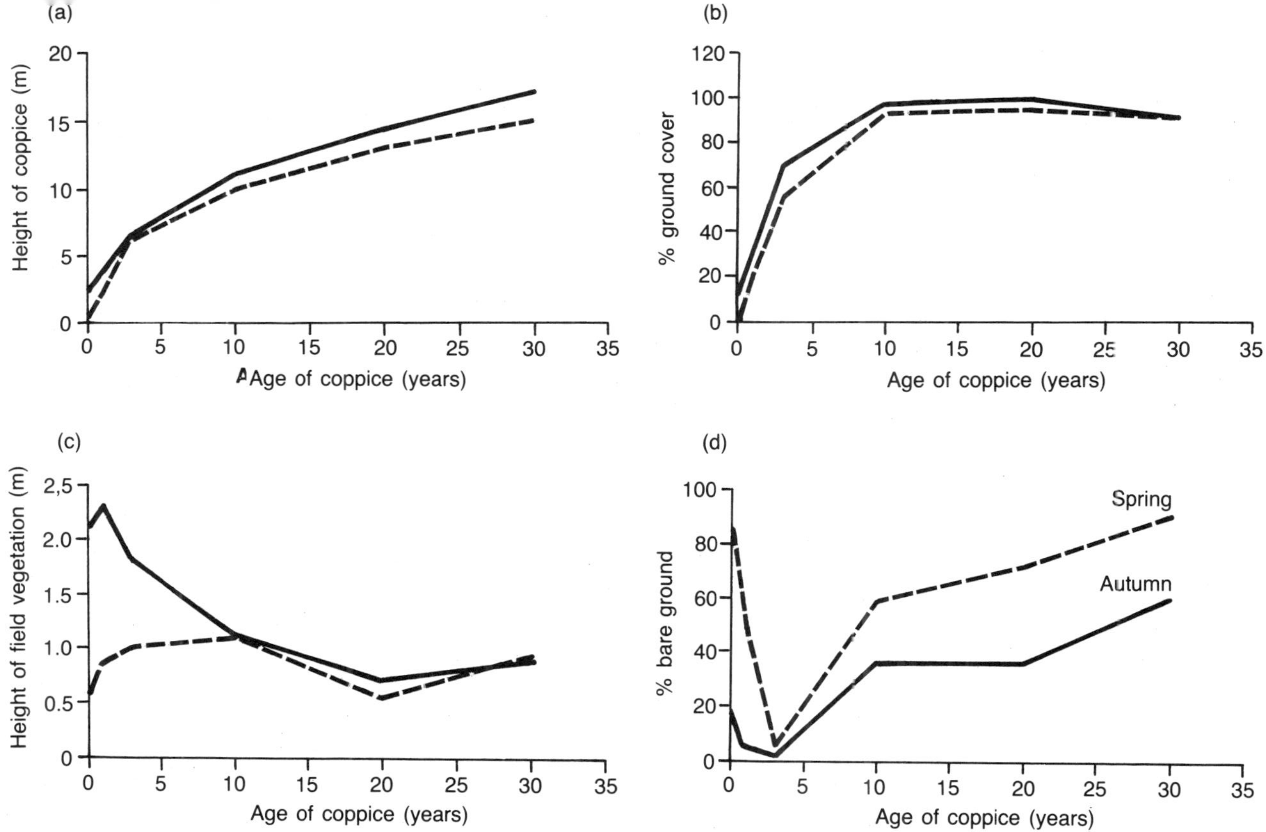

Figure 11.1 Examples of successional and seasonal changes in vegetation structure at Bradfield Woods, Suffolk. Changes are shown against age of coppice for: (a) height of coppice; (b) percentage ground cover of coppice canopy; (c) height of ground and field vegetation; (d) percentage bare ground. ———, autumn; — —, spring.

soil type and age of coppice, and is an important structural component of the habitat as well as a valuable source of food for mammals (including deer).

The time taken for a coppice species to regrow and produce flowers and fruits can also be an important influence on small mammals. For example, hazel (*Corylus avellana*) first produces seed at four to five years and sweet chestnut (*Castanea sativa*) probably at four to seven years. However, the age of maximum seed production varies among species and sites. For example, good seed crops may not occur on sweet chestnut until they are 20 years of age, even though rotation length is usually 15–17 years. In addition to the temporal changes in cover and food availability within a stand, there is the added complexity of spatial changes resulting from the juxtaposition of stands of different ages and perhaps species composition.

11.3 MICE, VOLES AND SHREWS

The vegetation profiles for coppice woodland depicted in Figure 11.1 are based on studies carried out at Bradfield Wood, Suffolk, a site continuously coppiced for over 700 years. Bradfield is best described as a mixed coppice-with-standards wood; dominant species include hazel, ash (*Fraxinus excelsior*), birch (*Betula pendula*), field maple (*Acer campestre*), sallow (*Salix caprea*), oak (*Quercus robur*) and notably alder (*Alnus glutinosa*), and in fact it is classified as 'plateau alderwood' by Peterken (1981). Small mammal populations were studied at Bradfield between 1985 and 1988 in five age-classes of coppice, namely 1 year, 3 year, *c.* 10 year (short rotation), *c.* 20 year (long rotation) and *c.* 30 year (neglected). In the main, different stands of coppice were studied for a given age-class each year. To provide a comparison with an 'unmanaged high forest' woodland, the results from a small mammal live-trapping study carried out in a 170-year-old oak (*Q. robur*) wood at Alice Holt Forest, Hampshire, are also presented. The oak wood included an understorey of hazel, hawthorn (*Crataegus monogyna*) and holly (*Ilex aquifolium*) which had not been coppiced for more than 40 years. Many natural gaps had appeared in the canopy so that in places there was a considerable amount of ground and field vegetation in the form of herbs, grasses, bracken (*Pteridium aquilinum*) and bramble (see Ludolf, 1986).

The number of small mammals captured at Bradfield and Alice Holt varied among seasons and between years, but the seasonal variation was generally larger than the year-to-year variation – a pattern common to different ages of coppice woodland as well as the oak wood. The seasonal cycle of numbers was, however, particularly striking at Bradfield, and is linked with the equally striking seasonal cycle in vegetation (Figures 11.1 to 11.3).

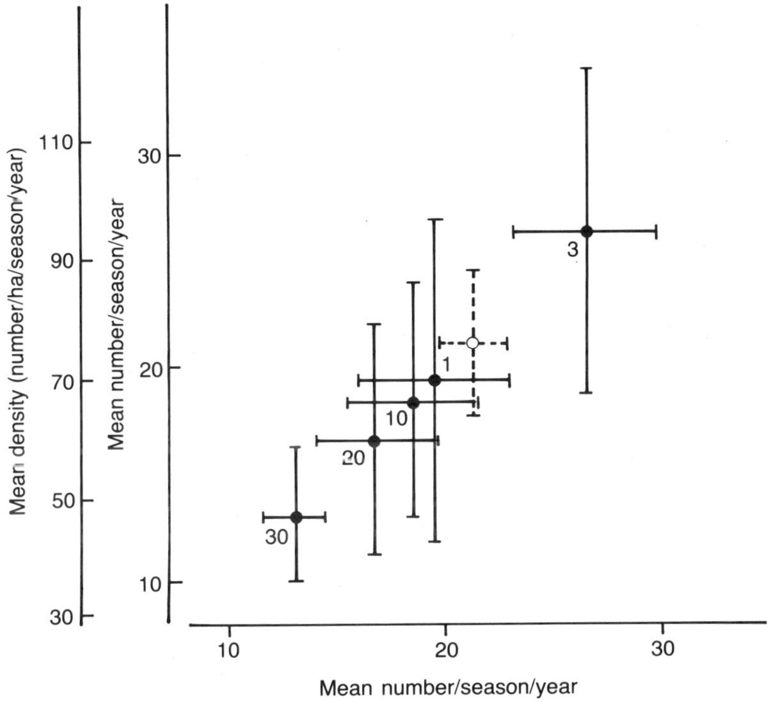

Figure 11.2 Mean number of mice, voles and shrews captured per year per season according to age of coppice (numbers on graph indicate age classes, years) in a four-year study, Bradfield Woods, Suffolk. Note that means are based on four different compartments of the same age, taken one per year. Horizontal bars represent the standard error (±1) of the mean number captured each year; vertical bars represent the standard error of the mean number captured each season. The additional scale on the left gives a crude indication of animal density. The open circle and dashed standard error bars represent the results from a four-year study in an old oak wood in Surrey (Gurnell, unpublished). This shows the seasonal and yearly patterns in changes in numbers of small mammals in mature broadleaved woodland for comparison with the coppice stands. However, its position with respect to numbers captured in the coppice stands is very approximate due to different sampling intensities used.

At Bradfield, the highest number of individuals captured was in 3-year coppice. The next largest number was in 1-year coppice and thereafter, on average, fewer small mammals were captured the older the coppice. In contrast to the oak wood, the 30-year coppice at Bradfield had very little field or shrub vegetation. While comparisons between different studies must be made with caution because of different site conditions and varying numbers and patterns of traps used, a number of observations can be made. Overall densities in the oak wood were considerably higher than those in the 30-year coppice, a reflection of the different amounts of ground and field vegetation. Also, relatively fewer bank voles (*Clethrionomys glareolus*)

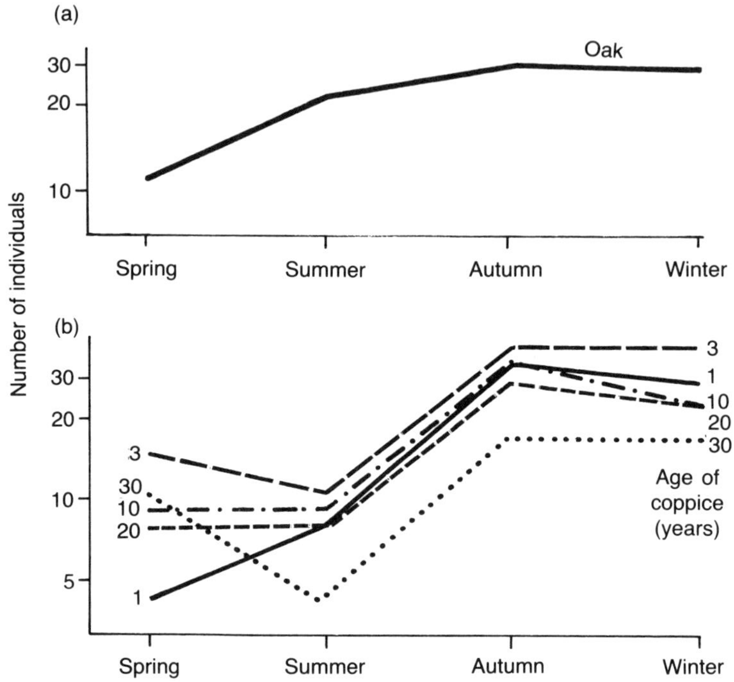

Figure 11.3 Seasonal changes in numbers of small mammals (average of four years) for: (a) the Alice Holt Forest; (b) different ages of coppice at Bradfield Woods, Suffolk.

were captured in the 30-year coppice than in the oak wood (see Table 11.1). The 30-year coppice showed a steep decline in numbers from spring to summer which is characteristic of woodland with an open, exposed floor. In contrast, there was an increase in numbers from spring to summer in the oak wood. One-year coppice also showed an increase from spring to summer but there was little change in other ages of coppice. With the exception of spring sampling, fewer small mammals were captured in the 30-year coppice than any of the four other coppice ages sampled at Bradfield.

The pattern of small mammal numbers at Bradfield and at Alice Holt is consistent with the accepted dynamics of small mammal populations generally (e.g. Alibhai and Gipps, 1985; Flowerdew, 1985; Montgomery, 1989; Churchfield, 1990), and coppicing does not appear to alter the fundamental processes underlying the seasonal cycles observed. Typically numbers build up in all ages of coppice during the autumn and early winter, and then decline over the winter to spring (Figure 11.3). These trends tend to reflect changes in the abundance of food such as seeds and invertebrates (e.g. Hansson, 1977).

The relative abundance of species in different ages of coppice and in the oak wood is shown in Table 11.1. Trap response factors may have affected the results obtained (e.g. Gurnell and Flowerdew, 1990; Gurnell and Gipps, 1989) but we believe that the figures presented give a fair reflection of relative abundance of species. Three species were widespread and abundant at both sites: wood mice (*Apodemus sylvaticus*), common shrews (*Sorex araneus*) and bank voles. At Bradfield, wood mice tended to be the dominant species, except in the 3-year coppice where bank voles and common shrews were most numerous. Wood mice outnumbered all the other species in the 30-year coppice whereas they were as numerous as bank voles in the oak wood – a reflection of the different accounts of field cover in these habitats. Yellow-necked mice were increasingly numerous in 10, 20 and 30-year coppice while bank voles showed a clear preference for 3-year coppice. Pygmy shrews (*Sorex minutus*) may be under-represented in the samples because they are very lightweight and may not activate the trap treadle mechanism and as a rule water shrews (*Neomys fodiens*) only appear occasionally in terrestrial habitats away from water. However, both species were consistently captured in 3-year coppice, an age at which common shrews were also most abundant.

Table 11.1 Relative abundance of species (% of total numbers) from live-trapping studies carried out in 5 ages of coppice from a four-year study at Bradfield Woods, Suffolk. Oak represents a comparable set of results from a four-year study in an old oak woodland in Surrey. Pygmy shrews are probably under-represented in the samples because they are often too light to activate the tripping mechanism of the traps. Water shrews are ephemeral animals in terrestrial habitats away from water

	Age of coppice (years)					
Species	*1*	*3*	*10*	*20*	*30*	*Oak*
Wood mouse	40	18	52	55	59	41
Common shrew	23	25	17	17	14	13
Bank vole	10	37	19	16	10	41
Yellow-necked mouse	2	2	10	11	14	4
Harvest mouse	20	5	<1	-	<1	<1
Pygmy shrew	<1	7	2	2	2	2
Field vole	3	3	<1	-	-	<1
Water shrew	<1	2	<1	-	-	<1
Number of species >1%	6	8	4	5	5	5

Field voles (*Microtus agrestis*) and harvest mice, both grassland specialists, were almost entirely restricted to 1- and 3-year coppice, and were captured in every year of the study. They rarely appear in mature high forest, as seen from the results from the oak wood, and then only where suitable patches of grass and scrub are found, along open rides or in glades. In general, more harvest mice were captured in 1-year coppice than 3-year coppice, although there may be some bias in the results because the *Deschampsia* tussocks were taller in the latter making it more difficult to catch these animals at ground level. Breeding has been confirmed in both ages and in 1988 the numbers of harvest mice and the density of breeding nests were exceptionally high in 1-year coppice. Some populations of *Microtus* exhibit multi-annual cycles in favourable years (Taitt and Krebs, 1985) when high populations may be reached, particularly in young forest plantations with a large amount of grass cover. Although populations of field voles in young coppice at Bradfield did not reach very high densities during the study period, they did show a clear preference for young coppice and a decline in numbers in older coppice as shade gradually eliminated the field cover (see Ferns, 1979a,b).

Perhaps the most significant result was that, overall, the greatest mix of species, as well as the highest densities, were found in the 3-year coppice at Bradfield. Moreover, taken together, the results demonstrate the benefits of having different macro-patches of habitat within one woodland site. Mice, voles and shrews are short-lived, few ever exceeding one year of age, and in general their populations are characterized by a high turnover of individuals. Coppicing, therefore, can produce a patchwork of discrete habitats with a high diversity of vegetation structure and composition, linked by well-established ride systems. This not only provides a wide range of opportunities that can be exploited quickly by a full complement of small mammals, but also facilitates their wide dispersal and possible colonization of new habitats.

11.4 DORMICE

Although most small woodland rodents and shrews readily ascend into vegetation above ground (e.g. Holisova, 1969; Montgomery, 1980) dormice are almost exclusively arboreal (Whitbread, 1986a,b). Their small size, agility and prehensile feet enable them to reach food items inaccessible to other species. By building up reserves of subcutaneous fat in the autumn (an adult may increase from 16–20 g in July to 25–32 g in October (Catzeflis, 1983)) dormice are able to hibernate through the long period of food scarcity in winter. In Britain, dormice are not known to breed until the summer following their birth, producing one or two litters of, on average, four young (Corbet and Southern, 1977), and they live for up to 4 years (Lozan, 1961; Catzeflis, 1984). Thus, dormice populations are likely to be at lower densities than those of other species of similar size and be less able to replace losses. Individual dormice are probably resident within an area for a comparatively long time. As a result of such population dynamics they are likely to be affected more by coppice management than any other woodland mammal. For dormice, the major consequences of coppicing are those concerned with the effects on the animals' food supply, the ability to locate their food, and on the availability of nest sites.

Dormice are strictly nocturnal and usually spend the day in nests of distinctive form (Hurrell, 1980). They maintain and use up to three nests within their home range at any given time. They also occasionally use bird-nests, nestboxes (Schulze, 1970), tree-holes and piles of brushwood. Nests may be situated at any height from below ground level to the top of the tree canopy according to the availability of cover in the different storeys, and there is no evidence to suggest that dormice numbers are limited by competition for available nesting sites. Nests may be in thick scrub, at ride edges, in the base of coppice stools, in tree-holes or in clumps of honeysuckle (*Lonicera periclymenum*) (Table 11.2). Coppice management results in uniform stands within which the range of potential nest sites may be reduced, yet adjacent areas of different ages provide a suitable diversity of habitat structure. Bramble thickets in which nests are commonly found (Hurrell and McIntosh, 1984) are present in younger regrowth, whereas mature stools may support thick bundles of honeysuckle. Nestboxes are most frequently used where there is little cover in the ground and field layers of a wood (Catzeflis, 1984; Morris *et al.*, 1990). At other sites, dormice will use nests located below ground or in tree-holes in the spring until suitable cover develops. They frequently cohabit outside the late-May to September breeding season (e.g. Holland, 1967; Catzeflis, 1984) and mixed-sex groups of three or more individuals are regularly reported. Reproductively active males and females may be found together for short periods during the summer. Winter hibernacula are sited close to or below ground level – where temperature fluctuations are at a minimum. Torpid dormice are often uncovered in the course of coppicing.

Dormice are unlikely to be found in areas of open woodland where it is not possible for them to travel between the widely spaced trees and shrubs without coming down to the ground as they are poorly adapted for travelling along the ground. A prime requirement is a dense shrub layer that allows free movement above ground, therefore coppice less than four years old or of an age at which the shrub layer becomes shaded out will be unsuitable for dormice. In overmature coppice reverting to high forest, dormice are largely restricted to the wood edges and ride margins where such shrubby conditions persist.

Dormice rely almost entirely on food sources above ground, many of which are highly seasonal and ephemeral in nature. Buds, leaves and flowers are most frequently consumed following emergence from hibernation in March–April. Aphids, caterpillars and other insect larvae, leaves and

Table 11.2 Dormouse nest locations in a hazel coppice, Isle of Wight, with oak standards, and a wide variety of other tree and shrub species, in 1985–1986

Vegetation layer	*Location*	*Coppice age class (year)* 1–7	25–30	35–40
Canopy	Oak tree – trunk/branch	2	17	3
	Ash tree	0	1	0
	Honeysuckle on tree	0	6	1
Shrub	Honeysuckle bundle	0	3	2
Field	Hawthorn	0	2	0
	Hazel: hollow pole	0	7	0
	Bramble	6	0	0
Ground	Habitat pile	2	0	0
	Tree/shrub base	3	5	0
	On ground	2	0	0
	Under stool/stump	1	2	0
Total (%)		16 (25)	43 (66)	6 (9)

honeysuckle flowers are major components of the diet in midsummer, and a variety of fruits, nuts and berries are taken increasingly from the end of July. It is in summer when bottlenecks are most likely to occur in the food supply, particularly in areas with a paucity of shrub species, and that dormice must rely on food sources of relatively poor quality such as leaf matter (Holisova, 1968). In autumn, coppice under appropriate management can produce heavy mast crops which enable more animals to fatten up prior to hibernation, resulting in greater overwinter survival and early breeding the following year. The value of the various parts of a coppice woodland, in terms of the variety, quantity and quality of food likely to be available at different times of year, depends on the species composition and age of the regrowth. Clearly, a rotation length that allows the various coppice species to regrow, flower and fruit in a sufficient proportion of the wood will be best suited to the needs of dormice. Therefore, in mixed hazel coppices in 10–25-year regrowth, conditions are particularly suitable for dormice.

Information collected about dormice by means of live-trapping (Morris and Whitbread, 1986) and radiotracking highlights the preference of dormice for thick, species-rich coppice with interconnected trees and shrubs (Table 11.3). It was found that animals in good quality habitat, such as mature hazel coppice stands, had smaller range areas, travelled shorter distances each night and were present in higher densities than in areas of either young coppice regrowth (less than five years old) or over-mature 'relict' coppice producing little fruit, with a patchy shrub layer and numerous dead poles. In fact, animals only used young or relict coppice to nest in, and spent most of their time in coppice that had good fruit production. Of particular interest was the seasonal use of neighbouring stands of different ages. Several animals were nesting in the bramble and young regrowth of 5–6-year coupes but fed entirely in adjacent, mature coppice stands with relatively little undergrowth, except when utilizing oak standards in May and hazel nuts from the end of August.

Table 11.3 Dormice in hazel coppice with oak standards on the Isle of Wight; no coppice of 8–24 years present

	Recent (1–7 years)	*Mature (25–30 years)*	*Relict (35–40 years)*
Number of shrub species	11	11	11
Continuity of vegetation above ground			
0–3.0 m	Good	Moderate	Patchy
3.0–6.0 m	Moderate	Good	Poor
6.0–9.0 m	Very little	Good	Patchy
9.0- 12 m	Moderate	Moderate	Patchy
Use of area	Very seasonal; nests, but majority of time spent in mature coppice	All year; nests; population density: 12–18/ha; home range sizes: 0.07–0.39ha* to 0.47–0.91ha†	In nests only, at edges; time spent mainly in mature coppice

*Range sizes in mature dense coppice.
†Range sizes in over-mature, patchy, dense/sparse coppice

In contrast to hazel coppice, sweet chestnut fruits later in the year, and as a commercial crop it is most frequently grown without standards (Evans, 1984). Such coppice supports few other shrub species and is cut in large coupes (e.g. 2.5 ha). Sweet chestnut is generally more upright in its growth form than hazel and is intolerant of shade: branches from neighbouring poles rarely interconnect; lower branches are shed and weaker poles and stools are shaded out entirely. Typically then, sweet chestnut coppice is only likely to be suitable for dormice where it occurs with other shrub species.

Table 11.4 compares captures of dormice and other small mammals above ground in coppice regrowth of different ages in (a) hazel coppice with standards and (b) sweet chestnut coppice. It illustrates the greater suitability of hazel, the effect of the age of regrowth, and the importance of other shrub species for dormice in sweet chestnut coppice. Toms (1990) examined small mammal communities in sweet chestnut coppice at the same site and encountered no dormice in age-classes of 1–8 years.

Table 11.4 Captures of dormice and other rodents 1–3 m above the ground. (a) Within different parts of a small coppice hazel with oak standards wood, restored to a ten-year rotation of 0.4 ha coupes on the Isle of Wight (1986). (b) Sweet chestnut coppice (April–November, 1989) cut in 2.5 ha coupes on a 17-year rotation in West Sussex. Note that no coppice of 8–24 years was present at the hazel site, and yellow-necked mice are absent from the Isle of Wight.

(a) Hazel coppice (*Corylus avellana*) with oak (*Querus*) standards

			Captures per 1000 trap-nights			
Coppice age-class	*Number of shrub species†*	*Number of trap-nights*	*Dormouse*	*Bank vole*	*Wood mouse*	*Yellow-necked mouse*
5–7	8(3)	320	52	175	0	–
25–30	8(3)	3136	96	18	1	–
35–40	7(4)	646	2	51	8	–

(b) Sweet chestnut (*Castanea sativa*) coppice

			Captures per 1000 trap-nights			
Coppice age-class	*Number of shrub species†*	*Number of trap-nights*	*Dormouse*	*Bank vole*	*Wood mouse*	*Yellow-necked mouse*
12	4(1)	1600	6	2	1	2
12	4(2)	960	1	0	0	0
14	1(1)*	640	0	0	0	0
15	1(1)*	640	0	0	0	0
19	7(3)	1600	28	8	2	0

*Areas with almost no stools other than of sweet chestnut.
†Brackets indicate other tree/shrub species of less than 5% cover.

Dormice will travel considerable distances out of their way to where branches meet across rides (even paths 2 m wide) in order to avoid travelling on the ground (S. Whitbread, personal observation). Wide rides, which may be introduced to encourage the colonization of new coppice by open-phase species, and large clearings will act as significant barriers to the free movement of dormice. Whilst their effects on dispersal within a wood are unknown, problems are more likely to arise where the areas bounded by rides and paths are small (e.g. less than 1 ha), necessitating frequent crossings. Such difficulties can normally be overcome with a well-planned rotation mosaic and the effective area of suitable dormouse habitat can be increased by linking small fragmented patches with 'corridors' of suitable habitat such as branches spanning suitably managed rides (e.g. with scallops and constrictions), or hedges connecting adjacent woods (e.g. at Stoke Woods in Northamptonshire). Such features not only increase the effective area of suitable dormouse habitat but also may assist dispersal (Bright and Morris, 1989).

11.5 SQUIRRELS

Whether or not squirrels are resident in coppice woodland will largely depend upon the presence or absence of standard trees. The habitat will generally be good for squirrels if the density of standards is about 25–50/ha, and particularly if they are large seeded trees such as oak or beech. In areas where there are red squirrels (*Sciurus vulgaris*), oak trees are particularly valuable as standards. Although red squirrels feed little on acorns (Kenward and Holm, 1989), oak trees can be vital for summer catkin feeding, for insects and for shelter and nest sites (J.L. Holm, personal communication). Hazel coppice is especially valuable to squirrels, hazel nuts being particularly favoured by red squirrels – their agility and small size enabling them to forage on the slender branches in coppice stands (see Kenward and Tonkin, 1986). On the Isle of Wight, it has been found that spring, summer and autumn densities of red squirrels, and juvenile survival to trapping age, were all positively related to the previous autumn's hazel crops but not acorns (Holm, 1990). Therefore, hazel coppice with oak standards is an especially good type of habitat for red squirrels, as long as they are not in danger from invasion by grey squirrelss (*Sciurus carolinensis*).

As indicated, hazel coppice with standards can also provide a good habitat for grey squirrels, although their spatial and population dynamics respond to the distribution and abundance of seeds from the standard trees, such as acorns, rather than to hazel nuts (see Gurnell, 1989; Kenward and Holm, 1989). Even so, grey squirrels, with their larger body size than red squirrels and much higher densities in broadleaved woodland (Gurnell and Pepper, 1988), can have a marked impact on the hazel crop. The median time for hazel nuts to fall in the oak wood at Alice Holt was by week 36 (Figure 11.4; this is equivalent to the week ending 26 August 1990). Few nuts survived until the end of September and, overall, 81% of the hazel nuts which fell had already been eaten, mostly by squirrels. Kenward and Holm (1989) also note that grey squirrels generally consume the hazel crop in Dorset and Hampshire woods by October. Additional studies at Bradfield and Alice Holt woods substantiate the fact that grey squirrels eat by far the majority of the hazel crop (Table 11.5). Thus, in stands in which the coppice regrowth is sufficiently mature physically to support the larger grey squirrel, they probably deplete the supply of hazel nuts much more quickly than red squirrels, so reducing the stock of this valuable food that is

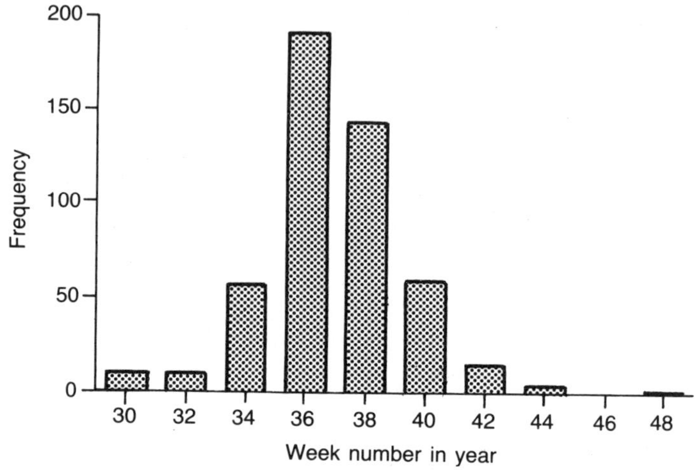

Figure 11.4 Time of fall of hazel nuts in an oak wood in Alice Holt Forest, Hampshire (81% of the nuts that fell had already been eaten, mainly by squirrels).

available for the latter. The consequences of such food exploitation for other seed-eaters, such as mice and dormice, also requires further detailed study.

Table 11.5 Proportion (%) of hazel nuts eaten by grey squirrels, dormice and other animals. Bradfield Wood, Suffolk data from ground quadrat counts under a range of coppice ages over a four-year period. Alice Holt Forest, Hampshire data from ground quadrat counts made during October 1983.

Seed predator	*Bradfield Wood (n=813)*	*Alice Holt Forest (n=180)*
Grey squirrels	85	71
Dormice	9	9
Other (mice, voles, birds)	7	20

Grey squirrels may also strip the bark from certain species of coppice (e.g. sycamore (*Acer pseudoplatanus*), hornbeam (*Carpinus betulus*) and sweet chestnut) when it is of sufficient size to support them: in some cases this may cause the tops of the branches to die back, which reduces the size of the coppice harvest.

11.6 RELATIONSHIPS WITH OTHER MEMBERS OF THE COMMUNITY

So far we have discussed in detail some of the most numerous and most studied small mammal species in coppice woodland, but others deserve mention. Rabbits (*Oryctolagus cuniculus*) are frequently present in short-grass areas such as along grassy rides or in young plantations. These may nibble back the bark on coppice regrowth when densities are very high (D. Saunders, personal communication). Moles (*Talpa europeae*) will also be found where the soil is suitable (i.e. not too stony, acidic or waterlogged), and hares (*Lepus capensis*) regularly use young coppice by day (e.g. at Bradfield). Furthermore, coppice woodlands, with their high structural diversity and insect diversity, can be good feeding grounds for woodland bats and standard trees, where present, provide suitable roosting sites (Steele, 1975). As a whole, the small mammal community of coppice woodland can attract a variety of resident and transient predators. Carnivores, such as fox (*Vulpes vulpes*), stoat (*Mustela erminea*), weasel (*M. nivalis*), mink (*M. vison*), polecat (*M. putorius*), pine marten (*Martes martes*) and wild cat (*Felis catus*) regularly hunt in woodland situations (King, 1985). Of these animals, weasels specialize more than the others in small mammal prey and stoats tend to feed on birds and lagomorphs as well as small mammals. The others are generalist and opportunistic predators, but some, e.g. the pine marten, eat proportionately more small mammal prey, such as the field vole, when they are abundant (Pulliainen, 1982). Very young coppice with little ground cover provides good hunting grounds for avian predators (e.g. owls and kestrels). However, standard trees may be required by some species for suitable nesting sites or perches. Owls, such as the barn owl, *Tyto alba*, may not be woodland residents, but young coppice and edge ecotones can provide suitable hunting grounds for them. Tawny owls, *Strix aluco*, however, are woodland resident specialists requiring an open forest floor to ambush their small mammal prey from a perch (King, 1985), thus, they may also hunt in the older coppice stands where undergrowth is lacking.

Predators can take large numbers of prey, but whether they control prey numbers is still a matter for debate (King, 1985). On the other hand, the distribution and abundance of prey has a strong influence on the occurrence and population dynamics of predators (e.g. Southern, 1970; Hirons, 1982). Active coppice woodlands, therefore, with its variety of habitat structure and diversity of small mammal prey, are especially suitable habitats for a wide range of predators.

11.7 IMPLICATIONS OF COPPICE MANAGEMENT FOR SMALL MAMMAL SPECIES

An actively managed coppice woodland provides a patchwork of habitats which, overall, is generally beneficial to small mammal populations. Structural diversity, especially of the undergrowth (field and shrub layers), probably influences the distribution of small mammal species more than plant species composition (Hansson, 1977; Rozenzweig and Winakur, 1969). However, the species of coppice in turn influences the diversity and persistence of undergrowth and the form of the shrub layer. Coppice of between one and seven years old supports the greatest diversity of small mammals, but a mosaic of coppice stands up to 30 years old ensures the greatest diversity of mice, voles and shrews. Such a managed woodland, therefore, provides habitats for both 'grassland' and 'mature woodland' species. Sweet chestnut coppice, found on species-poor, acidic sites, is the commonest type of coppice now found (Chapter 2), much of which is still managed. As discussed in the dormouse section above, it is probably not such a good habitat for woodland small mammals as is hazel or mixed coppice with standards. More information is needed about the effects of managing different pure coppice species (e.g. hornbeam, sweet chestnut, sycamore, ash) on small mammal populations – although we feel that the patterns discussed in this paper are generally applicable.

The length of the rotation cycle and the size and juxtaposition of coupes of different age are important management considerations. Only about a third of the worked coppice in Britain is coppice with standards (Evans, 1984) but, and although in need of specific study, we believe that a coppice with a low to moderate density of standard trees (25–50/ha) offers both a wider range and greater quantity of food for small mammals than simple coppice, especially in younger regrowth. Broadleaved woodland with hazel, such as the oak wood at Alice Holt, is a common type of habitat in certain parts of the country (e.g. Surrey and Hampshire). Restoring the hazel to coppice would be beneficial to the wildlife for some of these woods. When either restoring management to a stored coppice, or planting up a new area to manage as coppice, consideration should be given to the size and isolation of the wood, the species composition and density of trees and shrubs in those parts to be coppiced, and the width of paths and rides which separate the stands. Some small mammals, such as bank voles and dormice, may be reluctant to cross wide, open rides (e.g. Bakowski and Kozakiewicz, 1988; Geuse *et al.*, 1985), whereas appropriate ride and edge management may encourage movement between stands (Mayle and Gurnell, 1991).

Coppice managed primarily for commercial purposes can be enhanced by interplanting even a small number of other trees and shrubs and leaving some to develop as maidens, even if these are restricted to compartment corners or margins where they will provide connections with neighbouring stands. Boundary strips of coppice with standards around both broadleaved and coniferous plantations could encourage a greater diversity of small mammals and may be particularly beneficial to species such as red squirrels in areas less at risk from grey squirrel invasion, such as the north of England and Scotland.

The effects of the size of stand or felling coupe has not been fully investigated but from the available evidence it appears that several small coupes (0.5–1 ha) provide a better patchwork of habitats than few large ones, depending of course on the overall area of woodland available for coppicing. Furthermore, mosaics containing a wide range of age classes provide food and habitat structures that are beneficial to a wide range of small mammal species. For example, long-rotation stands of 20–30 years would be beneficial to squirrel and yellow-necked mice, while very young stages would be most suited to harvest mice, voles and shrews. Perhaps with the exceptions of very young or very old coppice, the full range of ages could be exploited at various times by dormice, while the ubiquitous wood mouse is likely to be widespread throughout. In a small (e.g. 5 ha), isolated wood, typical of many coppice woodlands (Chapter 1), a regime which retains 30–50% of the coppice at 20–30 years growth at any one time would appear to suit the needs of all these species.

In summary, active management of coppice woodland that produces a fairly even mix of age classes is beneficial to small mammal communities. This, in turn, should be beneficial to predatory species (e.g. owls, weasels). The diversity of small mammals will be enhanced if the stands are disturbed as little as possible between cutting so that the three-dimensional structure of the shrub layers can develop fully. Ride edges, hedgerows and ecotones between neighbouring stands increase the overall structural diversity of the woodland and facilitate the movement of small animals within the woodland. Coppice woodland may also play a part in encouraging particular species, e.g. harvest mice and water shrews, including some, such as dormice and red squirrels, which are important for their conservation interest.

ACKNOWLEDGEMENTS

We would like to thank the Forestry Commission for permission to work at Alice Holt Forest, the Suffolk Wildlife Trust for permission to work at Bradfield Woods and the support of the warden, Pete Fordham. We are grateful to the many people who have discussed various aspects of this chapter with us including Jessica Holm, Rob Fuller and Robert Kenward, and we are very grateful to Phil Ratcliffe and Peter Buckley for reading the manuscript. The studies were supported by grants from the Central Research Fund, University of London in 1975, British Ecological Society, 1985–1986, and 1985, and the Mammal Conservation Trust, 1985–1989, and 1986, and the Mammal Society, 1988. Part of the work carried out in 1986 was funded under a Nature Conservancy Council research contract to P.A. Morris.

REFERENCES

Alibhai, S.K. and Gipps, J.H.W. (1985) The population dynamics of bank voles, in *Symposium of the Zoological Society of London*, 55, Oxford University Press, Oxford, pp. 277–313.

Bakowski, C. and Kozakiewicz, M. (1988) The effect of a forest road on bank vole and yellow-necked mouse populations. *Acta Theriologica*, **33**, 345–53.

Bright, P. and Morris, P.A. (1989) *A Practical Guide to Dormouse Conservation.* Occasional publication of the Mammal Society, 11, The Mammal Society, London.

Catzeflis, F. (1983) Le poids du Muscardin (*Muscardinus avellanarius* Kaup, 1829) dans la nature (Gliridae, Rodentia). *Bull. Soc. Vaud. Sci. Nat.*, **76**, 295–8.

Catzeflis, F. (1984) Étude d'une population de Muscardins (*Muscardinus avellanarius*) lors du répos journalier (Mammalia, Gliridae). *Revue Suisse de Zoologie*, **91**, 851–60.

Churchfield, S. (1990) *The Natural History of Shrews*, Christopher Helm, London.

Corbet, G.B. and Southern, H.N. (1977) *The Handbook of British Mammals*, 2nd edn, Blackwell, Oxford.

Corbet, G.B. and Harris, S. (1990) *The Handbook of British Mammals*, 3rd edn, Blackwell, Oxford.

Evans, J. (1984) *Silviculture of Broadleaved Woodland.* Forestry Commission Bulletin, 62, HMSO, London.

Ferns, P.N. (1979a) Successional changes in a small mammal community of a young larch plantation in south-west Britain. *Mammalia*, **43**, 439–52.

Ferns, P.N. (1979b) Growth, reproduction and residency in a declining population of *Microtus agrestis. Journal of Animal Ecology*, **48**, 739–58.

Flowerdew, J.R. (1985) The population dynamics of wood mice and yellow-necked mice, in *Symposium of the Zoological Society of London*, 55, Oxford University Press, Oxford, pp. 315–38.

Flowerdew, J.R., Gurnell, J. and Gipps, J.H.W. (1985) The ecology of woodland rodents: bank voles and wood mice, in *Symposium of the Zoological Society of London*, 55, Oxford University Press, Oxford.

Geuse, P., Bauchau, V. and Le Boulenge, E. (1985) Distribution and population dynamics of bank voles and wood mice in a patchy woodland habitat in central Belgium. *Acta Zoologica Fennica*, **173**, 65–8.

Gurnell, J. (1985) Woodland rodent communities, in *Symposium of the Zoological Society of London*, 55, Oxford University Press, Oxford, pp. 377–411.

Gurnell, J. (1989) Demographic implications for the control of grey squirrels, in *Mammals as Pests* (Ed. R.J. Putman), Chapman and Hall, London, pp. 131–43.

Gurnell, J. and Pepper, H. (1988) Perspectives on the management of red and grey squirrels, in *Wildlife Management in Forests* (Ed. D.C. Jardine), Institute of Chartered Foresters, Edinburgh, pp. 92–109.

Gurnell, J. and Flowerdew, J.R. (1990) *A Trapping Guide to Small Mammals*, 2nd edn. Occasional Publication of the Mammal Society, London.

Gurnell, J. and Gipps, J.H.W. (1989) Inter-trap movement and estimating rodent densities. *Journal of Zoology, London*, **217**, 241–54.

Hansson, L. (1977) Spatial dynamics of field voles, *Microtus agrestis*, in heterogeneous landscapes. *Oikos*, **29**, 539–44.

Hirons, G.J.M. (1982) The effects of fluctuations in rodent numbers on breeding success in the tawny owl, *Strix aluco*. *Mammal Review*, **12**, 55–157.

Holisova, V. (1968) Notes on the food of dormice (Gliridae). *Zool. Listy*, **17**, 109–14.

Holisova, V. (1969) Vertical movements of some small mammals in a forest. *Zool. Listy*, **18**, 121–41.

Holland, S.C. (1967) Mammal notes and records. *Journal of the N. Gloucestershire Naturalists Society*, **18**, 225–6.

Holm, J. (1990) *The ecology and feeding behaviour of red squirrels (*Sciurus vulgaris* L.) in deciduous woodland*, PhD thesis, University of London.

Hurrell, E. (1980) *The Common Dormouse*, Blandford Press, London.

Hurrell, E. and McIntosh, G. (1984) Mammal Society dormouse survey, January 1975–April 1979. *Mammal Review*, **14**, 1–18.

Kenward, R.E. and Holm, J.L. (1989) What future for British red squirrels? *Biological Journal of the Linnean Society*, **38**, 83–9.

Kenward, R.E. and Tonkin, J.M. (1986) Red and grey squirrels: some behavioural and biometric differences. *Journal of Zoology*, London, **209**, 279–81.

King, C.M. (1985) Interactions between woodland rodents and their predators, in *Symposium of the Zoological Society of London, 55*, Oxford University Press, Oxford, 219–48.

Lozan, M.N. (1961) Age determination of *Dryomys nitedula* Pall. and of *Muscardinus avellanarius* L. *Zool. Zh*, **40**, 1740–3.

Ludolf, C.I. (1986) *Population dynamics of rodents in a mosaic of forest habitats*, PhD thesis, University of London.

Mayle, B. and Gurnell, J. (1991) Edge management and mammals, in *Edge Management in Woodlands* (Ed. R. Ferris-Kaan), Forestry Commission Occasional Paper, 28, pp. 42–8.

Mazurkiewicz, M. (1986) The influence of undergrowth distribution on the utilisation of space by bank vole populations. *Acta Theriologica*, **31**, 55–69.

Montgomery, W.I. (1980) The use of arboreal runways by the woodland rodents, *Apodemus sylvaticus* (L.), *A. flavicollis* (Melchior) and *Clethrionomys glareolus* (Schreber). *Mammal Review*, **10**, 189–96.

Montgomery, W.I. (1989) Population regulation in the wood mouse, *Apodemus sylvaticus*. I. Density dependence in the annual cycle of abundance. *Journal of Animal Ecology*, **58**, 465–75.

Morris, P.A. and Whitbread, S. (1986) A method for trapping the Common Dormouse (*Muscardinus avellanarius*). *Journal of Zoology, London*, **210**, 642–44

Morris, P.A., Bright, P. and Woods, D. (1990) Use of nestboxes by the dormouse *Muscadinus avellanarius*. *Biological Conservation*, **51**, 1–13.

Peterken, G.F. (1981) *Woodland Conservation and Management*, Chapman and Hall, London.

Pulliainen, E. (1982) Some characteristics of an exceptionally dense population of the red squirrel, *Sciurus vulgaris* L., on the south coast of Finland. *Aquilo Ser. Zool.*, **21**, 9–12.

Rozenzweig, M.L. and Winakur, J. (1969) Population ecology of desert rodent communities: habitats and environmental complexity. *Ecology*, **50**, 558–72.

Schulze, W. (1970) Beitrage zum Vorkommen und zur Biologie der Haselmaus (*Muscardinus avellanarius* L.) und des Siebenschlafers (*Glis glis* L.) im Sudharz. *Hercynia (N.F.)*, 7, 355–71.

Southern, H.N. (1970) The natural control of a population of tawny owls (*Strix aluco*). *Journal of Zoology, London*, **162**, 197–285.

Steele, R.C. (1975) Forests and wildlife. *Philosophical Transactions of the Royal Society*, *B*. 271, 163–78.

Taitt, M.J. and Krebs, C.J. (1985) Population dynamics and cycles, in *The Biology of New World* Microtus (Ed. R.H. Tamarin), Special Publication of the American Society of Mammalogy, 8, 267–720.

Toms, M. (1990) *An investigation into the distribution of small mammal species in different year classes of sweet chestnut coppice*, BSc Honours Project, Biology Department, University of Southampton.

Whitbread, S. (1986a), A radiotracking study of the common dormouse (*Muscardinus avellanarius*). *Bulletin of the British Ecological Society*, **17**, 66–8.

Whitbread, S. (1986b), The aerial dormouse. *BBC Wildlife Magazine*, **4**, 588–602.

– 12

The interaction of deer and vegetation in coppice woods

PHILIP R. RATCLIFFE

12.1 INTRODUCTION

Coppicing has been practised since medieval times to produce small roundwood for fencing, fuel, etc. (Chapter 2). More recently coppice management has been valued for its conservation benefits in terms of both generally increasing structural diversity of woodlands and species diversity, and supporting specific species groups, such as butterflies (Chapter 13) and birds (Chapter 9).

Additionally, coppice management may increase carrying capacity for gamebirds, especially pheasants (*Phasianus colchicus*) by increasing the availability of vegetation and insects (Hill and Robertson, 1988; Chapter 10). Deer will also benefit from the increased availability of food and the intimate mosaic of food, shelter and cover created. Although coppicing has seldom been practised purposely to increase deer densities in the UK, this has been the case in the USA (Cook, 1969; Shipman, 1972), France (Martinot-Lagarde, 1975) and Russia (Dunin, 1985). Historically, however, stock-proof boundaries were used in Britain so that domestic stock and deer could be excluded for the early part of the rotation and thereafter allowed in (Peterken, 1981).

Group selection felling, i.e. the periodic felling of small groups of trees, thereby creating canopy gaps (see Peterken, 1981), is probably the best silvicultural system for providing deer with an ideal interspersion of food, shelter (from weather) and cover (from disturbance) in space and time, but undoubtedly coppice management can produce similar results in providing a continuous succession of early growth stages. In such habitats deer can achieve high levels of performance (in terms of body weights, reproductive rates) and density.

Roe (*Capreolus capreolus*), muntjac (*Muntiacus reevesi*), and fallow deer (*Dama dama*) occur throughout much of the coppice area in Great Britain, and although their feeding ecology is different, all can have an impact on coppice woods. This can markedly affect plant and invertebrate species composition, birds and small mammals, as well as limiting the growth of the coppice crop itself.

12.2 HABITAT PREFERENCES OF DEER

Sequential changes in forest growth influence plant distribution and abundance, largely through changes in light transmittance to the forest floor (Hill, 1979; Chapter 3). Canopy closure is therefore an important influence on the carrying capacity of any animal dependent upon the availability of ground vegetation. Additionally, deer require shelter from bad weather and cover to hide from disturbance. Clearly, intimate mixtures of different temporal stages of woodland succession will provide ideal conditions for deer, providing food, shelter and cover in close proximity. The suitability of habitats for deer is indicated in Figure 12.1. Coppice management can provide these optimum conditions and support high numbers of deer (Prior, 1983), but rotation length, growth rate and coupe size have important influences.

Rotation lengths are usually of 5–30 years but have sometimes reached 50 years, allowing the stand to revert to high forest if the coppice is totally neglected. The terms of the management lease, the type of produce and the tree species involved have all influenced rotation length in historic times

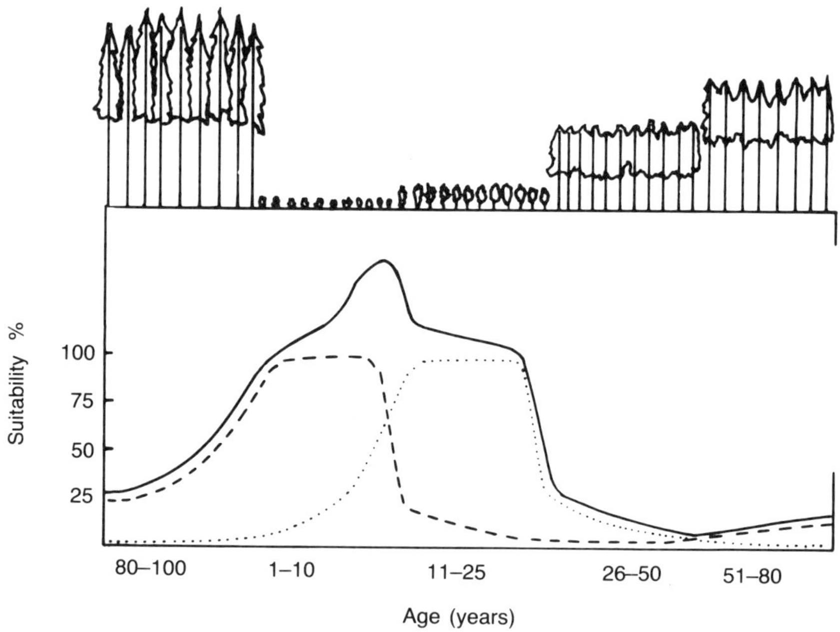

Figure 12.1 Schematic representation of the suitability of the forest habitat for deer during the course of a rotation (—, cover and food; — — —, available food; . . . . , cover). (Reproduced with permission from Koning and Gossow, 1979.)

(Peterken, 1981) but it is the perpetuation of the early growth stages (approximately five to ten years) which will support high numbers of deer.

All three deer species respond positively to high plant diversity and a mix of food, shelter and cover (Figure 12.1), but there are differences between species. Roe and muntjac deer are concentrate selectors (they select vegetation rich in nutrients), adapted to selecting the parts of plants that are high in nitrogen; fallow deer are less selective, taking relatively more fibre (Hofmann, 1985). As a consequence, fallow deer prefer rather open deciduous and mixed woodlands and open fields, arable crops or woodland glades, where they feed preferentially as grazers (Putman, 1988). Roe and muntjac deer prefer denser habitats with a wide variety of herbs and woody plants on which they can browse. Their demands for escape cover can be met by earlier successional stages due to their much smaller size, bramble (*Rubus fruticosus*) and bracken (*Pteridium aquilium*) thickets providing sufficient cover for these species.

12.3 USE OF COPPICE HABITATS

There are few studies of deer and coppice although inferences can be drawn from studies of habitat use and feeding ecology in woodlands of similar structure. Martinot-Lagarde (1975) suggested that abandoned coppice areas in France were unsuitable for deer, due to reduced cover and food availability, and advocated periodic cutting on 35–50-year rotations to provide suitable areas of young woody and herbaceous vegetation. Lozovoi and Sukhoroslov (1976) proposed a classification system for areas used for game based on plant biomass of annual shoots. Woodlands with coppice up to 10 years old were highly valued by fallow and roe deer. Cibien and Sempere (1989) compared the habitat use of roe deer in habitats with an abundant shrub layer with that of older coppice areas with low food-availability, and they showed that the population density of roe deer was higher in the shrubby areas with higher food-availability.

A brief, but valuable, study of habitat use by fallow deer in coppice woods and agricultural land in Suffolk, England, was conducted by Sackur (1984). The coppice woods were classified into four habitat types based on the age of coppice and the vegetation structure with the surrounding fields making the fifth habitat type (Table 12.1). Visual observations were recorded along fixed transects through a representative sample of all habitat types on a monthly basis from February to October (Table 12.2). Additionally, dung groups were counted on the same basis using quadrats in each habitat type which were counted and cleared each month (Table 12.3). Subsequent 'corrections' applied by Sackur (1984) to the data on visual observations, taking account of the representative proportions of each habitat in the study area, did not affect the pattern of habitat use, though it did substantially increase the number of deer using all of the woodland habitats relative to the fields.

Table 12.1 Habitat types used by fallow deer in Suffolk (After Sackur, 1984)

Type	*Description*
Recent open	Coppice newly cut with little regrowth; little ground vegetation
Recent closed	Extensive shrubby growth on coppice stools; dense ground vegetation
Old closed	Dense growth of mature coppice; little ground vegetation
Old open	Lack of shrub layer; grass-dominated ground vegetation
Agricultural fields	Mainly winter wheat crops

Table 12.2 Numbers of fallow deer seen in each habitat, percentages given in brackets (After Sackur, 1984)

Month	*Recent open*	*Recent closed*	*Old closed*	*Old open*	*Fields*	*Total*
February	–	2(2)	16(17)	5(5)	73(76)	96
March	–	5(13)	–	3(8)	32(80)	40
April	–	15(24)	3(5)	–	45(71)	63
May	–	–	5(6)	2(2)	81(92)	88
June	–	4(12)	–	14(41)	16(47)	34
July	7(19)	3(8)	2(6)	14(39)	10(28)	36
August	15(26)	3(5)	2(4)	37(65)	–	57
September	14(29)	–	4(8)	27(56)	3(6)	48
October	13(15)	11(12)	10(11)	35(39)	20(23)	89

There was relatively good agreement between habitat use as determined independently from visual observations and from dung accumulation, except in fields. Sackur (1984) suggests that this is largely due to very clumped use of fields by deer relative to rather more uniform use of woodlands. Figure 12.2 indicates habitat use by fallow deer of the woodland areas, based upon the dung accumulation data (Table 12.3).

Recent open areas with low food availability were little used by deer during the spring but there was a gradual increase in usage through summer to autumn (Figure 12.2). Recent closed areas with high food availability of both woody and herbaceous plants had a marked seasonal usage – it being particularly high in late spring and early summer, and winter. Old closed habitats with dense coppice growth but little herbaceous vegetation showed high usage in winter, and in late summer and autumn with a pronounced reduction in usage in spring and early summer. Old open areas

Table 12.3 Mean number of dung groups per 50 m² (standard deviations given in brackets) in each habitat. (After Sackur, 1984)

Month	*Recent open*	*Recent closed*	*Old closed*	*Old open*	*Fields*	*Total*
February	0.125 (0.35)	1.875 (1.46)	0.625 (0.74)	0.25 (0.71)	–	2.875
March	0.125 (0.35)	0.125 (0.35)	1.0 (1.41)	0.125 (0.35)	–	1.375
April	–	0.125 (0.35)	0.375 (0.74)	0.125 (0.35)	0.125 (0.35)	0.75
May	0.125 (0.35)	1.125 (1.73)	0.375 (0.52)	0.875 (0.83)	–	2.5
June	0.375 (0.74)	0.75 (1.16)	0.5 (0.76)	0.75) (1.49)	–	2.375
July	0.25 (0.46)	0.25 (0.46)	–	0.375 (0.74)	–	0.875
August	0.625 (0.74)	0.5 (0.76)	1.0 (0.93)	0.875 (0.83)	–	3.0
September	0.5 (0.76)	0.625 (0.92)	1.0 (0.93)	0.5 (0.76)	–	2.625
October	0.625 (0.74)	0.375 (0.52)	1.25 (1.17)	0.875 (0.83)	–	3.125

with a grass-dominated field layer supported high usage in summer and autumn with very low usage in spring.

These results support previous views on the importance of food and cover occurring in close proximity. Presumably, as recently cut areas become vegetated, deer begin to exploit them in the late summer. Maximum food availability of both woody browse and herbaceous vegetation is present in the recent closed areas. The old closed areas would offer an abundance of vernal species and some woody browse but shading during the summer would effectively reduce food availability on the woodland floor. The high usage of the old open areas in summer and autumn probably reflects the ability of fallow deer to graze efficiently on grasses. It is unlikely that roe or muntjac deer would make such high usage of these areas. Clearly, the arable field crops are of importance during the spring and presumably through the winter. It is important that this type of study is expanded to include all three deer species over several years and such a study is now in progress (C. Pringle and R.J. Putman, personal communication).

12.4 IMPACT OF DEER AND COPPICE HABITATS

The impact of deer on woodland regeneration and the species composition of ground vegetation is well documented (Peterken and Tubbs, 1965; Ratcliffe, 1988; Putman, 1986, 1988). Impact is closely related to palatability

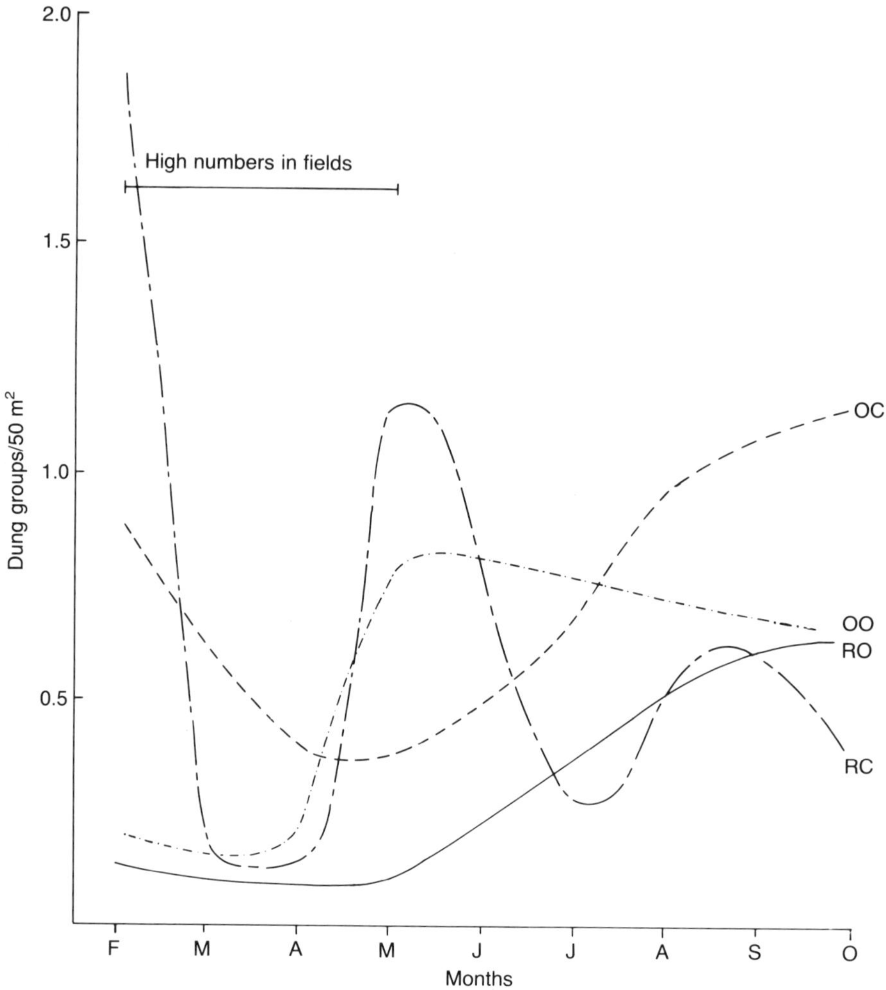

Figure 12.2 Habitat use of coppice woodland by fallow deer based on dung accumulation data (see Table 12.3) (after Sackur, 1984). See Table 12.1 for description of coppice types. RO, recent open; RC, recent closed; OC, old closed; OO, old open.

and many authors have listed preferences for certain browse species by deer (see Mitchell *et al.*, 1977). Unfortunately such lists are simplistic and difficult to interpret in different habitats because palatability is so often influenced by the availability of alternative foods. For example, it is of interest that aspen (*Populus tremula*) is considered to be highly preferred by deer, according to a number of European authors (Mitchell *et al.*, 1977) while in coppice areas in Great Britain it has been found to be unpalatable (Rackham, 1975).

Many factors may influence plant palatability and selection by herbivores. The presence of secondary plant compounds which protect plants from attack by herbivores can inhibit digestion or produce toxins (Rhoades and Gates, 1976). Toxic allelochemics are more characteristic of ephemeral plants or plant parts that usually contain high levels of nitrogen, while digestibility-reducing allelochemics are usually associated with long-lived plants or plant parts that are often low in nitrogen (Mattson, 1980). Clearly, growth form and phenology as well as species-type, and the presence of other more or less palatable species can all influence plant selection by deer. Willows (*Salix* spp.) are almost always highly palatable to deer, but for a given plant species voluntary intake and digestibility are lower in secondary growth than in primary growth (McCabe and Barry, 1988).

In New Zealand deer browsing is considered to have had an influence on the amount of coppice regrowth occurring (Smale, 1982) which was still obvious 23 years after logging. In Lady Park Wood, England, small-scale patchiness of tree regeneration, considered to be partly due to deer browsing, soon disappeared after fencing (Peterken and Jones, 1989). Picard (1976) compared the impact of roe and red deer in areas of montane conifers and lowland oak/hornbeam coppice stands in France. Deer browsing caused a reduction in the abundance of many plant species and an increase in unpalatable species such as great wood-rush (*Luzula sylvatica*) and *P. aquilinum* in the montane area. Much less effect was noticed in the lowland area which had a richer vegetation. In a coppice area in eastern England Rackham (1975) concluded that deer (mainly fallow) were incompatible with most woodland management objectives.

Perhaps the most useful study of grazing impact by fallow deer in lowland Britain was conducted in the New Forest. The study compared grazed (*c.* 1 deer/ha) and ungrazed plots over a 26-year period. It was initiated by the Forestry Commission in 1963 and has subsequently been studied by Mann (1978), How (1986) and Putman *et al.* (1989). In 1969, six years after establishment, there were 6440 young trees/ha (of <10 cm diameter at breast height) in the ungrazed plot compared with only 20 trees/ha in the grazed plot. The main species were sallow (*Salix cinerea*), hawthorn (*Crataegus monogyna*), blackthorn (*Prunus spinosa*), sycamore (*Acer pseudoplatanus*), oak (*Quercus robur*) and birch (*Betula* spp.). Holly (*Ilex aquifolium*), hawthorn, sallow, goat willow (*Salix caprea*) and blackthorn were only recorded in the ungrazed plot in the 1977 and 1985 surveys. Maple (*Acer campestris*), alder buckthorn (*Frangula alnus*) and rowan (*Sorbus aucuparia*) were recorded for the first time in the ungrazed plot in 1985.

By 1985 there was little change in the grazed plot, but in the ungrazed plot canopy closure had occurred and an inpenetrable thicket had developed (Figure 12.3). The main species in this thicket were birch, sallow

Figure 12.3 Part of the Denny enclosure, New Forest, 1985, demonstrating the effect of excluding fallow deer for a 22-year period.

and Douglas fir (*Pseudotsuga menziesii*) and sapling density had reached 7115/ha. The shrub layer had become sparse in places and gorse (*Ulex europaeus*) and heather (*Calluna vulgaris*) had disappeared, leaving some places dominated by bramble. Ivy (*Hedera helix*) and moss (*Thuidium tamariscinum*) dominated the field layer. The species composition of the ground flora in grazed and ungrazed plots is presented in Table 12.4.

After six years of continued grazing, *C. vulgaris*, *Epilobium montanum*, *Rosa arvensis*, *R. canina* had been eliminated from the grazed area and *R. fruticosus* had declined to a low level. Selective grazing had favoured an increase in abundance of *Digitalis purpurea*, *Juncus conglomeratus*, *Luzula campestris* and *Molinia caerulea* in the grazed area with all of these disappearing from the ungrazed area. The grasses *Agrostis capillaris* and *Deschampsia cespitosa* achieved dominance in the grazed area due to the selective grazing of the fallow deer.

In 1986, 66 flowering plants and ferns were found in the grazed plot, of which 26 were also in the woodland, with 40 species being found on rides. In the ungrazed plot 52 species were found, of which 27 were confined to

Table 12.4 Species composition of the ground flora at Denny Lodge, New Forest, in grazed (G) and ungrazed (U) plots. Frequency of occurrence of species in 20 quadrats in 1969, 1977 and 1985. (After Putman *et al.*, 1989)

Species	*1969*		*1977*		*1985*	
	G	*U*	*G*	*U*	*G*	*U*
Agrostis canina	2	2	–	1	–	–
A. capillaris	2	2	10	1	8	4
Anthoxanthum odoratum	–	–	–	1	–	–
Calluna vulgaris	–	6	–	6	–	–
Deschampsia cespitosa	2	–	10	3	6	1
Digitalis purpurea	1	–	3	–	1	–
Epilobium montanum	–	2	–	1	–	–
Euphorbia amygdaloides	2	1	–	–	–	–
Festuca rubra	2	1	–	–	–	1
Galium saxatile	2	1	–	1	1	–
Hedera helix	–	3	5	7	3	9
Hypericum perforatum	–	–	–	3	–	–
Juncus conglomeratus	10	9	7	–	2	–
Lonicera periclymenum	9	5	7	6	2	7
Luzula campestris	8	1	3	–	–	–
Molinia caerulea	10	10	4	–	8	–
Oxalis acetosella	–	1	1	–	–	–
Pteridium aquilinum	14	13	15	13	13	9
Rosa arvensis/canina	7	–	–	3	–	2
Rubus fruticosus agg.	13	16	6	19	1	20
Teucrium scorodonia	–	–	–	1	–	–
Ulex europaeus	2	4	–	2	–	–
Viola riviniana	4	2	3	–	3	1

rides. Many of the smaller, light-demanding species had been lost from the ungrazed plot. Thus, although species diversity was similar in grazed and ungrazed plots, species composition was markedly atered (Table 12.4).

Sackur (1984) compared the growth of hazel (*Corylus avellana*) coppice on unprotected stools with those on stools protected by a fence. By October shoots on the protected stools were 1.5–2.0 m in length; all unprotected shoots retained fewer than 20% of the original number of shoots, none of which was longer than 0.1 m. No follow-up work was possible, and the degree of regrowth following browsing is not known.

12.5 MANAGEMENT

Before the 'correct' management can be addressed it is necessary to define the objectives. Clearly, management for timber production, high wildlife diversity, conservation of one or a few rare or sensitive species, or deer or pheasant production for sport, will all make different demands upon the habitat. Presumably different woods will be managed for different reasons,

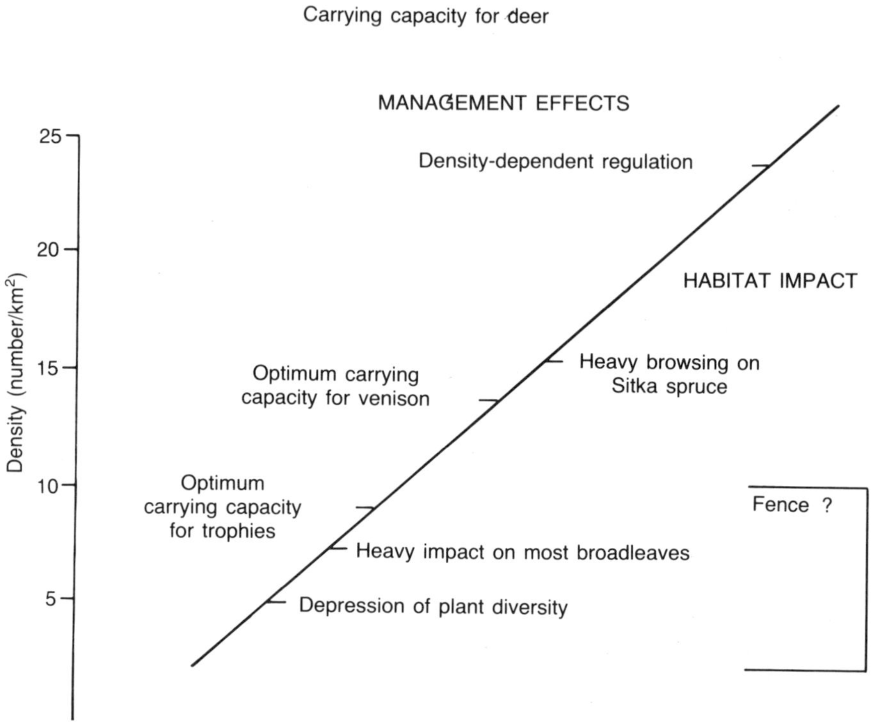

Figure 12.4 Management effects and habitat impact related to deer density (after Ratcliffe, 1988).

and some will be managed for a variety of reasons. If there are many objectives for one wood then at least some compatibility between objectives is necessary.

If the objectives demand that no browsing by deer can be tolerated then fencing is the only sure way to achieve long-term protection (Pepper and Tee, 1986). However, it is clear that no browsing or grazing pressure can result in very dense stands of tree saplings which will ultimately shade-out much of the ground vegetation. Traditional coppice management will prevent this occurring and perpetuate the early growth stages – ensuring that a rich ground flora is always present.

Ratcliffe (1988) summarized the management effects and habitat impact of deer related to deer density (Figure 12.4). It is clear that the optimum densities of deer for most management objectives are well below the level at which density-dependent regulation might occur. Additionally, deer densities in southern England frequently achieve levels exceeding 20 deer/km^2 and it is unlikely that shooting alone can either reduce population size or prevent impact on woodland regeneration, or plant diversity, sufficiently to ensure coppice regrowth and prevent the loss or reduction of sensitive plant species.

Individual tree protection (Pepper *et al.*, 1985) or protection of coppice stools using wire or polythene mesh guards will undoubtedly prevent damage to trees but obviously will have no effect on protecting the ground flora. Tree-shelters are not effective protection against fallow deer. Chemical repellents (Pepper, 1978) have only limited value, only one has been shown to be effective and this is phytotoxic to young growing shoots, and their application is labour intensive. Trials have so far failed to demonstrate the effectiveness of electric fences against deer. The decision to fence or use tree-guards depends upon the tree density and size of area covered but the cut-off point is about 2 ha, assuming a tree density of about 2000/ha (Ratcliffe and Pepper, 1987).

Where the objective is to encourage deer, then coppicing can be used to advantage, in maintaining an abundant supply of young vigorously growing shoots. Willows in particular are likely to be favoured by deer.

ACKNOWLEDGEMENTS

I am grateful to Dr J. Gurnell and Mr R.M.A. Gill for comments on earlier drafts.

REFERENCES

Cibien, C. and Sempere, A. (1989) Food availability as a factor in habitat use by roe deer. *Acta Theriol.*, **34**, 111–24.

Cook, J.L. (1969) Conservation by bulldozer. *Michigan Academician*, **1**, 179–91.

Dunin, V.F. (1985) Improving the food resource for deer. *Lesnoe Khozyaistvo*, **4**, 58–60.

Hill, D. and Robertson, P. (1988) *The Pheasant: Ecology, management and conservation*, Blackwell, Oxford.

Hill, M.O. (1979) The development of a flora in even-aged plantations, in *Ecology of Even-aged Forest Plantations* (eds E.D. Ford, D.C. Malcolm and J. Atterson), Institute of Terrestrial Ecology, Cambridge, 175–192.

Hofmann, R.R. (1985) Digestive physiology of the deer – their morphophysiological specialization and adaptation, in *Biology of Deer Production* (eds P.F. Fennessy and K.R. Drew), Proceedings of the Royal Society of New Zealand, Dunedin, 393–407.

How, R.C. (1986) *An investigation into the vegetational differences between an ungrazed and grazed areas of enclosed woodland in the New Forest*, BSc thesis, University of Southampton.

Konig, E. and Gossow, H. (1979) Even-aged plantations as a habitat for deer in Central Europe, in *Ecology of Even-aged Forest Plantations* (eds E.D. Ford, D.C. Malcolm and J. Atterson), Institute of Terrestrial Ecology, Cambridge, 429–52.

Lozovoi, A.D. and Sukhoroslov, M.S. (1976) Classifying hunting areas by quality on the basis of the total phytomass of the stands. *Lesnoe Khozyaistvo*, **6**, 72–5.

McCabe, S.M. and Barry, T.N. (1988) Nutritive value of willow, *Salix* spp., for sheep, goats and deer. *Journal of Agricultural Science*, **111**, 1–10.

Mann, J.C.E. (1978) *An investigation into the vegetational differences between an ungrazed and a grazed portion of the New Forest, with special emphasis on the species composition, species diversity and productivity of the areas.* BSc thesis, University of Southampton.

Martinot-Lagarde, P. (1975) The management of coppice kept for game. *Bulletin Technique, Office National des Forêts*, **7**, 41–4.

Mattson, W.J. (1980) Herbivory in relation to plant nitrogen content. A review. *Annual Review of Ecology and Systematics*, **11**, 119–62.

Mitchell, B., Staines, B.W. and Welch, D.W. (1977) *Ecology of red deer. A research review relevant to their management in Scotland*, Institute of Terrestrial Ecology, Cambridge.

Pepper, H.W. (1978) *Chemical repellents*, Forestry Commission Leaflet, 73, HMSO, London.

Pepper, H.W., Rowe, J.J. and Tee, L.A. (1985) *Individual tree protection*, Forestry Commission Arboricultural Leaflet, 10, HMSO, London.

Pepper, H.W. and Tee, L.A. (1986) *Forest fencing*, Forestry Commission Leaflet, 87, HMSO, London.

Peterken, G.F. (1981) *Woodland Conservation and Management*, Chapman and Hall, London.

Peterken, G.F. and Jones, E.W. (1989) Forty years of change in Lady Park Wood, UK. The young growth stands. *Journal of Ecology*, 77, 401–29.

Peterken, G.F. and Tubbs, C.R. (1965) Woodland regeneration in the New Forest, Hampshire since 1650. *Journal of Applied Ecology*, **2**, 159–70.

Picard, J.F. (1976) Feeding preferences of deer and their consequences: first conclusions from two years experimentation. *Revue Forestière Française*, **28**, 106–14.

Prior, R. (1983) *Trees and Deer*, Batsford, London.

Putman, R.J. (1986) *Grazing in Temperate Ecosystems: Large herbivores and the ecology of the New Forest*, Croom Helm, London.

Putman, R.J. (1988) *The Natural History of Deer*, Christopher Helm, London.

Putman, R.J., Edwards, P.J., Mann, J.C.E. *et al.* (1989) Vegetational and faunal changes in an area of heavily grazed woodland following relief of grazing. *Biological Conservation*, **47**, 13–32.

Rackham, O. (1975) *Hayley Wood: its history and natural history*, Cambridgeshire and Isle of Ely Naturalists' Trust, Cambridge.

Ratcliffe, P.R. (1988) The management of red deer populations resident in upland forests, in *Wildlife Management in Forests* (Ed. D. Jardine), Institute of Chartered Foresters, Edinburgh.

Ratcliffe, P.R. and Pepper, H.W. (1987) The impact of roe deer, rabbits and grey squirrels on the management of broadleaved woodlands, in *Proceedings of the National Hardwoods Programme*, Oxford Forestry Institute Occasional Paper, 34, Department of Plant Sciences, University of Oxford, pp. 39–50.

Rhoades, D.F. and Gates, R.G. (1976) Toward a general theory of plant anti-herbivore chemistry. *Recent Advances in Phytochemistry*, **10**, 166–213.

Sackur, V. (1984) *A study of the patterns of habitat use by fallow deer* (Dama dama) *in south-west Suffolk (with particular reference to coppiced woodlands*, BSc thesis, University of Southampton.

Shipman, R.D. (1972) Stump sprouting studied for timber, deer browse. *Science in Agriculture*, **19**, 3–4.

Smale, M.C. (1982) Vegetative regrowth of *Beilschmiledia tawa* after selective logging at Pureora and Rotoehu. *New Zealand Journal of Forest Science*, **12**, 442–7.

Part Seven

EFFECTS OF COPPICING ON INVERTEBRATE POPULATIONS

13

Butterfly responses to coppicing

M.S. WARREN and J.A. THOMAS

13.1 INTRODUCTION

Almost three-quarters of Britain's 59 resident butterflies regularly breed in woodland and about one-third are confined to this biotope through a large part of their British range. Most species breed in open woodland habitats such as rides, glades and clearings, where their larvae feed on herbs or grasses growing in the field layer. Relatively few species breed on shrubs and trees (Thomas, 1986; Warren and Fuller, 1990).

Although most British butterflies have declined over the last 150 years, the greatest losses have been among species that breed in the earliest successional stages of woods and grasslands (e.g. Heath *et al.*, 1984). In woods, nearly all losses have been attributed to the decline in coppicing as the traditional form of woodland management. Moreover, the same species have often become extinct on nature reserves which were established partly to conserve them (Thomas, 1984). The plight of woodland butterflies has therefore aroused much concern amongst conservationists, particularly as butterflies are thought to be good indicators of other, less conspicuous insect groups (e.g. Warren and Key, 1991). Conservation efforts have tended to focus on restoring coppice regimes or on managing woodland rides and glades but, until recently, there have been few quantitative data on which to base conservation programmes.

The aim of this chapter is to review the autecological research to date that has been conducted on several woodland butterfly species, and to consider the implications for their survival and conservation. We also hope to highlight the considerable gaps that remain in our knowledge of these vital topics, and to identify some of the problems that may result from future changes in woodland management.

13.2 THE HABITAT REQUIREMENTS OF BUTTERFLIES IN WOODLANDS

Within woodland, butterflies breed in four broad categories of habitat: grassland, fresh woodland clearings, the shady interior of woods, and the woodland canopy (after Warren and Fuller, 1990). The vast majority of British species that regularly breed in woods use predominantly grassy areas, such as rides and permanent glades, and all but four of these require very open, sunny conditions with less than 20% direct shade (e.g. Warren and Fuller, 1990). The wood white (*Leptidea sinapis*) and ringlet (*Aphantopus hyperantus*) prefer lightly shaded conditions in grassland with 10–40% direct shade; and only the speckled wood (*Pararge aegeria*) and green-veined white (*Pieris napi*) can tolerate heavier shade up to 90% (Warren, 1985a).

The second category contains three species – (the pearl-bordered fritillary (*Boloria euphrosyne*), the high brown fritillary (*Argynnis adippe*) and the heath fritillary (*Mellicta athalia*)) that also breed on the ground flora, but generally avoid the permanent grassland within woods – being restricted, instead, to food-plants growing in the open, warm situations that are created briefly after coppicing or clearing. At least four of the open grassland species also thrive in newly cut woodland: the grizzled skipper (*Pyrgus malvae*), the Duke of Burgundy (*Hamearis lucina*), the small pearl-bordered fritillary (*Boloria selene*) and the dark-green fritillary (*Argynnis aglaja*). Only two woodland species breed in dappled shade within woodland interiors, the white admiral (*Ladoga camilla*) and the silver-washed fritillary (*Argynnis paphia*). The final category contains five species which are more or less confined to the canopy where they breed on tree species and feed on honeydew as adults.

13.3 THE RESPONSE OF BUTTERFLIES TO COPPICING

13.3.1 Population changes through the coppice cycle

There have been few studies of butterflies in coppice woodland and these involve only a small number of species. This paucity of data is partly due to the current rarity of actively coppiced woods where studies can be undertaken but also to the fact that the importance of coppicing to butterfly conservation has only recently been recognized.

The most comprehensive data available are for the nationally endangered heath fritillary, which is now confined to the Blean Woods complex in Kent, and a few sites in south-west England (Warren *et al.*, 1984). In the Blean, it breeds solely on common cow-wheat, *Melampyrum pratense*, which is a common constituent of the ground flora of deciduous woodland on some acid soils (Figure 13.1). However, eggs are laid only on plants growing under large canopy gaps, shaded plants beneath closed canopies are never used (Warren, 1987a). As a result, heath fritillary population size is correlated not with the absolute number of cow-wheat plants in its woods, but with the abundance of plants growing in open, sunny situations.

Figure 13.1 Heath fritillary egg-laying on common cow-wheat (*Melampyrum pratense*), a plant that flourishes after coppicing. (Photograph: M.S. Warren.)

For example, populations in coppice woodland build up rapidly following clearance and usually reach a peak in year classes 1–3 (Figure 13.2). Very high densities can be reached at this stage (up to 3000 adults/ha) but populations decline rapidly as the canopy closes and usually die out within five years in vigorous chestnut coppice. The decline may be slower in poorly stocked or less vigorous coppice where the canopy sometimes takes ten years to close after cutting (Figure 13.2b). Young conifer plantations can also provide suitable breeding habitats for up to ten years, but only when they are planted on deciduous woodland sites as the food-plant is thought to be parasitic mainly on the roots of deciduous trees (Warren, 1987c).

Equally large and rapid responses to coppicing were recorded in three other butterfly species during a preliminary study of one of the last commercially-managed hazel coppices in Hampshire (Figure 13.3). Here, adult densities of the pearl-bordered fritillary, Duke of Burgundy and grizzled skipper were greatest in year-classes 1 and 2, and none were recorded after canopy closure in year-class 4. All three species breed on ground flora plants growing in open situations (respectively violets (*Viola* sp.), primroses (*Primula vulgaris*), wild strawberry (*Fragaria vesca*)) and will not breed in

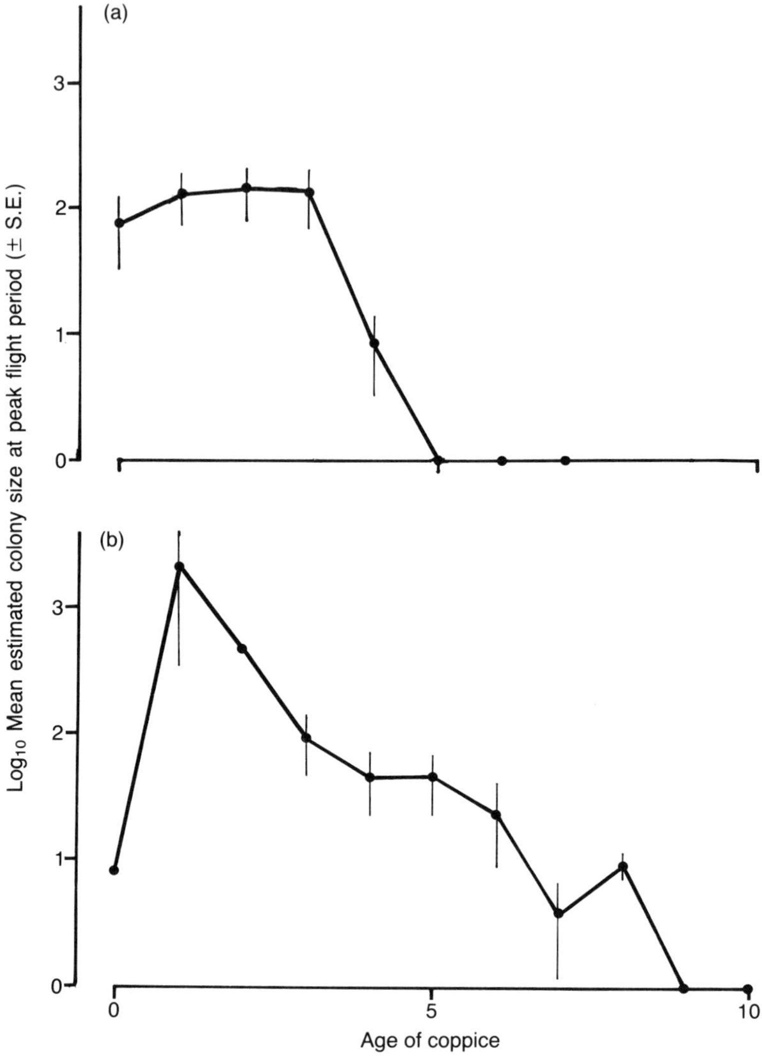

Figure 13.2 The size of heath fritillary populations (note log. scale) in predominantly sweet chestnut coppice of different ages, Kent, 1980–1984. (a) Vigorous coppice (16 sites); (b) poor coppice (8 sites). (After Warren, 1987c.)

densely shaded conditions. Although sample sizes are small, this and other population studies (Thomas, unpublished) show that populations of the pearl-bordered fritillary and grizzled skipper peak under the most open conditions provided by year-class 1, and thereafter decline rapidly, whereas those of Duke of Burgundy peak later in year-class 2 (Figure 13.3). It also seems that the Duke of Burgundy is more tolerant of shading by standard

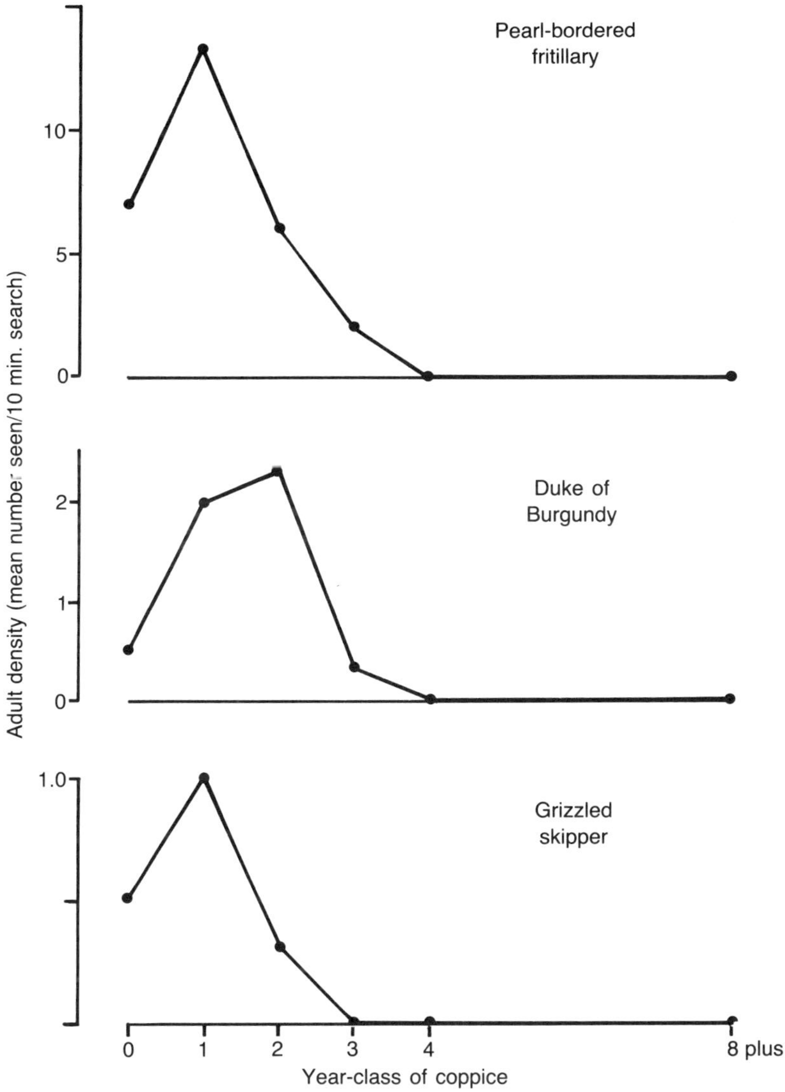

Figure 13.3 The density of three butterfly species in vigorous hazel coppice (with oak standards) of different ages. (Data collected by M.S. Warren, Hampshire, 1987.)
(a) pearl-bordered fritillary; (b) Duke of Burgundy; and (c) grizzled skipper

trees than the other two species (see below). These observations are supported by the known egg-laying preferences of the species: the pearl-bordered fritillary and grizzled skipper preferring short, sparse vegetation, while the Duke of Burgundy prefers taller vegetation (Thomas, 1986). In the case of the Duke of Burgundy, M.R. Oates (personal communication) has suggested that food-plants exposed to too much direct sunshine are less suitable because they are prone to desiccation before the larvae can complete development – usually by the end of August or September. The Duke of Burgundy may therefore avoid the most open coppice panels because the food-plants are liable to die back prematurely, particularly on the thin chalky soils of the study site. A year or two later, the same panels will be more shaded and the food-plants more likely to survive until late summer.

Several other species are most abundant in the earliest stages of the coppice cycle and probably have similar responses to those described above. They include the small pearl-bordered fritillary and dark-green fritillary, which both occur in most worked hazel coppices in Dorset and Hampshire, and notably the high brown fritillary which has increased markedly in abundance on Gait Barrows National Nature Reserve in Cumbria since coppicing was resumed (Pollard *et al.*, 1986). The silver-washed fritillary is also abundant in worked hazel coppices, but breeds on violets growing in the edges where recently cut panels adjoin older ones, providing a mixture of sun and shade. However, the adults of this and many other woodland butterflies are often seen more frequently in newly cut panels because these provide a particularly abundant supply of flowers.

Many of the butterflies associated with woodland rides and glades rarely breed within the coppice panels themselves because the latter develop a very different flora and seldom contain their larval food-plants. No study has been carried out in actively coppiced woodland but, as most prefer very open, sunny grassland, their abundance would probably increase greatly when adjacent panels are coppiced – as they do when rides are widened (e.g. Pollard, 1982; Pollard *et al.*, 1986; Warren *et al.*, 1986).

The only butterflies that can breed in the dense thicket stage after canopy closure are the few canopy-dwelling species, but again there has been no study of these species in worked coppice. As few of the commonly coppiced trees are butterfly food-plants, most canopy species will probably rely on the standard trees and are unlikely to be greatly affected by the coppice cycle beneath them. However, there may be exceptions such as the purple hairstreak (*Quercusia quercus*) which lives as an egg, larva and adult on oak canopies for 11 months of the year but descends to pupate in the ground during June. The pupae experienced 80% mortality during one population study (Thomas, 1975), suggesting that this may be a key period affecting the population size of this insect. Purple hairstreak pupae have been found only inside the brood chambers of *Myrmica* ant nests (Thomas and Snazell, unpublished) and it is likely that their survival is correlated with the status of these ants. If so, coppicing probably affects purple hairstreak numbers, for *Myrmica* ant nests are often abundant in the early stages of the coppice cycle but are very rare in deeply shaded woodland.

Once a coppice becomes neglected, the canopy usually begins to thin and develop a high forest structure, allowing small patches of sunlight to penetrate through to the ground floor. Such conditions may be suitable for a few shade-tolerant species such as the speckled wood, which breeds on grasses, and the white admiral, which breeds on clumps of honeysuckle (*Lonicera periclymenum*) in sparse woodland.

13.3.2 The effect of standard density

The density of standard trees in coppice woodland has a profound effect on ground-breeding butterflies, largely because of the shading of their food-plants. Populations of the heath fritillary can tolerate canopy cover of up to about 50% (e.g. in heavily thinned high forest), but no colony has been recorded above this level, even though the larval food-plant can be abundant under a canopy cover of 80–90% (Warren *et al.*, 1984).

Figure 13.4 shows an analysis of the effects of standard density on the abundance of pearl-bordered fritillaries in hazel coppice, where the adult density recorded per panel has been expressed as a proportion of the mean for each age class of coppice. The results show that the 11 panels studied in year classes 0–3 (i.e. all those which contained adults, see Figure 13.3) had a wide range of standard densities ranging from 20–80% cover. In general, far higher adult densities were found where the cover was 25% or less and densities declined significantly as the cover increased (Kendall rank coefficient = −38, n=11, p<0.01). Only one panel had a cover of greater than 60% but this contained no adults despite being in age class 1, which was otherwise the most suitable class for the species (Figure 13.3).

A similar trend was noticed in the abundance of the grizzled skipper and the Duke of Burgundy, but too few data were obtained for this to be statistically significant. However, the relationship appeared to be somewhat less strong for the Duke of Burgundy and, in year-classes 0 and 1, this species seemed to avoid panels with very low standard cover. As mentioned above, this may be because it avoids breeding in very open areas where the food-plants may desiccate early. In contrast, in year-class 2 and 3, it was often more abundant where there was low standard cover, which suggests that the combined cover of standards and coppice may be critical because both affect the microclimate (see section 13.3.3).

The cover of standard trees recorded in many of the panels on this study site was unusually high for a commercially-managed coppice as more than

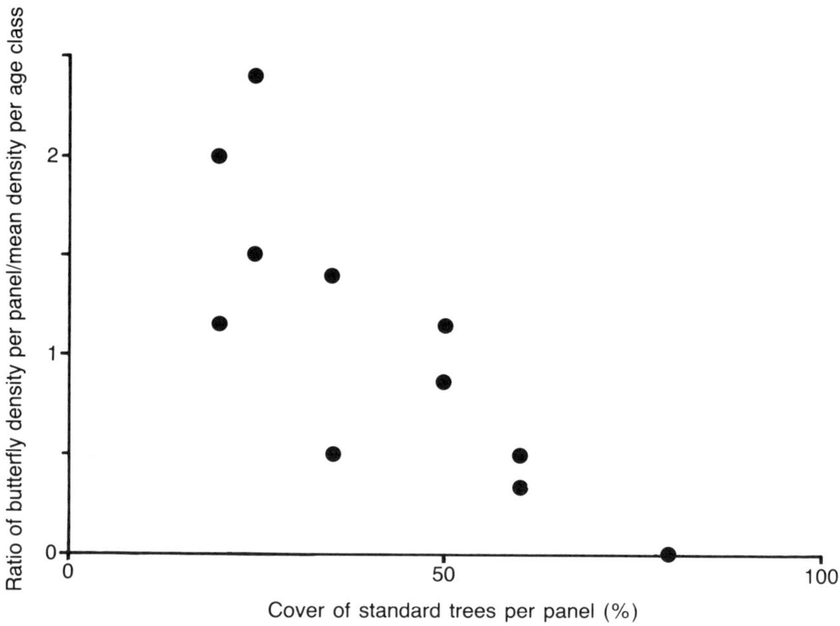

Figure 13.4 The influence of standard tree cover on the density of pearl-bordered fritillaries in recently cut hazel coppice. (Data collected by M.S. Warren, Hampshire, 1987.)

20–30% canopy cover can significantly suppress the coppice regrowth. In this case, the high density of standards occurred only because of the peculiar history of this wood, and will probably be reduced in the future. However, it is clear that more than about 60% cover is unsuitable for nearly all ground-dwelling butterflies in coppice, although it may favour a few shade-tolerant species. In general, a standard tree cover of less than about 25% appears to support the highest population densities, and covers of 30–60% support much smaller populations which may not be viable in the long term.

13.3.3 The role of microclimate

The preference by butterflies for the early stages of coppice appears to be due to their requirement for a particularly warm microclimate. This factor is most important for the developmental stages, which have limited mobility and require warmth, and in some cases direct sunlight, in order to develop successfully. For example, the larvae of most coppice butterflies are active only during the daytime and several fritillary larvae spend much of their time basking, particularly in cooler weather (e.g. Emmet and Heath, 1989; Warren, 1987c). Porter (1982) has shown that similar behaviour among the gregarious larvae of the marsh fritillary (*Eurodryas aurinia*) allows them to raise their body temperatures by up to 25°C above the ambient, enabling them to develop quickly even in cool weather. The less well-developed basking behaviour of the other fritillary larvae, most of which are solitary, probably means that the selection of a warm microclimate by egg-laying females is crucial.

Recent research on the violet-feeding fritillaries has shown that they select plants growing in different types of vegetation, corresponding to different microclimates, although there is considerable overlap between each species (Thomas and Snazell, 1989). For example, the pearl-bordered fritillary lays its eggs mainly in sparse patches of vegetation where the microclimate is at least 1°C warmer than that surrounding the bushier violets chosen by the small pearl-bordered fritillary. This explains why populations of the former species peak at about two years after coppice clearing even though the biomass of violets is usually much greater in the third and fourth years. In Brackett's Coppice, Dorset, the effect of an extra year's regrowth in recently cut pure coppice was to reduce the ground temperature by 1–2°C during spring and summer. When even low densities of standard trees were present above the coppice (e.g. 10–20% canopy cover) the ground temperature was reduced by a further 1–2°C (Thomas *et al.*, unpublished data).

13.4 MOBILITY OF COPPICE SPECIES

Because the habitat needed by most butterflies in coppice woodland is ephemeral, it might be expected that the adults are highly mobile so that they can colonize newly cut areas rapidly. However, the opposite appears to be the case: research has shown that several species are in fact sedentary (Thomas, 1991). Most species form closed colonies within discrete areas (Thomas, 1984), and most individuals rarely move outside the habitat patch where they emerge.

Mark–recapture experiments on three adjacent colonies of the heath fritillary, covering over 1 km^2 in the Blean Woods, have shown that the mean range of adults was about 120 m for males and 90 m for females (Warren, 1987b). Although there was some movement between colonies this was very small and estimated to involve less than 2% of the combined population. The rate of dispersal also appeared to vary a little with age, with older females moving greater distances between captures than younger ones, probably because the latter were slowed down by a heavier egg-load. The opposite was true for adult males – they became more sedentary with age.

Further information on the mobility of the heath fritillary has been obtained by monitoring the colonization rate of suitable clearings created in a large block of continuous woodland, covering 8 km^2 in the Blean Woods complex, Kent (Warren, 1987c). Clearings within about 300 m of the nearest colony were colonized rapidly, usually within the first year after cutting, but colonization was delayed, and often did not occur at all, in sites created further than 600 m from the nearest colony. Given the ephemeral nature of its woodland habitat it was considered unlikely that the species would colonize a newly coppiced panel if it is further than 1 km from the nearest colony.

The mobility of the small pearl-bordered fritillary has also been studied in Brackett's Coppice after coppicing was resumed in 1988 (Thomas and Snazell, 1989). Four coppice panels were cut at varying distances from the surviving colony, which had become restricted to a small field within the wood, and 20 adults were transferred to each new panel. Subsequent mark-recapture experiments showed that the transferred adults stayed (and bred) within the new panels, but movement from the original population of about 1000 adults was very small, except to the panel that was cut alongside (Figure 13.5).

Although it is likely that greater dispersal occurs from sites where the habitat deteriorates, the main conclusions of both these studies are that small fritillaries not only need a continual supply of new clearings in which they can breed but also that clearings must be created close together so that colonies can move easily from one to another. If there is even a short break in the supply of new clearings, or if the distance between them is too great, then extinction is likely to be rapid. Even strong-flying species may be more sedentary than most conservationists suppose. Thomas and Snazell (1989) showed that the silver-washed fritillary flew freely throughout Brackett's Coppice, but recorded no interchange with its nearest neighbouring population which bred in a wood about 1 km away.

It is easy to underestimate the number of individuals that occasionally disperse long distances from a source colony (Dempster, 1989). Nevertheless, recent evidence suggests that many locally-distributed butterflies frequently fail to colonize isolated patches of new habitat during the period that these remain suitable for breeding (e.g. Oates and Warren, 1990; Thomas, 1991; Thomas and Snazell, 1989; Warren, 1987c). This is a serious problem for conservationists since ecological theory predicts that species with ephemeral habitats (such as most that breed in coppice woodland) are likely to have high dispersal rates (e.g. Southwood, 1977). However, Thomas (1991) argues that reduced dispersal is a likely outcome of a combination of changing woodland management and climate cooling 2000–4500 years ago. In addition, Dempster (1991) has demonstrated that two species of butterfly underwent morphological changes associated with reduced dispersal (e.g. reduced wing length and/or reduced thoracic size) once their populations became more fragmented and isolated during the

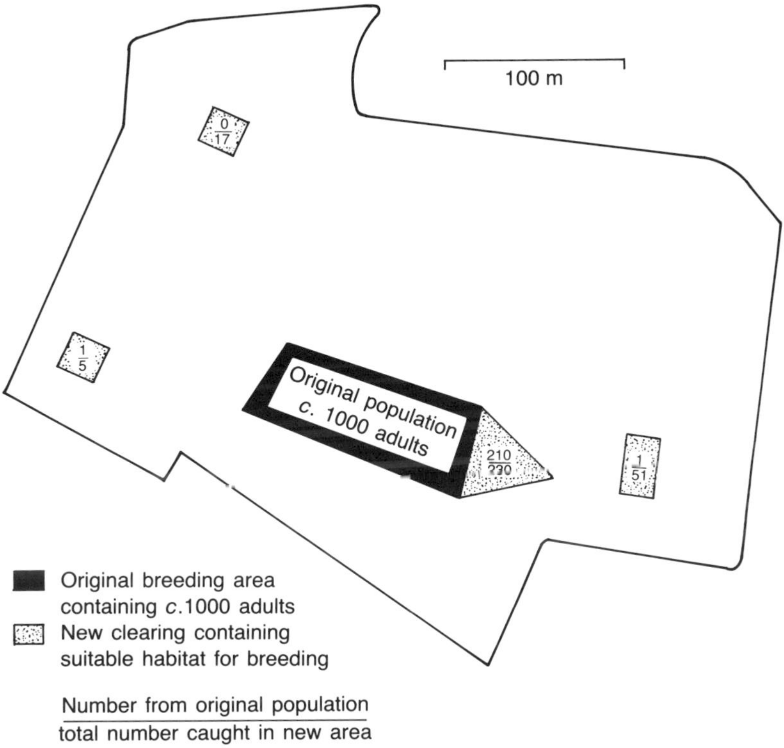

Figure 13.5 Movements of the small pearl-bordered fritillary between its breeding areas within Brackett's Coppice, Dorset. The original population was confined to a field (shown in black) and the four newly cut coppice panels (stippled) contained abundant violets suitable for egg-laying during the experiment. The figures represent the number of captures made in the two weeks following the release of 20 marked adults in each new panel, and show the proportion that were immigrants from the original population. (After Thomas and Snazell, 1989.)

present century; these changes evolved quite quickly over about 20 years (i.e. 20 generations).

13.5 RESPONSE TO HIGH FOREST MANAGEMENT

Virtually no data are available on the response of coppice butterflies to high forest management, but enough is now known of their ecology to make informed guesses of how they might respond to different treatments. Models of the rate of canopy gap creation under five forestry regimes are shown in Figure 13.6. Figure 13.6a simulates coppice rotations, the first part representing the short rotations that were prevalent throughout lowland Britain, at least since the Middle Ages (Rackham, 1980), and typical

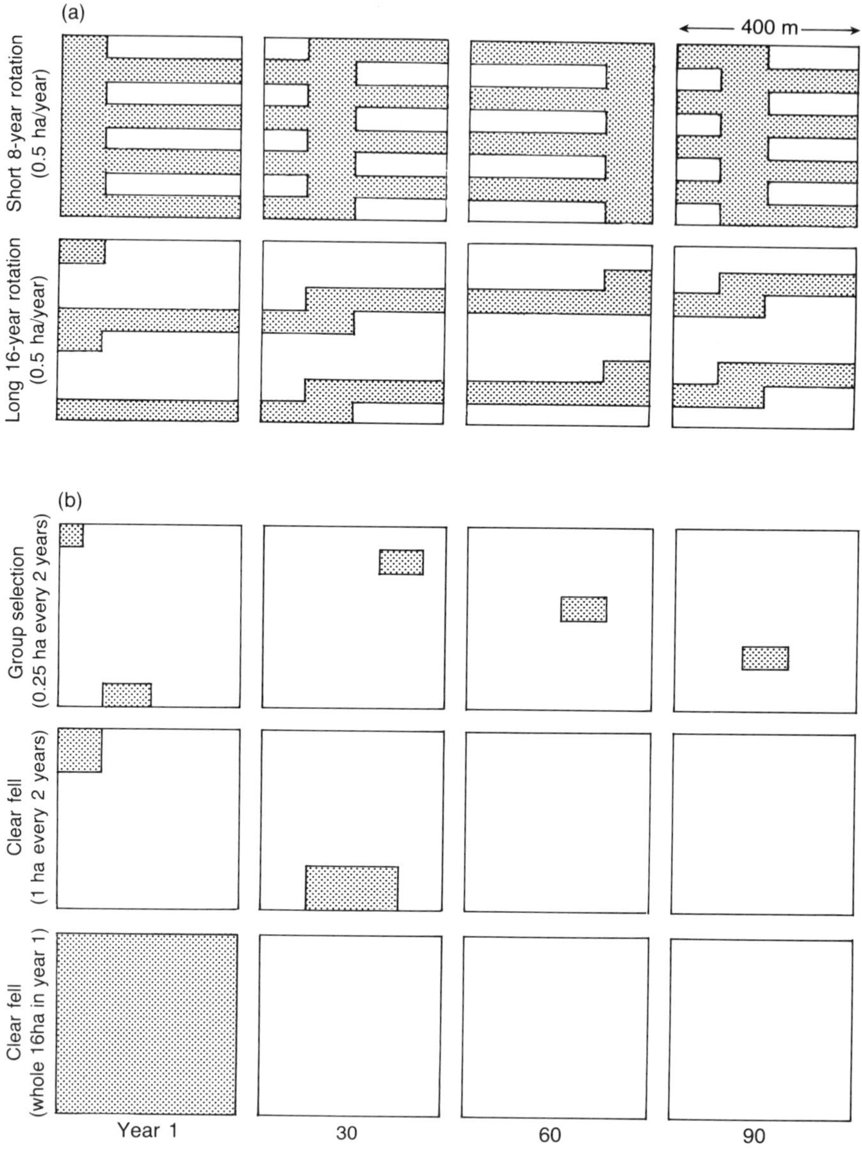

Figure 13.6 Simplified model of gap creation under different forestry regimes in a woodland block of 16 hectares. (a) coppice systems; (b) high forest systems (120-year cycle for broad-leaves). (Adapted from Mitchell and Kirby, 1989.) Shaded areas indicate stands less than five years old that are potentially suitable for coppice butterflies. The cutting cycle starts in the top left-hand corner of each block in year 1.

of the small number of woods that are still managed as hazel coppice today. The longer rotation is more representative of modern coppices (e.g. of sweet chestnut (*Castanea sativa*), where the main produce is used for woodchip or firewood. Both systems have a large proportion of the total area (*c.* 30–60%) under five years of age, which is potentially suitable for coppice butterflies. Moreover, the usual practice of cutting panels on rotation ensures that the distance between successive habitats is very small. In short, these systems are highly suitable for species that require newly cleared woodland.

The creation of canopy gaps under high forest systems is quite different from coppicing (cf. Figures 13.6a and b). Under group selection systems such as those now encouraged by the Forestry Commission's Broadleaves Policy (Forestry Commission, 1985), less than 4% of the woodland area will normally be under five years of age. There may be some continuity of new clearings if successive groups are cut adjacent to each other, as in a very long coppice rotation, but this is unlikely to occur in practice for any length of time. Clear-felling systems usually generate large clearings but these are created infrequently and are widely separated in space. Again, there may be some continuity if the clear-fells are cut in sequence but a very large wood would be needed to accommodate a full and regular rotation.

Not only do high forest systems differ from coppice in the quantity and continuity of gap creation, they also differ significantly in the quality of habitat that they provide. The vegetation that develops in coppice clearings, after a short period of intense shade, is usually very different from that which develops in clear-felled or group-felled clearings, which have undergone a long period of moderate shade during successive thinnings (see the review by Mitchell and Kirby, 1989). Moreover, most modern high forests are of introduced conifers which rarely develop a suitable flora after felling. In general terms, the vegetation of newly coppiced woodland is likely to be more suitable for butterflies than cleared high forest, although there is considerable variability in the response of the flora under each system (see Chapter 7). Long-established ancient woodlands are also likely to have potential for a greater range of butterflies because they usually contain a wider variety of food-plants than recent secondary woods. Broadleaved high forests may provide suitable conditions for some coppice butterflies in the short term but these are unlikely to be perpetuated for long unless special measures are taken (see below).

The response to high forest systems of butterflies that breed in grassy habitats, such as rides, is more clearly known, although it is largely outside the scope of this chapter. Briefly, however, the response is determined by the degree of shading in the ride, which is itself determined by ride width and orientation, and by the height of the surrounding trees (e.g. Warren,

1985a). Other factors include the structure of the ground vegetation, which varies according to its management (Warren and Fuller, 1990).

13.6 THE DECLINE OF COPPICE AND ITS ASSOCIATED BUTTERFLIES

Coppicing was the traditional method of woodland management throughout lowland Britain until the 19th century when it became increasingly uneconomic and began to decline (e.g. Peterken, 1981; Rackham, 1980). Actively coppiced woodland has declined by 82% over the last 50 years alone (Warren and Key, 1991) and today less than 2% of British woods are managed by coppicing (Forestry Commission, 1984). Nearly all the remaining coppice is in Kent and Sussex but this mostly consists of planted sweet chestnut, which casts an extremely dense shade. As a consequence, chestnut coppices generally have a poor ground flora and support few butterflies, although they are one of the last major strongholds of the heath fritillary (Figure 13.7).

Figure 13.7 Habitat of the endangered heath fritillary butterfly in recently cut sweet chestnut coppice. Note the flush of its larval food-plant, common cow-wheat (*Melampyrum pratense*) amongst the coppice stools. (Photograph: M.S. Warren.)

The decline of native (i.e. non-introduced) species of coppice, which provide the best habitats for most butterflies, has been about 94% over the last 50 years (Warren and Key, 1991). As a result of this massive loss of open woodland habitat, the heath fritillary and high brown fritillary, which were once fairly widespread in England, have become highly endangered and have been listed as 'vulnerable' in the *British Insect Red Data Book* (Shirt, 1987). The pearl-bordered fritillary and Duke of Burgundy are now considered to be nationally scarce in Britain, their distribution being reduced to fewer than 100 10-km-grid squares (Nature Conservancy Council, 1989).

These and other species have been declining steadily for most of this century, not because their woods have disappeared, but because most former coppices became too shady. However, their populations received a temporary boost during the 1950s and early 1960s when many lowland broadleaved woods were cleared and replanted, usually with conifers (see Forestry Commission, 1984). When this phase of planting ceased during the 1970s, most of them underwent a severe and sudden decline, which continues to the present day. The habitat of the heath fritillary in the Blean Woods complex increased significantly during the 1970s, due to replanting with conifers, but these areas are now too shady and the area of potential habitat is far smaller than before (Warren, 1985b).

Most fritillaries have now virtually disappeared from eastern Britain, but have survived rather better in western regions (and sometimes in Scotland) where they also use certain types of non-woodland habitats which are seldom present further east. However, throughout Britain as a whole, a large proportion of colonies occur in young conifer plantations, and these are expected to become unsuitable very soon, leading to many more extinctions over the next decade. For example, in central-southern Britain, half the surviving colonies of the pearl-bordered and high brown fritillaries, and a third of the small pearl-bordered fritillary bred mainly in young conifer plantations. Most of the other colonies were small and survived in the rides and glades of large, commercially-managed woods.

The decline of coppicing is also thought to be responsible for the decline of several butterflies associated with open grassland in woods, notably the wood white (Warren, 1984) and the chequered skipper (*Carterocephalus palaemon*) which is now extinct in England (Collier, 1986). In contrast, the white admiral and speckled wood have spread during this century probably because the shady habitats that they prefer have increased (e.g. Pollard, 1979; Emmet *et al*., 1985; Mendel and Piotrowski, 1986).

13.7 HABITAT MANAGEMENT FOR COPPICE BUTTERFLIES

Many other animals and plants thrive in the early growth stages of woodland development and share with butterflies many of the same general requirements for the coppice habitat (Fuller and Warren, 1990). However, unlike some other wildlife groups, butterflies and many other insects need to complete their life-cycle at least once a year. They cannot survive temporary periods of adversity by entering a dormant stage, such as plants with buried seeds or with dormant rootstocks, and they may not be able to fly away to return a few years later.

Butterflies therefore not only need the regular cutting of panels of woodland – preferably at least one sizeable panel each year – but also need the panels to be cut close together, preferably in a regular sequence. One way to encourage adults to move between coppice panels is to connect them with a network of wide, sunny rides. Ideally, these should be maintained as permanently open habitats using an appropriate cutting regime (e.g. Warren and Fuller, 1990).

This principle has been implemented on the Blean Woods National Nature Reserve, Kent, since the late 1970s, with the hope of encouraging the heath fritillary population. At first numbers were very small, with less than 50 adults in 1980, but the population increased greatly following the improved management and, since 1984, has been consistently above 1500 adults (Warren, 1991). Figure 13.8 shows how the colony has shifted substantially over the last seven years in response to the coppice cycle, and how the network of wide rides and glades has ensured the rapid colonization of each newly cut panel. The widest rides (and some of the glades) briefly contained some breeding habitat, but steadily lost the larval foodplant as the vegetation became dominated by a grassland, as opposed to a woodland, flora. They are still valuable as flight paths for the heath fritillary and also provide an abundant supply of nectar, which is often lacking within cut panels of sweet chestnut. Similar forms of management are now being practised on all of this butterfly's remaining woodland sites in Kent, including nature reserves owned by the Royal Society for the Protection of Birds and the Kent Trust for Nature Conservation (Warren, 1991).

One positive feature of most coppice butterflies is that large populations can be supported by small patches of habitat – provided these are suitably managed. Most coppice species can have viable populations within 1–2 ha of habitat (Thomas, 1984) although a larger area is clearly needed if the habitat is to be rotated through the full coppice cycle. For example, the heath fritillary has been successfully re-established in a woodland in Essex where coppicing was reinstated in 1981. Here a population of several hundred adults has been maintained since 1984 in a clearing which has never exceeded 0.75 ha and where the total area to be coppiced is no

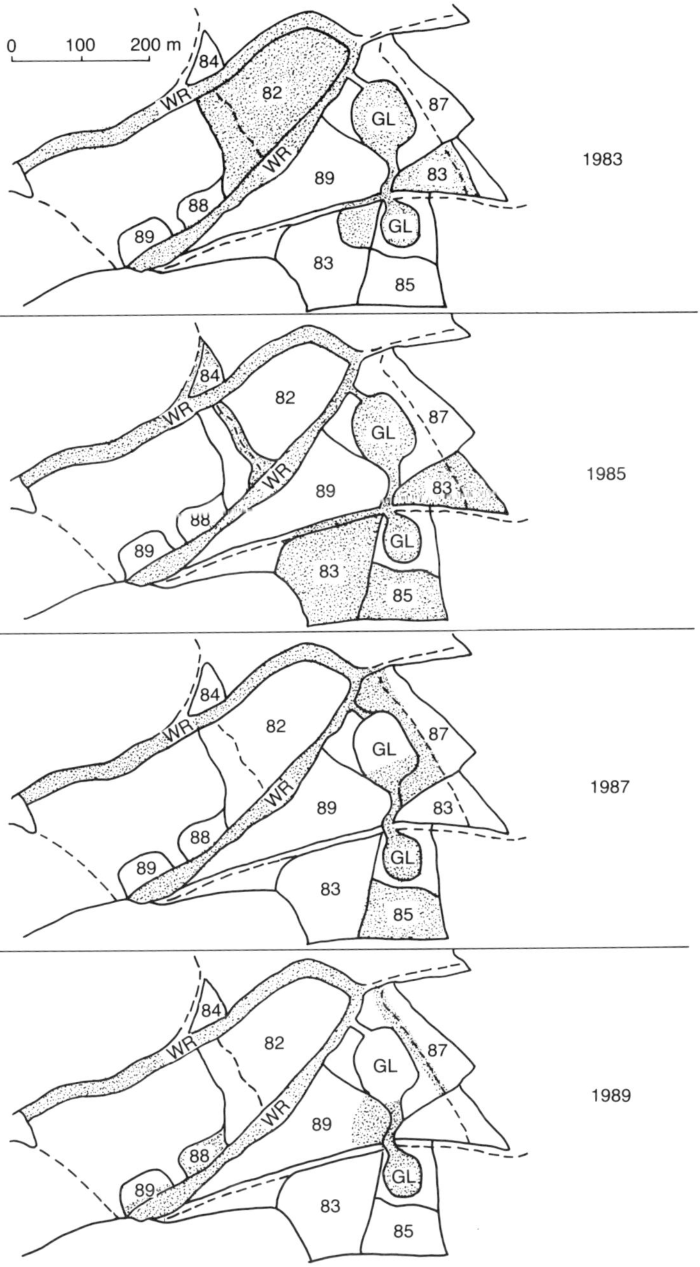

Figure 13.8 Changes in the distribution of the heath fritillary colony on the Blean Woods National Nature Reserve, Kent, in response to the coppice cycle, 1983–1989. Shaded areas indicate distribution of adults; numbers refer to the year of coppicing; WR, wide rides; and GL, permanent glades. (After Warren, 1991).

greater than about 4 ha. However, the scale of cutting required to support a population in the long term may not be the same for each species. The high brown fritillary probably needs larger breeding areas than other coppice butterflies and may require cutting on a far greater scale. Also, there is considerable variation in the response of the field layer to coppicing as not every newly cut panel will develop a suitable vegetation for any one species. Consequently, it is better to cut more than the minimum area needed by a population each year, preferably as several panels in different parts of the wood, so as to increase the chances of continuity of suitable habitat.

In general, short rotations are usually better for most butterfly species, but it is essential that the shady phase is long enough to cause the vigorous light-demanding ground vegetation to die well back. This is important because it ensures that some bare ground will be present after cutting, favouring a ground flora dominated by uncompetitive food-plants such as violets (*Viola* spp.), rather than by a sward of coarse grasses or brambles (*Rubus fruticosus*). It also results in the unusually warm microclimate required by species such as the pearl-bordered fritillary. Open, sunny conditions are also favoured by keeping the canopy cover of any standard trees among a coppice to less than about 20% of the ground area. If longer rotations are more practical for economic reasons, it is essential that the panels are connected by wide rides.

Because full-scale coppicing is uneconomic in many regions, conservationists have often concentrated on rides and other existing openings in woods that might be managed to encourage coppice species. Various cutting regimes have been suggested, based around the division of rides into parallel zones, cutting each on a different length of rotation (e.g. Warren and Fuller, 1990). The principle is to create a variety of habitats ranging from a strip of short grassland in the centre, taller grassland and herbs on either side, culminating in scrub on the margins of the tree crop, which is usually managed as high forest. The scrub margins can be treated as coppice and cut on rotation to create a succession of linear clearings. Variations on this general theme include the creation of a series of sheltered bays or scallops along the ride edges, and the creation of glades or 'box junctions' at the intersection of rides. Both types can either be allowed to develop scrub and managed on a coppice rotation, or cut more frequently by machine to create permanent grassland habitats.

The width and orientation of the rides are also crucial in order to let in sufficient sunshine for the most light-demanding species. In most of southern Britain rides orientated east–west receive the most sunshine: it is preferable to select these if only a few rides can be widened or have their margins coppiced (Warren, 1985a). Rides must be widened to at least 30–40 m (and possibly more) to let in enough light when the surrounding crop is mature, although as much as two-thirds of this width can be taken up by the coppice zone itself (Warren and Fuller, 1990).

Many of these recommendations have yet to be fully tested but initial results are encouraging (e.g. Pollard, 1982; Pollard *et al.*, 1986; Warren *et al.*, 1986). For example, we still know little about the scale of management needed for each species, the ideal length of the rotation, and the variation in response of different types of vegetation on different soils. Another major problem in southern woods is that increasing deer populations may completely suppress coppice regrowth, especially along narrow rideside strips. There are also many practical problems such as whether the cut material should be left, stacked, burnt or removed, and whether certain machinery could be used to reduce the expense and labour. Another possibility that has scarcely been examined is that the sale of the rideside coppice as firewood could offset some, or all, of the costs of ride management. For effective conservation in the long term it is of the utmost importance that such management should be carried out as cheaply as possible otherwise there is a high risk of it being discontinued in the future. What is needed is the development of practical systems of maintaining early successional conditions that are sustainable in the future.

13.8 CONCLUSIONS: THE FUTURE OF COPPICE BUTTERFLIES IN BRITISH WOODS

Coppice woodlands have been an important element of the British landscape for hundreds, and probably thousands, of years. The recent decline of their characteristic butterflies suggests that a similar decline has occurred in many other insects of early successional woodland (Warren and Key, 1991; Chapter 14). Butterflies are still declining rapidly as woodlands develop into high forest and become too shady, few colonies now rely for their survival on active coppicing. Clearly, the conservation of these last remaining examples of traditionally managed woodland is of the utmost priority.

Woodland butterfly populations in most regions of Britain can only be conserved if coppice rotations are resumed or if coppice conditions are simulated in other ways. In most commercially managed woodlands, the only practical way of conserving these threatened species is to adapt high forest management to maximize the continuity of young growth stages, and to combine this with the appropriate management of rides, glades and wood edges. We are only beginning to explore these subjects, and further research is urgently needed if some butterfly species are not to become confined to a handful of nature reserves (e.g. the heath fritillary, Warren, 1991). The onus of conserving viable populations falls on the managers of large woods or complexes involving several woods in close proximity where there are good prospects for maintaining sufficient suitable habitat in the long term. A national strategy is urgently needed to ensure that such woods are managed to create as much early successional habitat as possible, so that at least some coppice butterflies and other insects might survive.

In conclusion, it seems likely that coppice butterflies will become increasingly rare in the foreseeable future. The only chance that this situation might be reversed is if (a) the climate warms (thereby expanding the current narrow woodland niches of some butterfly species), (b) there is a major revival of coppicing, or (c) if high forest managers are successful in creating suitable habitats on a large enough scale along ride edges. Even then, the poor dispersal powers of most butterflies will severely curtail any natural spread because of the present degree of woodland fragmentation. There is therefore an urgent need for renewed action to conserve all woods that either still contain substantial populations of early successional species, or have the potential to maintain suitable habitats in the future.

ACKNOWLEDGEMENTS

We thank the World Wide Fund for Nature and the Vincent Wildlife Trust for funding various surveys of the heath fritillary that were conducted by the Joint Committee for the Conservation of British Insects 1980–1989; R. Snazell and I. Moy for unpublished field data; and the Nature Conservancy Council and Institute of Terrestrial Ecology for jointly funding research on the heath fritillary, 1982–1984, and the research on the violet feeding fritillaries (by JAT), 1986–1989.

REFERENCES

Collier, R.V. (1986) *The conservation of the chequered skipper in Britain*, Focus on Nature Conservation, 16, Nature Conservancy Council, Peterborough.

Dempster, J.P. (1989) Insect introductions: natural dispersal and population persistence in insects. *The Entomologist*, **108**, 5–13.

Dempster, J.P. (1991) Fragmentation, isolation and mobility of insect populations, in *The Conservation of Insects and their Habitats* (eds N.M. Collins and J.A. Thomas), Symposium of the Royal Entomological Society, No. 15, pp. 143–53, Academic Press, London.

Emmet, A.M. and Heath, J. (1989) *The Moths and Butterflies of Great Britain and Ireland, Volume 7(1): The butterflies*, Harley Books, Colchester.

Emmet, A.M., Pyman, G.A. and Corke, D. (1985) *The Larger Moths and Butterflies of Essex*. Essex Field Club (Passmore Edwards Museum), London.

Forestry Commission (1984) *Census of Woodland Trees, 1979–82: Great Britain*, Forestry Commission, Edinburgh.

Forestry Commission (1985) *Guidelines for the management of broadleaved woodland*, Forestry Commission, Edinburgh.

Fuller, R.J. and Warren, M.S. (1990) *Coppiced Woodland: its conservation for wildlife*, Nature Conservancy Council, Peterborough.

Heath, J., Pollard, E. and Thomas, J.A. (1984) *Atlas of Butterflies in Britain and Ireland*, Viking, Middlesex.

Mendel, H. and Piotrowski, S.H. (1986) *The Butterflies of Suffolk: An atlas and history*, Suffolk Naturalists' Society, Ipswich.

Mitchell, P.L. and Kirby, K.J. (1989) *Ecological Effects of Forestry Practices in Long-established Woodland and their Implications for Nature Conservation*, Oxford Forestry Institute Occasional Paper, 39, Department of Plant Sciences, University of Oxford.

Nature Conservancy Council (1989) *Guidelines for the Selection of Biological SSSIs*, Nature Conservancy Council, Peterborough.

Oates, M.R. and Warren, M.S. (1990) *A Review of Butterfly Introductions in Britain and Ireland*, World Wide Fund for Nature, Godalming.

Peterken, G.F. (1981) *Woodland Conservation and Management*, Chapman and Hall, London.

Pollard, E. (1979) Population ecology and change in range of the white admiral, *Ladoga camilla*, in England. *Ecological Entomology*, **4**, 61–74.

Pollard, E. (1982) Monitoring butterfly abundance in relation to the management of a nature reserve. *Biological Conservation*, **24**, 317–28.

Pollard, E., Hall, M.L. and Bibby, T.J. (1986) Monitoring the abundance of butterflies, 1976–1985. *Research and Survey in Nature Conservation*, No. 2, Nature Conservancy Council, Peterborough.

Porter, K. (1982) Basking behaviour in larvae of the butterfly *Euphydryas aurinia*. *Oikos*, **38**, 308–12.

Rackham, O. (1980) *Ancient Woodland*, Edward Arnold, London.

Shirt, D.B. (ed.) (1987) *British Red Data Books: 2. Insects*, Nature Conservancy Council, Peterborough.

Southwood, T.R.E. (1977) Habitat, the templet for ecological strategies? *Journal of Animal Ecology*, **46**, 337–65.

Thomas, J.A. (1975) Some observations on the early stages of the purple hairstreak butterfly, *Quercusia quercus* (Linnaeus). *Entomologist's Gazette*, **26**, 224–6.

Thomas, J.A. (1984) The conservation of butterflies in temperate countries: past efforts and lessons for the future. *Symposium of the Royal Entomological Society of London*, **11**, Academic Press, London 333–53.

Thomas, J.A. (1986) *The RSNC Guide to Butterflies of the British Isles*, Country Life Books, Twickenham.

Thomas, J.A. and Snazell, R.G. (1989) Declining fritillaries: the next challenge in the conservation of Britain's butterflies. *Annual Report of the Institute of Terrestrial Ecology 1988–89*, Cambridge, 54–6.

Thomas, J.A. (1991) Rare species conservation: Case studies of European butterflies, in *The Scientific Management of Temperate Communities for Conservation* (eds I.F. Spellerburg, F.B. Goldsmith and M.G. Morris), 31st Symposium of the British Ecological Society, Southampton, 1989. Blackwell, Oxford, 149–97.

Warren, M.S. (1984) The biology and status of the wood white butterfly, *Leptidea sinapis* (L.), in the British Isles. *Entomologists's Gazette*, **35**, 207–23.

Warren, M.S. (1985a) The influence of shade on butterfly numbers in woodland rides, with special reference to the wood white, *Leptidea sinapis*. *Biological Conservation*, **33**, 147–64.

Warren, M.S. (1985b) The status of the heath fritillary butterfly, *Mellicta athalia* Rott., in relation to changing woodland management in the Blean Woods, Kent. *Quarterly Journal of Forestry*, **79**, 174–82.

Warren, M.S. (1987a) The ecology and conservation of the heath fritillary butterfly, *Mellicta athalia*. I. Host selection and phenology. *Journal of Applied Ecology*, **24**, 467–82.

Warren, M.S. (1987b) The ecology and conservation of the heath fritillary butterfly, *Mellicta athalia*. II. Adult population structure and mobility. *Journal of Applied Ecology*, **24**, 483–98.

Warren, M.S. (1987c) The ecology and conservation of the heath fritillary butterfly, *Mellicta athalia*. III. Population dynamics and the effect of habitat management. *Journal of Applied Ecology*, **24**, 499–513.

Warren, M.S. (1991) The successful conservation of an endangered species, *Mellicta athalia* (the heath fritillary butterfly), in Britain. *Biological Conservation* (in press).

Warren, M.S. and Fuller, R.J. (1990) *Woodland Rides and Glades: their management for wildlife*, Nature Conservancy Council, Peterborough.

Warren, M.S. and Key, R.S. (1991) Woodlands: past, present and potential for insects, in *The Conservation of Insects and their Habitats* (eds N.M. Collins and J.A. Thomas), Symposium of the Royal Entomological Society, No. 15, Academic Press, London, 155–212.

Warren, M.S., Pollard, E. and Bibby, T.J. (1986) Annual and long-term changes in a population of the wood white butterfly, *Leptidea sinapis*. *Journal of Animal Ecology*, **55**, 707–19.

Warren, M.S., Thomas, C.D. and Thomas, J.A. (1984) The status of the heath fritillary butterfly, *Mellicta athalia* Rott., in Britain. *Biological Conservation*, **29**, 287–305.

14

The quality of coppice woods as habitats for invertebrates

J.N. GREATOREX-DAVIES and R.H. MARRS

14.1 INTRODUCTION

The relative merits of coppicing for many woodland invertebrates other than butterflies is poorly understood, this is probably due to the large number of species involved. However, some studies have been published (Welch, 1968, 1969, 1978; Duffey, 1975; Steel and Mills, 1988; Sterling and Hambler, 1988; Waring, 1988; Hill *et al.*, 1990). Information is also available for some species that seem to be particularly associated with recent coppice or other open areas within woodland (Massee, 1965; Hambler, 1987; Waring, 1989b; Waring and Haggett, 1991; Warren and Key, 1991; Kirby in preparation). Very little has been published on the leaf litter or the soil fauna of coppice, and consequently these aspects have been largely ignored in this chapter.

The advantages and disadvantages of coppicing for invertebrates can to some extent be predicted. Coppicing is likely to benefit those species that occur in the early growth stages (*sensu* Chapter 3) (Tables 14.1–14.4; Waring and Haggett, 1991; Warren and Key, 1991). In contrast species associated with the dead wood of mature old trees will tend to be reduced or eliminated in coppice woodland (Harding and Rose, 1986; Warren and Key, 1991; Chapter 6). Between these two extremes are many invertebrates which might find mature trees more favourable than coppice (Welch, 1968, 1969; Sterling and Hambler, 1988; Waring, 1988). Thus, one factor that should be considered is the proportion of the fauna of a tree or shrub host species that can be supported on different stages of coppice regrowth. The available data suggest that a high proportion can be found on young trees (Welch, 1969; Fowler, 1985; Waring and Haggett, 1991).

More importantly in terms of conservation are the species that benefit particularly from the coppice cycle, by exploiting either the vigorous young coppice regrowth or the abundant ground flora. Many of these species have declined in Britain and are now rare, this rarity often being attributed to the decline of coppicing (Massee, 1965; Fuller and Warren, 1990; Waring and Haggett, 1991; Warren and Key, 1991; Kirby in preparation), see Tables 14.1–14.4 for examples.

Particular features of coppice woods affect the quality of their invertebrate fauna and should be considered when planning conservation management. These include: plant architecture, length of the coppice rotation, density and species of standards, nutrient status of the young coppice regrowth, microclimate of the fresh clearings, composition of the field layer, edge effects, and the historical continuity of coppice management at the site – particularly in relation to site isolation and invertebrate mobility. With respect to invertebrates, the quality of coppice woods can be defined in terms of the comparative biomass or density of invertebrates supported, species richness, species diversity, or in terms of assemblages of invertebrates associated with coppice woodland, particularly those that are considered to be scarce or rare.

14.2 PLANT ARCHITECTURE AND THE LENGTH OF THE COPPICE ROTATION

Plant architecture is believed to be an important factor in both ecological and evolutionary terms in determining the number of invertebrate species associated with any particular plant species. For example trees, with their large size, structural complexity and diversity, may be expected to host more (perhaps one order of magnitude more) species of invertebrate than herbaceous plants (Figure 14.1), which are relatively small and simple in structure and diversity (Lawton and Schroder, 1977; Strong and Levin, 1979; Southwood *et al.*, 1979; Lawton, 1983; Strong, Lawton and Southwood, 1984).

In the first few years after coppicing, plant species richness is at its highest. As the coppice grows the ground flora is increasingly shaded out – particularly the more light-demanding species (Adamson, 1921; Salisbury, 1924; Ash and Barkham, 1976). However, the taxonomic diversity of the invertebrate fauna decreases less as regrowth proceeds than that of the plant species – due to the increasing structural diversity of the trees and shrubs (Southwood *et al.*, 1979).

14.2.1 Young coppice

Although there will be a substantial reduction in the abundance and species richness of the canopy fauna in the first few years after coppicing, there is evidence that a tree or shrub species can support a considerable proportion of the available fauna within a few years of growth. In addition some invertebrates will tend to be more abundant on or more or less restricted to young plants (Massee, 1965; Welch, 1969; Fowler, 1985; Godfray, 1985; Waring and Haggett, in press), particularly if their nutrient status is high (section 14.3).

Table 14.1 Some Red Data Book (RDB) (Shirt, 1987) and Nationally Scarce (NS) (Nature Conservancy Council, 1989) Macrolepidoptera thought to have declined due to lack of recent woodland management, especially coppicing. (Data from Massee, 1965; Skinner, 1984; Waring, 1989b)

Species	*Host-plant*	*Habitat*	*Distribution and status*
Synanthedon vespiformis Yellow-legged clearwing (Sesiidae)	Under bark of *Quercus* and stumps, sometimes other trees	Often seen in areas of felled woodland containing 1–3-year-old oak stumps	Generally distributed southern England–Yorkshire (NS)
Synanthedon culciformis Large red-belted clearwing	*Betula*, especially stumps	Light woodland and heathland	Widespread (NS)
Eupithecia plumbeolata Lead-coloured pug (Geometridae)	*Melampyrum pratense* (flowers), occasionally *Rhinanthus minor*	Open woodland occasionally sandhills	Local (NS)
Eupithecia expallidata Bleached pug	*Solidago virgaurea* (flowers)	Woodland rides and clearings; also grassy hillsides, waste ground and road verges	Local (NS)
Minoa murinata Drab looper	*Euphorbia amygdaloides*	Woodland rides and clearings	Southern England, south-west Midlands and south-east Wales; local (NS)
Hemaris fuciformis Broad-bordered bee hawk-moth (Sphingidae)	*Lonicera periclymenum*	Woodland rides and clearings	Southern England (NS)
Cucullia asteris Star-wort (Noctuidae)	*Solidago virgaurea* and *Aster tripolium*	Woodland rides and clearings and coastal marshes	England and Wales; local (NS)
Cucullia gnaphalii The cudweed	*Solidago virgaurea*	Woodland rides and clearings	East Sussex and south-east Kent; very local and uncommon (RDB1)
Acosmetia caliginosa Reddish buff	*Serratula tinctoria*	Heathland and coppice	Isle of Wight; very rare (RDB1)
Minucia lunaris Lunar double-stripe	Coppiced *Quercus* with mildewed leaves	Coppice	Kent and Sussex; very rare, transient resident (RDB3)
Colobochyla salicalis Lesser belle	*Populus tremula* (regrowth)	Woodland clearings and coppice plots	Kent; very local (RDB1)

Table 14.2 Some potential Red Data Book (pRDB) (Ball, 1986) and Nationally Scarce (NS) (Nature Conservancy Council, 1989) Microlepidoptera which may have declined in Britain due to lack of recent woodland management, especially coppicing (Data from Parsons, 1984; Bradley *et al.*, 1973, 1979 and Goater, 1986)

Species	*Host-plant*	*Habitat*	*Distribution and status*
Coleophora wockeella (Coleophoridae)	*Stachys officinalis*	Coppice rides	South-eastern England (pRDB2)
Phalonidia curvistrigana (Cochylidae)	*Solidago virgaurea*	Glades, clearings and coppiced woodland	South-eastern England–Cumbria, also in Wales; local (NS)
Cochlydia subroseana	*Solidago virgaurea*	Woodland, especially where coppiced	Southern and south-eastern England; local (NS)
Olethreutes arcuella (Tortricidae)	Decaying leaves on the ground	Woodland glades and wooded heaths	England–Yorkshire, also in Wales; generally distributed but scarce (NS)
Lobesia occidentis	*Euphorbia amygdaloides* and *E. paralias*	Damp woods, especially where recently coppiced; also coastal cliffs, sands and dunes	Southern and south-eastern England; very local (NS)
Gibberifera simplana	*Populus tremula*, particularly new growth	Coppice woodland	South-eastern England; rare and local (pRDB1)
Eucosma aemulana	*Solidago virgaurea* (unripe seeds)	Open woodland, especially where recently coppiced	South-eastern England; local (NS)
Eucosmomorpha albersana	*Lonicera* spp.	Woodland clearings and rides, especially oak woods	Southern England; widely distributed but local (NS)
Anania funebris	*Solidago virgaurea*	Glades and woodland margins, also rough hillsides and cliffs on the coast, especially on limestone	Widely distributed but local (NS)

Table 14.3 Red Data Book (RDB) (Shirt, 1987) and Nationally Scarce (NS – Nature Conservancy Council, 1989) Heteroptera thought to have declined due to lack of recent woodland management, especially coppicing. (Information summarized from P. Kirby, in press)

Species	*Host-plant*	*Habitat*	*Distribution and status*
Sehirus biguttatus (Cydnidae)	*Melampyrum pratense* (flowers and seeds)	Mainly open, sunny woodland rides and clearings	Mainly southern England and southern Wales; very local (NS)
Eurydema dominulus (Pentatomidae)	*Cardamine pratensis* and other crucifers (especially the developing fruits)	Recent coppice but also elsewhere	Southern England; very local (NS)
Dicranocephalus medius (Stenocephalidae)	*Euphorbia amygdaloides* (flowers and developing fruits)	Open, sheltered and sunny woodland clearings, rides and margins	Southern England; very local (NS)
Tingis reticulata (Tingidae)	*Ajuga reptans*	Open rides and sunny clearings	Local but widely distributed
Onochila simplex	*Euphorbia amygdaloides*	Sheltered, sunny woodland clearings, rides and edge	Southern England; very local (NS)
Charagochilus weberi Wagner (Miridae)	*Melampyrum pratense* (may be others)	Woodland in light shade, probably a species of coppice and woodland rides	Southern England; very local (RDB3)
Capsodes flavomarginatus	*Lotus uliginosus*	Damp woodland rides and recent coppice	Southern England; very local (NS)

Table 14.4 Some Red Data Book (RDB) (Shirt, 1987) and Nationally Scarce (NS) (Nature Conservancy Council, 1989) Coleoptera associated with the young growth of woody plants which may have declined in Britain due to lack of recent woodland management, especially coppicing (Data from Massee, 1965; Warren and Key, in press; Hyman, in preparation and R. Key, personal communication)

Species	*Host-plant*	*Habitat*	*Distribution and status*
Cryptocephalus bipunctatus (Chrysomelidae)	Particularly *Corylus avellana*, *Salix cinerea* and *Betula* spp.	Woodland rides, clearings and edges, also scrub and heathland	Widespread but uncommon (NS)
Cryptocephalus coryli	Various, including *Betula* spp., *Quercus* spp. and *Crataegus* spp.	Woodland rides, clearings and edges	Formally widespread, now very rare (RDB1)
Cryptocephalus punctiger	Mainly *Betula* spp.	Woodland rides, clearings and edges	Widespread but very scarce (NS)
Cryptocephalus sexpunctatus	Various, including *Corylus avellana*, *Betula* spp., *Populus tremula*, *Salix fragilis* and young *Quercus* spp.	Woodland rides, clearings and edges	Mainly S. England; formally locally common (RDB2)
Chrysomela tremula	*Populus* spp., particularly young *Populus tremula*	Woodland clearings and young regenerating aspen (*Populus tremula*)	Formally widespread in southern England, now extremely rare, possibly extinct in Britain (RDB1)
Byctiscus betulae (Attelabidae)	Polyphagus, but particularly on *Corylus avellana* and *Betula* spp.	Woodland, especially hazel coppice	Scarce, widespread but mainly in southern England (NS)
Byctiscus populi	*Populus tremula* and *P.* × *canescens*	Woodland	Southern England; very scarce (NS)
Rutidosoma globulus (Curculionidae)	*Populus tremula* (also other *Populus* spp. and *Salix* spp.)	Woodland	Mainly southern England (NS)

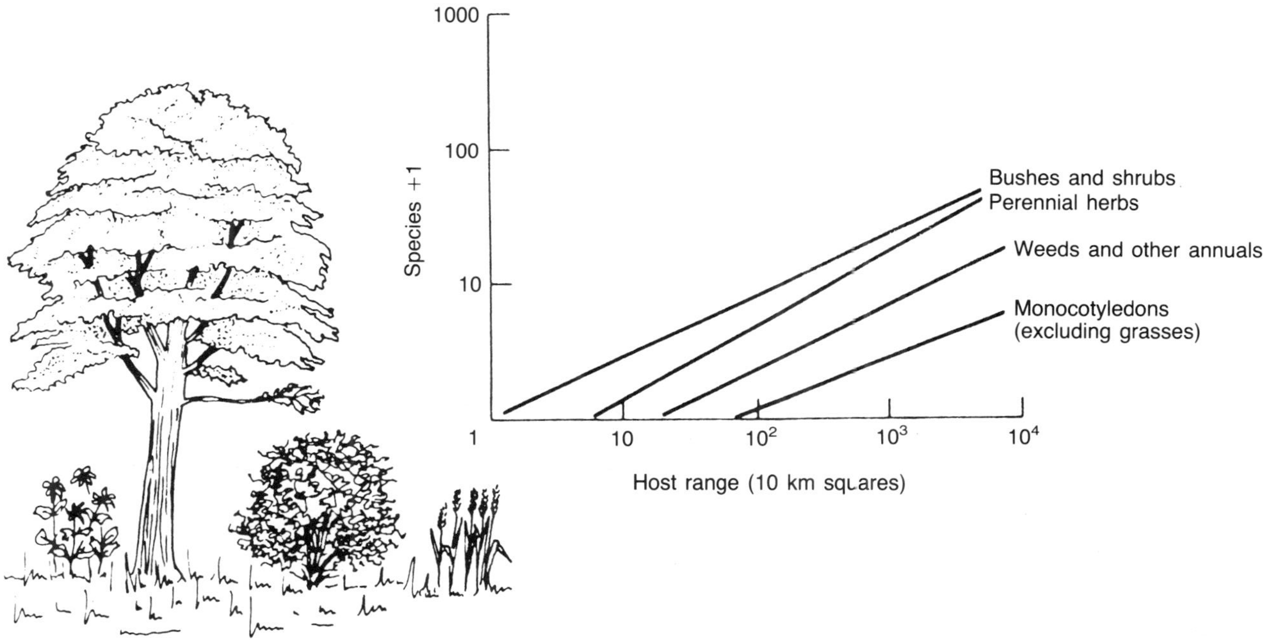

Figure 14.1 The relationship between plant architecture host-plant range and the number of insect species associated with British plants. (Adapted from Strong *et al.*, 1984.)

Fowler (1985) found no overall difference in the richness of associated insects between seedling (1–2-years old; 10–30 cm high) sapling (1–2 m high) and mature (5–10 m high) birth (*Betula pendula* and *B. pubescens*). However, no data were available for seedlings and saplings growing in the absence of mature birch, which may have acted as an invertebrate reservoir: the early spring lag in the number of species on immature birch suggested that some of the colonizing species came from mature trees. Nevertheless, at least 90% of the phytophagus insect fauna of birch fed on leaves, buds or shoots rather than on resources found mainly on mature birch (e.g. catkins, fruit, wood, bark or algae and lichen growing on the bark). Similarly, Welch (1969) concluded from a study of the fauna of hazel (*Corylus avellana*) coppice that the vast majority of associated insects could be maintained on a 10-year coppice cycle.

Waring and Haggett (1991) have pointed out that of about 530 Macrolepidoptera that regularly occur in British woodland over 60% feed on woody perennials. Within 2–3 years the young coppice growth may support over 100 species of Macrolepidoptera, depending on the number of tree and shrub species being coppiced.

Sterling and Hambler (1988) studied the abundance (mean number per stool under 2 m high) of both spiders (Araneae) and lepidopteran leaf miners on 0–8-year and 40-year-old hazel coppice in Brasenose Wood, Oxfordshire. They found that spiders and leaf miners increased in abundance with the age of coppice regrowth over the first eight years. Densities of leaf miners also increased with the age of coppice. In some cases leaf miners and spiders were more abundant in seven or eight-year-old coppice than in the 40-year-old coppice. However, as the bushes age an increasing proportion of their growth is above 2 m. They concluded that the abundance of these invertebrates would be somewhat greater on older bushes, particularly in the case of 40-year-old hazel.

14.2.2 Neglected coppice, high forest and mature standard trees

As trees and shrubs mature they increase in architectural complexity and so the variety of available niches increases. Saproxylic species, surface and sub-cortical bark dwellers, wood borers, catkin, flower and seed feeders, lepidopteran leaf miners and other leaf-feeding invertebrates increase in abundance and number of species on older trees and shrubs (Welch, 1969; Fowler, 1985; Godfray, 1985; Harding and Rose, 1986; Sterling and Hambler, 1988; Waring, 1988, Waring and Haggett, 1991).

Many wood-boring insects need mature timber. Welch (1969) pointed out that some hazel-dependent species would not be catered for in a ten-year coppice cycle and that mature hazel of perhaps 30–40 years would need to be present to maintain the wood-boring beetles such as *Synchita humeralis*, *Hypulus quercinus* and *Lissodema quadripustulatum*.

Bark-dwelling invertebrates may prefer, or only occur, on older trees and bushes. Welch (1969), sampling hazel for Psocoptera, found that some algae and lichen-feeding species only occurred on bushes of more than ten years growth. Other invertebrates may require the deeply-fissured bark of mature trees as refuges, e.g. the spider *Zygiella stroemi* which appears to be largely confined to large trees with tall, exposed trunks, such as the standards in actively coppiced woodland (Hambler, 1987). The larvae of the crimson underwing moths (*Catocala promissa* and *C. sponsa*) appear to require the presence of very mature oak and some moth collectors have suggested that the larvae hide by day in crevices in the deeply-fissured bark, (P. Waring, personal communication). These species and the heart moth (*Dicycla oo*), have disappeared from woods after all the large, mature oaks have been felled (Hadley, 1982; Waring, unpublished).

Waring (1988, 1989a) found that many macrolepidoptera associated with trees and shrubs were much more abundant in neglected coppice than in young coppice (1–2 years growth). In addition, Lepidoptera species were present that needed mature trees of species not present in the young coppice plot (as the trees had been felled during coppicing), e.g. the light-orange underwing moth (*Archiearis notha*) whose larvae feed on aspen (*Populus tremula*) – initially on the catkins. However, some species associated with standard trees were probably not affected and may even have benefited from coppicing of the underwood. These results were consistent with expectations based on the availability of the food-plant.

Coppice stools provide an important habitat for spiders and other litter-dwelling invertebrates. The multi-stem growth of the coppice stools acts as a trap for leaf and twig litter which can accumulate to a depth of 30 cm or more in long rotation or neglected coppice and will provide a habitat for an increasing number of invertebrates (Jones, 1970; Duffey, 1973, 1975). The cool shade under abandoned coppice may be important for other litter-dwelling invertebrates. Several specimens of a spider new to Britain, *Walckenaeria (Wideria) mitrata*, were taken in pitfall traps under sweet chestnut (*Castanea sativa*) coppiced 60 years previously where no ground vegetation was present (Russell-Smith and Swann, 1972).

14.2.3 Over-mature timber

Many saproxylic invertebrates require the dead wood of mature old trees to breed in. In many cases these saproxylic species are restricted to a few areas of parkland and woodland that have little known history of coppicing, such as Windsor Great Forest and the New Forest (Harding and Rose,

1986; Warren and Key, 1991). Many of these species appear to have poor powers of dispersal and despite the increase in neglected coppice and high forest habitats they have apparently not colonized these areas (Warren and Key, 1991; Chapter 6).

14.2.4 Tree and shrub species

The species of tree or shrub that is coppiced or grown as standards amongst coppice will profoundly affect not only the species of invertebrates present, e.g. Macrolepidoptera (Waring, 1988) but also the abundance of invertebrates in general, with some trees supporting much richer faunas and greater densities of invertebrates (Southwood *et al.*, 1982; Hill *et al.*, 1990). Many invertebrates are associated with one or a limited number of closely related tree species. The greater the number and diversity of tree and shrub species present in a wood the greater the number and diversity of invertebrates there will be.

Some invertebrates are associated with the flowers and fruits of the understorey shrubs. Many of these shrubs, e.g. blackthorn (*Prunus spinosa*), hawthorn (*Crataegus monogyna*), privet (*Ligustrum vulgare*) and spindle (*Euonymus europaeus*), are less likely to flower in overgrown coppice except if present on the edges of wide rides. However, many understorey shrubs will flower within a few years after coppicing and provide both nectar and food for a variety of invertebrates with the added benefit of the shelter and warmth of the coppice glades.

14.2.5 Density of standards

A high density of standards will cause shading of the ground flora and the benefit of the coppice rotation to species that exploit the early growth stages will be lost. It has been suggested that when managing coppice for invertebrates, the density of standards should not exceed 15/ha (Fuller and Warren, 1990; Waring and Haggett, 1991). A short rotation, e.g. of ten years, will increase the amount of the early stages available to these invertebrates at any one time. However in view of the comparative abundance of abandoned coppice and high forest, providing for species that prefer young coppice should be considered a priority in sites where they occur.

To summarize: mature trees and shrubs can be expected to host more species of invertebrate than young trees or coppice regrowth. However, many invertebrates (possibly the majority) can be catered for within a coppice cycle of about 10–15 years and many others on the standards growing amongst the coppice. Furthermore, there are invertebrates that feed preferentially on young woody plants, while others feed on the flowers or fruits of the understorey shrubs which may flower more profusely in a coppice situation than in abandoned coppice or high forest. The continuation or the re-introduction of coppicing on sites with a long history of coppice management is unlikely to be detrimental to species of very mature woodland.

14.3 NUTRIENT STATUS

14.3.1 Coppice regrowth

Coppice regrowth is generally very vigorous in the first few years after coppicing and the new leaves are often particularly large and have a higher photosynthetic rate than leaves on the shoots of mature plants (Kramer and Kozlowski, 1979; Chapin *et al*., 1985). Coppice plants have an extensive root system that is able to supply large amounts of both nutrients and water to the new shoots after coppicing. No studies relating the nutrient status of coppice regrowth and its relative merits for invertebrate herbivores have been found. However, from what is known about the nutrient status of young versus old leaves, and young versus mature plants of trees and shrubs some inferences may be drawn.

Young leaves are considered to be more nutritious for many phytophagus insects than mature leaves because they are relatively rich in available nitrogen, high in water content, less tough and less fibrous (Feeny, 1970, 1976; Lawton, 1976; Rhoades and Cates, 1976; McNeill and Southwood, 1978; Rauscher, 1981; Coley, 1983). As a consequence they tend to be grazed more heavily by invertebrates than mature leaves (Feeny, 1970, 1976; Coley, 1980, 1983). As leaves age, both the available nitrogen and water content decrease and they become less suitable as food for many invertebrate species (Scriber, 1977; Scriber and Slansky, 1981; Lawson *et al*., 1984). There is also an increase in quantitative toxins (digestibility-reducing herbivore-defence substances) such as tannins and resins, which are thought to inhibit the growth of many phytophagus insects (Feeny, 1970, 1976; Rhoades and Cates, 1976; Lawton and McNeill, 1979). However, these compounds may not be as important as originally thought in limiting many invertebrate herbivores, it is rather the availability of succulent foliage with a high nitrogen content that is the limiting factor, e.g. for spring-feeding Lepidoptera (Fox and Macauley, 1977; Bernays, 1981; Coley, 1983; Lawson *et al*., 1984). Conversely, some species feed preferentially on older leaves and appear to be less able to cope with the higher levels of qualitative phenolic herbivore-defence substances present in young leaves (Meyer and Montgomery, 1987) or, in the case of lepidopteran leaf miners, wound-induced toxins produced by young leaves in response to feeding macrophages such as large Lepidoptera larvae (West, 1985).

As young coppice regrowth contains a high proportion of young leaves, it can be expected to support larger overall populations of invertebrate herbivores (on a per leaf or area basis) than older regrowth. However, at present there appears to be little data to support this. In a recent study Hill *et al.* (1990) sampled all invertebrates from three age-classes (1–3, 5–6 and >7 years) of sweet chestnut and birch (*Betula pendula*) coppice in a woodland in Kent on three sampling dates (May, June and July 1988). The Hemiptera, Diptera and Arachnida were also analysed separately. They concluded that age of the coppice regrowth had little effect on the density and sample biomass of invertebrates, although there were large differences in both density and biomass of invertebrates between coppice plant species. Sterling and Hambler (1988) found that the density of five common species of leaf miner on hazel increased in the first few years after coppicing.

Godfray (1985) tested the hypothesis that the foliage of seedling trees supported higher densities of invertebrate herbivores than mature trees. He found that in his study there were nearly twice as many leaf-mining species on mature birch (*Betula pubescens*) (24 species) as on seedling (1–5-year-old) birch (13 species). However, densities of leaf miners were significantly greater on the seedling birch. Furthermore, five species were found only on seedling birch and all but one of the remaining species that occurred on the seedling birch were in higher densities on seedlings than on mature trees. It was also noted that the commonest species on the seedling birch, *Lyonetia clerkella*, showed a marked preference for the youngest foliage.

Several scarce or rare macrolepidoptera are particularly associated with young coppice regrowth (Table 14.1; Waring and Haggett, 1991). Other macrolepidoptera also appear to favour coppice regrowth, such as the purple clay (*Diarsia brunnea*) and the dotted clay (*Xestia baja*) (Waring and Haggett, 1991). In a study on the fauna of southern beech (*Nothofagus* spp.), the larvae of several noctuid species, mostly common, were taken only on young coppiced regrowth (Welch and Greatorex-Davies, unpublished data). The larvae of the scallop shell (*Rheumaptera undulata*) seems to be particularly associated with the young leaves of the regrowth of recently cut goat willow (*Salix caprea*), on which it can be found feeding on the top-most leaves of young shoots (P. Sterling, personal communication). Waring (1986) found five larvae of the bordered sallow (*Pyrrhia umbra*) on young hazel coppice. Restharrow (*Ononis repens*) is the normally stated host-plant of this species (e.g. Skinner, 1984). However, there are several old records of the species feeding on trees and shrubs, including young sweet chestnut and 'stool oak' (see Waring, 1986). The nationally scarce argent and sable moth (*Rheumaptera hastata*) is thought to prefer young birch (*Betula* spp.), and may benefit particularly from birch regrowth in young coppice (Harper, 1990; P. Waring, personal communication).

Many Coleoptera exploit young coppice growth and some of them appear to benefit particularly from it (Massee, 1965; Waring and Haggett, 1991; Warren and Key, 1991). Some of the rarer species are listed in Table 14.4. The current rarity of these Coleoptera in Britain is likely to be largely due to the decline in coppicing. They include the chrysomelid beetle *Chrysomela tremula* (now possibly extinct in Britain) and the weevils *Byctiscus populi* and *Rutidosoma globulus*, all of which are associated with the young regrowth or regeneration of aspen.

The comparison between coppice and other new growth cannot be made too strictly, as the nutrient status of new coppice regrowth may be rather different from the new growth produced by seedling regeneration or the new shoots of mature trees. However, it is clear that many insects exploit young coppice regrowth, including species that normally feed on herbaceous plants and the seedling foliage of woody plants. Other tree-feeding and shrub-feeding species may also feed preferentially on coppice regrowth. A major feature of young coppice is the abundance of young foliage. It is likely that this foliage has a high water content and nutrient status and is poorly defended by quantitative toxins and may be similar to seedling foliage; however, this needs to be substantiated by further study.

14.3.2 Field layer

Although herbs and grasses of the coppice field layers generally host fewer species of invertebrate they usually support greater overall populations (on a per leaf or area basis) than the trees and shrubs of the later growth stages (Lawton and McNeill, 1979; Godfray, 1985). For example, Godfray (1985) found that densities of leaf miners (both lepidopteran and those of other orders) were significantly greater on early-successional plants than on birch (*Betula pubescens*). Early-successional plants tend to be defended against herbivore attack by qualitative toxins, whereas late-successional plants tend to be defended in addition by quantitative toxins, particularly as the leaves mature (Feeny 1976; Rhoades and Cates, 1976; Lawton and McNeill, 1979; Coley *et al.*, 1985). Herbivore species which are adapted to feed on one host-plant species, or on a small range of related host plants (i.e. monophages or oligophages), are more likely to succeed on early-successional species which specialize in qualitative rather than quantitative herbivore-defence compounds, because by implication they detoxify or neutralize and even sequester these compounds for their own use as toxins (Lawton and McNeill, 1979; Strong *et al.*, 1984).

In a study on the invertebrate fauna of hazel coppice in Oxfordshire, Steel and Mills (1988) took replicated samples of a wide variety of invertebrates in 0–5 and 40-year-old coppice under oak standards throughout the summers of 1980 and 1981. Samples from the low vegetation and coppice regrowth (not separated) showed early peaks in abundance, but species richness increased gradually over the five years of coppice regrowth (Figure 14.2).

14.3.3 Standard trees

Standard trees amongst coppice often put on extra growth after the underwood is coppiced due to the reduced competition from the coppice regrowth. This was confirmed by Steel and Mills (1988) who found that the annual rings of standard trees were widest immediately after coppicing. Correlated with this increased growth was a peak in abundance of invertebrates present in the oak canopy. This may have been due to the larger amounts of young foliage on the canopy trees and their increased water and available nutrient content. The oak canopy foliage was relatively uniform in species richness throughout the sampling, and Steel and Mills concluded that the canopy fauna of the standard trees was not markedly influenced by the growth of the underwood. However, the species richness of the lower vegetation was greater than the oak foliage relative to the total numbers of individuals sampled. This produced a substantially greater index of biodiversity, indicating the importance of the contribution of the lower vegetation fauna to the overall diversity of coppice woodland.

14.4 CLEARINGS AND RIDES

14.4.1 Microclimate

Several aspects of the microclimate are likely to have an important impact on invertebrates in actively coppiced woodland. These include the amount of light reaching the field layer, wind speed, temperature and humidity (Chapter 3).

There is an enormous increase in the amount of direct sunlight reaching the ground and field layer in the first few years after coppicing (Salisbury, 1924). In addition, adjacent older coppice will provide considerable shelter to the newly cleared areas. These factors result in higher temperatures in new clearings than in closed-canopy coppice (Ash and Barkham, 1976; Mitchell and Woodward, 1987). Ash and Barkham (1976) found that the temperature at ground level in August in 0–1-year-old coppice was up to 10°C warmer than under closed-canopy coppice, and about 3°C warmer in 2–3-year-old coppice after the development of a field layer, but only a metre above the ground the temperatures were very high (Figure 14.3).

These higher temperatures will undoubtedly result in faster development of the young stages of invertebrates and may result in an increase in the numbers and species of ground-dwelling and herb-dwelling invertebrates (Figure 14.2) (Welch, 1969; Steel and Mills, 1988). In addition, the warm

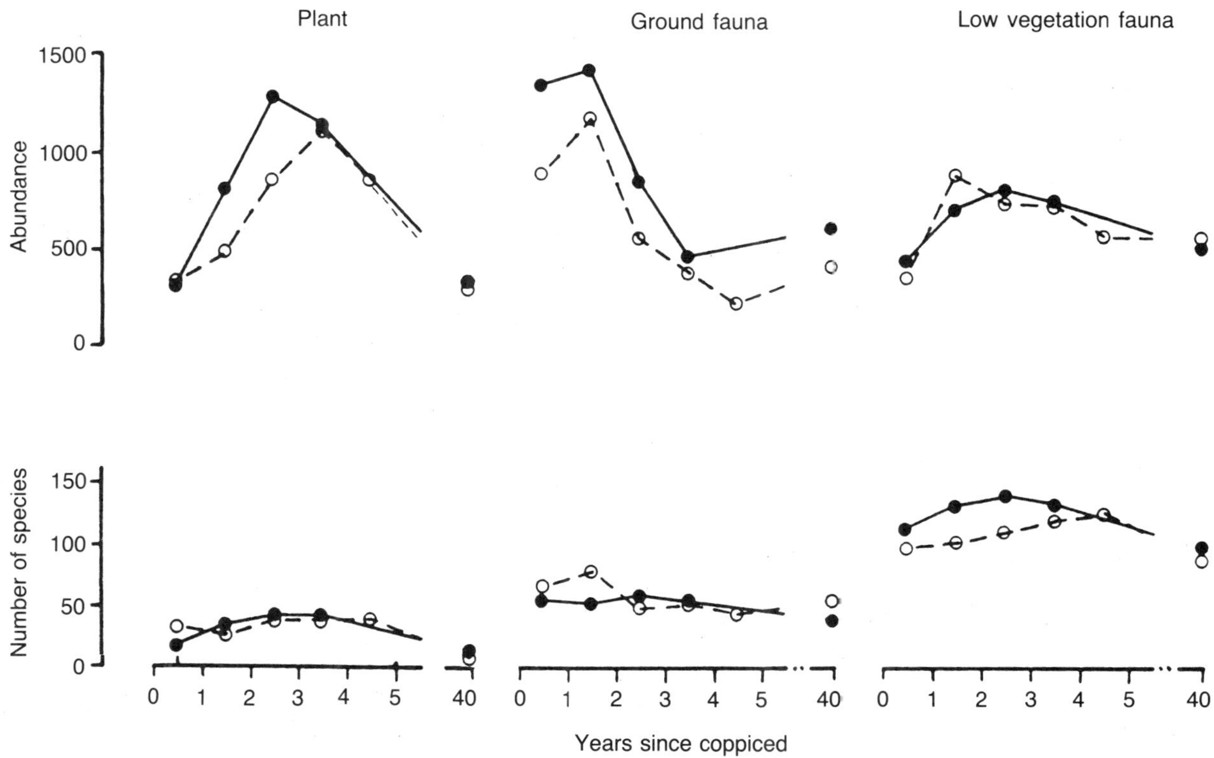

Figure 14.2 The abundance (number of individuals or plant hits in point quadrats) and number of species of plants and invertebrates in samples taken in hazel coppice under oak standards in Brasenose Wood, Oxfordshire, in 1980–1981 (●-●, 1980; ○–○, 1981). (Adapted from Steel and Mills, 1988.)

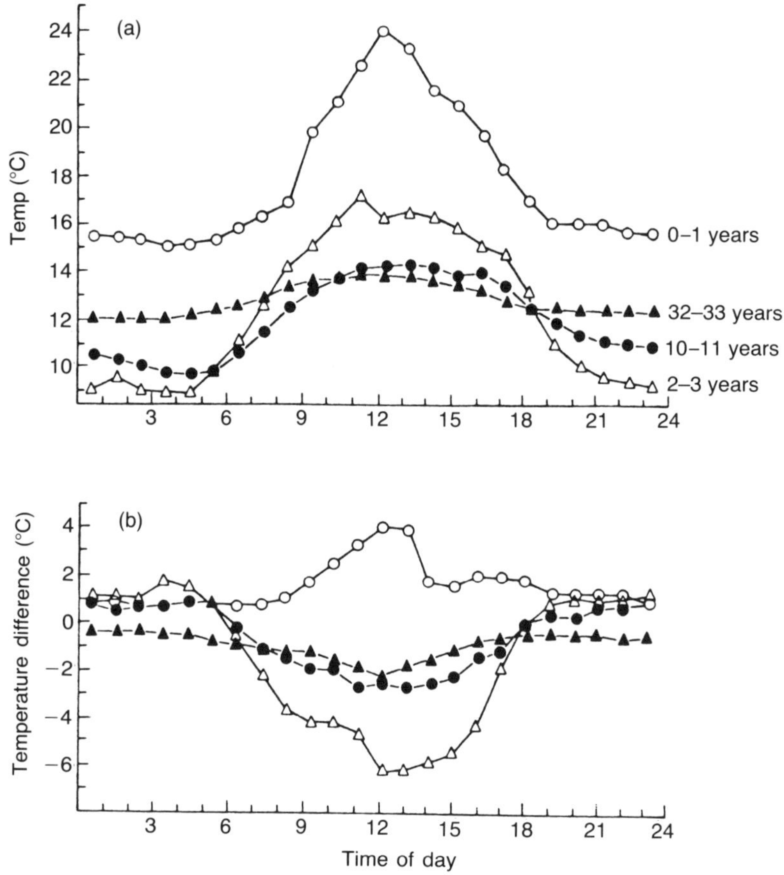

Figure 14.3 Soil surface temperature variation in four coppice sites in Foxley Wood, Norfolk, in August 1971. (a) Mean temperature at 0–1 cm depth: ○, exposed soil in 0–1-year-old coppice; △, under field layer only in 2–3-year-old coppice; ●, under field layer and tree canopy in 10–11-year-old coppice; ▲, under field layer and tree canopy in 32–33-year-old coppice. (b) Mean difference between soil temperature (at 0–1 cm depth) and air temperature (at a height of 100 cm), symbols as in (a). (From Ash and Barkham, 1976.)

microclimate of young coppice glades is likely to be particularly important for some woodland species currently on the northern edge of their range in Britain.

Open areas within woodland are certainly very rich in both species and numbers of invertebrates – as has been demonstrated in several studies. For example, Heydemann and Muller-Karch (1980, quoted in Niemeyer, 1986), working in the north of West Germany found that open glades, of at most a few hectares, had by far the greatest number of species compared with their surroundings. Clearings were particularly important for Aculeata, their abundance probably being due to the shelter of the surround-

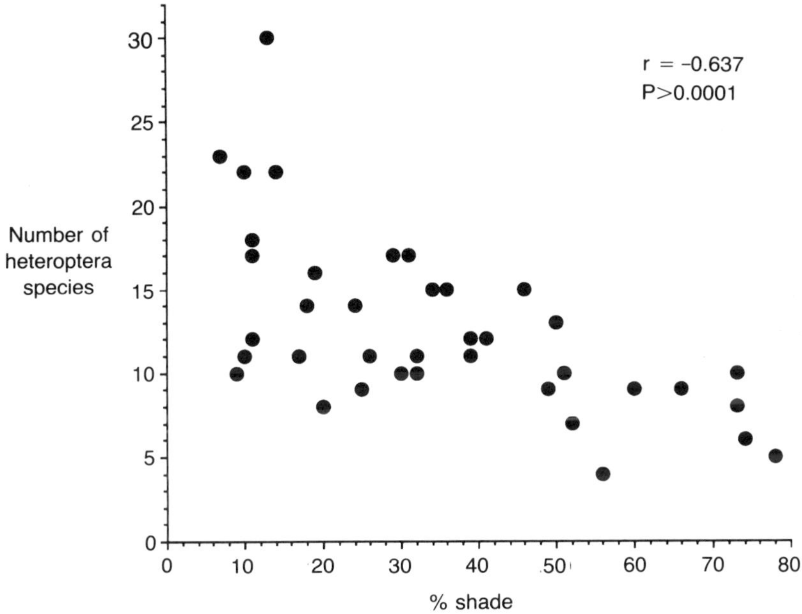

Figure 14.4 The relationship between shade (100% minus percentage open sky as calculated from hemispherical photographs), and the number of species of Heteroptera in samples from herbs and shrubs on rides in eight plantation woods on ancient woodland sites in southern England. (From Hall and Greatorex-Davies, 1989.)

ing trees and to the relatively higher temperature and humidity that resulted.

Hall and Greatorex-Davies (1989) examined the relationship between shade, and several groups of invertebrates in rides of coniferous plantations on ancient woodland sites. Open and sunny rides were much richer in numbers and species of butterflies and heteropteran bugs than shaded and closed rides (Figure 14.4). In the case of butterflies this has already been demonstrated by Warren (1981, 1985). When data for the shrub-feeding and herb-feeding Heteroptera were analysed separately there were still significantly more species and greater numbers of both groups in the more open rides. Chrysomelid beetles were also significantly more abundant in open rides. However lepidopteran leaf miners on hazel appeared to be generally unaffected by shade, although this was not suprising in view of the fact that hazel is relatively shade tolerant. In contrast, hawthorn has been found to host fewer macrolepidopteran larvae (in both number of individuals and species) in shady situations (P. Waring, personal communication). This is likely to be true of other shrubs of more open habitats, such as blackthorn and sallow.

At least 60% of 126 Red Data Book (Shirt, 1987) and nationally scarce

(Nature Conservancy Council, 1989) macro-moths associated with woodland, appear to be associated with open woodland, rides, clearings and coppice rather than dense or mature woodland stands (Waring, 1989b; see Table 14.1). Young (1986) noted that larvae of many woodland Lepidoptera feed on herbs, grasses and fungi and that the majority of these species appear to prefer to feed in clearings, rides or woodland margins.

Ground-dwelling invertebrates may be particularly abundant in the first years after coppicing (Figure 14.2; Welch, 1969; Steel and Mills, 1988). However, increased numbers trapped in these studies may at least partly reflect increased invertebrate activity because of the higher temperatures and the openness of the plots (Duffey, 1972; Southwood, 1978). Nevertheless, overgrown coppice may also be open at ground level and yet far fewer invertebrates be trapped (Steel and Mills, 1988).

The wood ant (*Formica rufa*) is a species of open woodland which requires some direct sunlight on its nest site. In a study by Welch (1978) the number of wood ant nests increased markedly in the two years following the cutting of neglected coppice. However, over the following ten years the number of nests had declined to below the original number. A large invertebrate fauna is associated with the nests of wood ants in Britain, including about 40 Coleoptera species (Donisthorpe, 1927) – a few of which are nationally scarce. Open woodland created by coppice management should result in an increased density of wood ants and hence in the associated myrmecophilus invertebrate fauna.

14.4.2 Field layer

One of the greatest benefits of coppice to insects is likely to be in the field layer of the newly cleared coppice and the rides. Not only is there considerable richness and diversity in the true woodland herbaceous plants, but also in the numbers of open-site perennial and ruderal species in the early years after coppicing (Salisbury, 1924; Ash and Barkham, 1976; Barkham, 1984). A number of insect species are particularly associated with woodland ground flora, but the ecology of many of them is still very poorly known. At least some are known to require their food-plants to be in open sunny situations, and have apparently declined considerably during the last century and are now rare (Tables 14.1–14.4). This decline can almost certainly be attributed to the decline of coppicing (Kirby, in preparation; Waring, personal communication).

14.4.3 Rides and edges

The value of any woodland is enhanced by its rides. The transition zone between different types of habitat (ecotone), such as the edges of rides, tends to be both structurally and compositionally diverse. Organisms of both overlapping communities tend to be present together in addition to species characteristic of and often restricted to the ecotone. Consequently the ecotone is likely to be much richer in species than the adjoining habitats (Odum, 1971).

Unlike the coppice areas themselves, the rides are not subject to such dramatic changes in light at the field layer, remaining more open throughout the coppice cycle. In addition, the rides are unlikely to become as shady as most rides in abandoned coppice or high forest, as the relatively short rotation length of most coppice cycles will mean that a high proportion of the rides will remain open at any one time.

Rides are often surviving fragments of herb-rich grassland and act as refuges for many grassland and woodland edge invertebrates – especially if they are open and sunny (Warren, 1985; Hall and Greatorex-Davies, 1989; Marshall and Haes, 1988). Those which contain abundant flowers are particularly important for nectar-feeders such as hoverflies (Diptera: Syrphidae) (Stubbs and Falk, 1983). The rides can also provide food-plants for uncommon invertebrate species that are often, or particularly, associated with open woodland rides and clearings. These include: the mere wainscot moth (*Photedes fluxa*), the larvae of which feed on wood small-reed (*Calamagrostis epigijos*), the bleached pug (*Eupithecia expallidata*) and the cudweed (*Cucullia gnaphalii*), the larvae of which feed on golden-rod (*Solidago virgaurea*) (Table 14.1).

A well-managed coppice woodland provides a patchwork mosaic of different successional stages, each with different structures, diversity and abundance of species. The increased amount of edge created by the patch-work will help to maximize the diversity of invertebrate species. This is likely to be particularly important along the ride edges.

14.5 INVERTEBRATE MOBILITY, ISOLATION AND CONTINUITY OF MANAGEMENT

Besides butterflies, many other invertebrates probably form more or less discrete colonies and tend to be closely associated with the areas in which they breed, avoiding travelling long distances or crossing areas of unsuitable habitat (Mader, 1984; Thomas, 1984, 1991; Harding and Rose, 1986; Warren, 1987; Warren and Key, 1991; Mitchell and Kirby, 1989). As pointed out in Chapter 13, butterflies and other invertebrates require the continual availability of both food-plants and suitable habitats.

In coppice woods there is a much larger proportion of early-successional habitats suitable for butterflies and other invertebrates than in unmanaged woodland (see Figure 13.6). Because coppicing was done on a regular cycle in the past, there would have been a continuous supply of early-successional stages which probably accounts for the survival of many invertebrates associated with these habitats in woodland into the present century. As the landscape of earlier centuries was also less intensively managed for agriculture than it is today (Chapter 1), at least some of the species that are now normally associated with open habitats within woodland would have been less restricted to woodland than they are now. This would have facilitated the recolonization of sites following extinctions that occurred during the periods when woodland management was neglected. Unfortunately, the increasing fragmentation of coppice woods and their tendency to become shady through lack of management means that the chance of natural colonization, or recolonization, of suitable sites by many invertebrate species is remote (Chapter 13).

At sites where there has been a break in the continuity of coppice management, at least some of the species of the early successional habitats are likely to have become extinct (see Tables 14.1–14.4). For example, Bernwood Forest Nature Reserve, Oxfordshire, is an ancient wood that was largely (90%) converted to conifer plantation in the late 1950s. More than 30 nationally scarce macro-moths have been recorded from the wood in the past (Waring, unpublished) and approximately two-thirds of these are associated with open habitats within woodland. Despite extensive ride management over the last decade and the re-introduction of coppicing in the abandoned coppice area, more than half of these species have not been recorded in the wood for at least 20 years.

14.6 CONCLUSIONS

Invertebrates associated with woodland exploit a wide range of habitats, but individual species often have very precise habitat requirements. The range of habitats provided in actively coppiced woodland is obviously limited but will depend on a variety of factors including the length of the coppice rotation, the tree and other plant species present and the number, species and age of standard trees.

Although it is clear that some invertebrate species will not be catered for in coppice woodland, e.g. those requiring mature, dead wood (Harding and Rose, 1986; Warren and Key, 1991; Chapter 6), it can support a rich and diverse fauna, including much of the canopy fauna (Welch, 1969; Fowler, 1985; Sterling and Hambler, 1988; Waring and Haggett, 1991). The majority of invertebrates present in British woods have apparently survived despite, and in some cases because of, regular coppicing for many centuries.

Coppice woods will be particularly important for invertebrate species associated with the early seral stages of woodland succession (e.g. butterflies), particularly those which are now rare and are thought to be so as a result of the decline in coppicing (Waring, 1989b; Waring and Haggett, 1991; Warren and Key, 1991; Kirby, in preparation). These include species which are on the northern limit of their range in Britain. Some of these species can undoubtedly be maintained by appropriate ride and edge management, such as the provision of wide, sunny rides where both the herbaceous edges of the ride and the scrub fringes are managed on separate rotations (Anderson and Carter, 1987; Carter and Anderson, 1987; Hall and Greatorex-Davies, 1989; Greatorex-Davies, 1991; Warren and Fuller, 1990; Fuller and Warren, 1991). But for some species that are particularly associated with coppice, e.g. the now rare high brown fritillary (*Argynnis adippe*), this is apparently not sufficient (Warren and Key, 1991).

It is important to clarify and establish management priorities before reintroducing coppice management to benefit invertebrates. It may be that many species of conservation interest particularly associated with coppice have already become extinct at the site, but others may be present whose ecology is still very poorly known. The main invertebrate interest may lie in the fauna of the mature canopy trees, in which case it would be desirable to leave at least some areas unmanaged or managed as high forest. However, because of the greater richness of invertebrates in open areas within woodland many commoner species would undoubtedly benefit from coppice management.

Actively coppiced woodland is a relatively rare habitat in Britain today and as such its value is enhanced. These woods provide small islands of habitat in areas where suitable young growth stages of woodland are rare. Coppice woodland on ancient woodland sites, with its rich and diverse plant communities, warm microclimate within the newly cleared areas and complex structural diversity, is undoubtedly of benefit to many invertebrates. On sites where it is known that there has been a long continuity of management or where species of conservation interest associated with the coppice rotation are still present the continuation of coppicing should be strongly encouraged.

ACKNOWLEDGEMENTS

The authors would like to thank Mr Paul Harding, Dr Paul Waring and Dr Colin Welch for their critical comments on earlier drafts of this chapter and would also like to thank Mr Brian Eversham, Dr Roger Key, Dr Peter Kirby, Dr Mike Morris, Dr Keith Porter, Dr David Steel, Dr Philip Sterling, Dr Jeremy Thomas, Dr Lena Ward and Dr Martin Warren for valuable discussions and for providing information and ideas during the preparation of this chapter.

REFERENCES

Adamson, R.S. (1921) Studies of the vegetation of the English chalk. I. The woodlands of Ditcham Park, Hampshire. *Journal of Ecology*, **20**, 152–6.

Anderson, M.A. and Carter, C.I. (1987) shaping ride sides to benefit wild plants and butterflies, in *Wildlife Management in Forests* (ed. D.C. Jardine) Institute of Chartered Foresters, Edinburgh, 66–80.

Ash, J.E. and Barkham, J.P. (1976) Changes and variability in the field layer of a coppiced woodland in Norfolk, England. *Journal of Ecology*, **64**, 697–712.

Ball, S.G. (1986) *Terrestrial and freshwater invertebrates with Red Data Book, Notable or Habitat Indicator Status.* Invertebrate Site Register, 66, Nature Conservancy Council, Peterborough.

Barkham, J. (1984) Management for ground flora in Norfolk woodlands. *Annual Report of the Norfolk Naturalists' Trust*, **58**, 35–9.

Bernays, E.A. (1981) Plant tannins and insect herbivores: an appraisal. *Ecological Entomology*, **6**, 353–60.

Bradley, J.D., Tremewan, W.G. and Smith, A. (1973) *British Tortricoid Moths. Cochylidae and Tortricidae: Torticinae*, The Ray Society, London.

Bradley, J.D., Tremewan, W.G. and Smith, A. (1979) *British Tortricoid Moths. Tortricidae: Olethreutinae*, The Ray Society, London.

Carter, C.I. and Anderson, M.A. (1987) Enhancement of lowland forest roadsides to benefit wild plants and butterflies. *Research Information Note 126*, Forestry Commission, Farnham.

Chapin, F.S. III, Bryant, J.P. and Fox, J.F. (1985) Lack of induced chemical defense in juvenile Alaskan woody plants in response to simulated browzing. *Oecologia*, **67**, 457–9.

Coley, P.D. (1980) Effects of leaf age and plant life history patterns on herbivory. *Nature*, **284**, 545–6.

Coley, P.D. (1983) Herbivory and defensive characteristics of tree species in a lowland tropical forest. *Ecological Monographs*, **53**, 209–33.

Coley, P.D., Bryant, J.P. and Chapin, F.S. III (1985) Resource availability and plant anti-herbivore defense. *Science*, **230**, 895–9.

Donisthorpe, H. St. J.K. (1927) *British Ants: their life-history and classification*, George Routledge, London.

Duffey, E. (1972) Ecological survey and the arachnologist. *Bulletin of the British Arachnological Society*, **2**, 69–82.

Duffey, E. (1973) Araneae, in *Monks Wood. A nature reserve record* (eds R.C. Steele and R.C. Welch), The Nature Conservancy, Abbots Ripton, Huntingdon, 234–40.

Duffey, E. (1975) Habitat selection by spiders in man-made environments. *Proceedings of the 6th International Arachnological Congress, 1974*, 53–67.

Feeny, P. (1970) Seasonal changes in oak leaf tannins and nutrients as a cause of spring feeding by winter moth caterpillars. *Ecology*, **51**, 565–81.

Feeny, P. (1976) Plant apparency and chemical defense. *Recent Advances in Phytochemistry*, **10**, 1–40.

Fowler, S.V. (1985) Differences in insect species richness and faunal composition of birch seedlings, saplings and trees: the importance of plant architecture. *Ecological Entomology*, **10**, 159–69.

Fox, L.R. and Macauley, B.J. (1977) Insect grazing on *Eucalyptus* in response to variation in leaf tannins and nitrogen. *Oecologia*, **29**, 145–62.

Fuller, R.J. and Warren, M.S. (1990) *Coppiced Woodlands: their management for wildlife*, Nature Conservancy Council, Peterborough.

Fuller, R.J. and Warren, M.S. (1991) Conservation management in ancient and modern woodlands: responses of fauna to edges and rotations, in *Scientific Management of Temperate Communities for Conservation* (eds I.F. Spellerberg, F.B. Goldsmith and M.G. Morris), Blackwell, Oxford 445–71.

Goater, B. (1986) *British Pyralid Moths: a guide to their identification*, Harley Books, Colchester.

Godfray, H.C.J. (1985) The absolute abundance of leaf miners on plants of different successional stages. *Oikos*, **45**, 17–25.

Greatorex-Davies, J.N. (1991) Woodland edge management for all invertebrates, in *Edge Management in Woodlands* (Ed. R. Ferris-Kaan), Forestry Commission Occasional Paper, 28, Forestry Commission, Edinburgh, pp. 25–30.

Hadley, M. (1982) The decline of the indigenous macrolepidoptera of Abbots Wood, East Sussex. *Entomologist's Record and Journal of Variation*, **94**, 92–6.

Hall, M.L. and Greatorex-Davies, J.N. (1989) *Management guidelines for the conservation of invertebrates, especially butterflies, in plantation woodland.* Unpublished Nature Conservancy Council/Natural Environment Research Council contract report (HF3/08/12).

Hambler, C. (1987) *Zygiella stroemi* on oak. *British Arachnological Society Newsletter*, **50**, 2–3.

Harding, P.T. and Rose, F. (1986) *Pasture woodlands in lowland Britain.* Institute of Terrestrial Ecology, Abbots Ripton, Huntingdon.

Harper, M.W. (1990) *Rheumaptera hastata* ssp. *hastata* Linn. (Lep.: Geometridae): A welcome return in Herefordshire. *Entomologist's Record and Journal of Variation*, **102**, 89.

Heydemann, B. and Muller-Karch, J. (1980) *Biologisher Atlas Schleswig-Holstein*, Karl Wachholtz, Neumunster.

Hill, D., Roberts, P. and Stork, N. (1990) Densities and biomass of invertebrates in stands of rotationally managed coppice woodland. *Biological Conservation*, **51**, 167–76.

Hyman, P.H., and Parsons, M. *A national review of the scarce and threatened beetles of Great Britain. Part 1. Joint Nature Conservation Committee*, Peterborough (in preparation).

Jones, P.E. (1970) The occurrence of *Chthonius ischnocheles* (Hermann) (Chelonethi: Cheloniidae) in two types of hazel coppice litter. *British Arachnological Society Bulletin*, **1**, 77–9.

Kirby, P. (in press) *A National Review of the Rarer British Hemiptera*, Nature Conservancy Council, Peterborough.

Kramer, P.J. and Kozlowski, T.T. (1979) *Physiology of Woody Plants*, Academic Press, London.

Lawson, D.L., Merritt, R.W., Martin *et al.* (1984) The nutritional ecology of larvae of *Alsophila pometaria* and *Anisota senatoria* feeding on early- and late-season oak foliage. *Entomologia Experimentalis et Applicata*, **35**, 105–14.

Lawton, J.H. (1976) The structure of the arthropod community on bracken. *Botanical Journal of the Linnean Society*, **73**, 187–216.

Lawton, J.H. (1983) Plant architecture and the diversity of phytophagus insects. *Annual Review of Entomology*, **28**, 23–9.

Lawton, J.H. and Shroder, D. (1977) Effects of plant type, size of geographical range and taxonomic isolation on number of insect species associated with British plants. *Nature*, **265**, 137–40.

Lawton, J.H. and McNeill, S. (1979) Between the devil and the deep blue sea: on the problem of being a herbivore, in *Population Dynamics* (eds R.M. Anderson, B.D. Turner and L.R. Taylor), Blackwell Scientific, Oxford, pp. 223–44.

Mader, H.-J. (1984) Animal habitat isolation by roads and agricultural fields. *Biological Conservation*, **29**, 81–96.

Marshall, J.A. and Haes, E.C.M. (1988) *Grasshoppers and Allied Insects*, Harley Books, Colchester.

Massee, A.M. (1965) Some features of conservation interest arising from surveys, in *The conservation of invertebrates. Monks Wood Experimental Station, Symposium No. 1* (eds E.A.G. Duffey and M.G. Morris), Nature Conservancy, Abbots Ripton, Huntingdon, pp. 77–82.

McNeill, S. and Southwood, T.R.E. (1978) The role of nitrogen in the development of insect plant relationships, in *Biochemical Aspects of Plant and Animal Coevolution* (Ed. J.B. Harborne), Academic Press, New York, pp. 79–88.

Meyer, G.A. and Montgomery, M.E. (1987) Relationships between leaf age and food quality of cottonwood foliage for the gipsy moth, *Lymantria dispar*. *Oecologia*, **72**, 527–32.

Mitchell, P.L. and Kirby, K.J. (1989) *Ecological Effects of Forestry Practices in Long-established Woodland and their Implications for Nature Conservation.* Oxford Forestry Institute Occasional Paper, 39, Department of Plant Sciences, University of Oxford.

Mitchell, P.L. and Woodward, F.I. (1987) Instrument for measuring temperature, photosynthetically active radiation and daylength, and its use in the measurement of daylength and temperature in coppice. *Journal of Applied Ecology*, **24**, 239–49.

Nature Conservancy Council (1989) *Guidelines for the selection of biological SSSIs*, Nature Conservancy Council, Peterborough.

Niemeyer, H. (1986) Managing forests for wildlife in Germany, in *Trees and Wildlife in the Scottish Uplands* (Ed. D. Jenkins), Institute of Terrestrial Ecology, Abbots Ripton, Huntingdon, pp. 158–65.

Odum, E.P. (1971) *Fundamentals of Ecology*, Saunders, Philadelphia.

Parsons, M. (1984) *A provisional review of the status of British microlepidoptera*, Invertebrate Site Register, 53, Nature Conservancy Council, Peterborough.

Rauscher, M.D. (1981) Host plant selection by *Battus philenor* butterflies: The roles of predation, nutrition, and plant chemistry. *Ecological Monographs*, **51**, 1–20.

Rhoades, D.F. and Cates, R.G. (1976) Toward a general theory of plant anti-herbivore chemistry. *Recent Advances in Phytochemistry*, **10**, 168–213.

Russell-Smith, A. and Swann, P. (1972) The activity of spiders in coppiced chestnut woodland in southern England. *Bulletin of the British Arachnological Society*, **2**, 99–103.

Salisbury, E.J. (1924) The effects of coppicing as illustrated by the woods of Hertfordshire. *Transactions of the Hertfordshire Natural History Society*, **18**, 1–21.

Scriber, J.M. (1977) Limiting effects of low leaf-water content on the nitrogen utilisation, energy budget, and larval growth of *Hyalophora cecropia* (Lep.: Saturnidae). *Oecologia*, **28**, 269–87.

Scriber, J.M. and Slansky, F. (1981) The nutritional ecology of immature insects. *Annual Review of Entomology*, **26**, 183–211.

Shirt, D.B. (ed.) (1987) *British Red Data Books: 2. Insects*, Nature Conservancy Council, Peterborough.

Skinner, B. (1984) *Colour Identification Guide to the Moths of the British Isles*, Viking, Harmondsworth.

Southwood, T.R.E. (1978) *Ecological Methods*, Chapman and Hall, London.

Southwood, T.R.E., Brown, V.K. and Reader, P.M. (1979) The relationship of plant and insect diversities in succession, *Biological Journal of the Linnean Society*, **12**, 327–48.

Southwood, T.R.E., Moran, V.C. and Kennedy, C.E.J. (1982) The assessment of arboreal insect faunas: comparisons of knockdown samples and faunal lists. *Ecological Entomology*, **7**, 331–40.

Steel, D. and Mills, N. (1988) A study of plants and invertebrates in an actively coppiced woodland (Brasenose Wood, Oxfordshire), in *Woodland Conservation and Research in the Clay Vale of Oxfordshire and Buckinghamshire* (eds K.J. Kirby and F.J. Wright), Research and Survey in Nature Conservation, 15, Nature Conservancy Council, Peterborough, pp. 116–122.

Sterling, P.H. and Hambler, C. (1988) Coppicing for conservation: Do hazel communities benefit? in *Woodland Conservation and Research in the Clay Vale of Oxfordshire and Buckinghamshire* (eds K.J. Kirby and F.J. Wright), Research and Survey in Nature Conservation, 15, Nature Conservancy Council, Peterborough, pp. 69–80.

Strong, D.R., Lawton, J.H. and Southwood, T.R.E. (1984) *Insects on Plants*, Blackwell Scientific, Oxford.

Strong, D.R. and Levin, D.A. (1979) Species richness of plant parasites and growth form of their hosts. *American Naturalist*, **114**, 1–22.

Stubbs, A.E. and Falk, S.T. (1983) *British Hoverflies: an illustrated guide*, British Entomological and Natural History Society, London.

Thomas, J.A. (1984) The conservation of butterflies in temperate countries; past effects and lessons for the future, in *The Biology of Butterflies* (eds R.I. Vane-Wright and P.R. Ackery), Academic Press, London, pp. 333–53.

Thomas, J.A. (1991) Rare species conservation; case studies of European butterflies, in *Scientific Management of Temperate Communities for Conservation* (eds I.F. Spellerberg, F.B. Goldsmith and M.G. Morris), Blackwell, Oxford, pp. 149–97.

Waring, P. (1986) The larva of *Pyrrhia umbra* (Hufn.), the bordered sallow, on young hazel coppice. *Entomologist's Record and Journal of Variation*, **98**, 80.

Waring, P. (1988) Response of moth populations to coppicing and the planting of conifers, in *Woodland Conservation and Research in the Clay Vale of Oxfordshire and Buckinghamshire* (eds K.J. Kirby and F.J. Wright), Research and Survey in Nature Conservation, 15, Nature Conservancy Council, Peterborough, pp. 82–113.

Waring, P. (1989a) Comparison of light-trap catches in deciduous and coniferous woodland habitats. *Entomologist's Record and Journal of Variation*, **101**, 1–10.

Waring, P. (1989b) *Moth conservation project news bulletin 2: April 1988 – August 1989*, Unpublished report, Nature Conservancy Council, Peterborough.

Waring, P. (unpublished) Hell Coppice/Shabbington Wood complex, Bernwood Forest Buckinghamshire, including the Bernwood meadows. Provisional annotated list of macro-moths including all known historical records.

Waring, P. and Haggett, G. (1991) Coppicing in woodlands, in *Insects in Habitat Conservation* (eds R. Fry and D. Lonsdale), Amateur Entomological Society, Feltham (in press).

Warren, M.S. (1981) *The ecology of the wood white butterfly* Leptidea sinapis (Lepidoptera: Pieridae), PhD thesis, University of Cambridge.

Warren, M.S. (1985) The influence of shade on butterfly numbers in woodland rides, with special reference to the wood white *Leptidea sinapis*. *Biological Conservation*, **33**, 147–64.

Warren, M.S. (1987) The ecology and conservation of the heath fritillary *Mellicta athalia*. II. Adult population structure and mobility. *Journal of Applied Ecology*, **24**, 483–8.

Warren, M.S. and Fuller, R.J. (1990) *Woodland Rides and Glades: their management for wildlife*, Nature Conservancy Council, Peterborough.

Warren, M.S. and Key, R.S. (1991) Woodlands: past, present and potential for insects, in *The Conservation of Insects and their Habitats*, 15th Royal Entomological Society Symposium, September 1989 (eds N.M. Collins and J.A. Thomas), Academic Press, London, pp. 155–212.

Welch, R.C. (1968) Some Psocoptera from Monks Wood National Nature Reserve, Huntingdonshire. *The Entomologist*, **101**, 173–4.

Welch, R.C. (1969) Coppicing and its effects on woodland invertebrates. *Devon Trust for Nature Conservation*, **22**, 969–73.

Welch, R.C. (1978) Changes in the distribution of the nests of *Formica rufa* L. (Hymenoptera: Formicidae) at Blean Woods National Nature Reserve, Kent, during the decade following coppicing. *Insectes Sociaux, Paris*, **25**, 173–86.

West, C. (1985) Factors underlying late seasonal appearance of the lepidopterous leaf-mining guild on oak. *Ecological Entomology*, **10**, 111–20.

Young, M.R. (1986) The effects of commercial forestry on woodland lepidoptera, in *Trees and wildlife in the Scottish uplands* (Ed. D. Jenkins), Institute of Terrestrial Ecology, Abbots Ripton, Huntingdon, pp. 88–94.

Part Eight

IMPLICATIONS OF COPPICING FOR CONSERVATION MANAGEMENT

– 15

Coppicing for nature conservation – the practical reality

JOHN BOOKER and RUTH TITTENSOR

15.1 INTRODUCTION

The arguments and proposals put forward below stem from 15 years of management of the 16 ha West Dean Woods Nature Reserve in west Sussex. This reserve is an ancient woodland and a Site of Special Scientific Interest designated by the Nature Conservancy Council, representing just 7% of the area of coppice or coppice with standards present in the parish of West Dean in 1950. The Reserve consists predominantly of hazel (*Corylus avellana*) coppice with a little sweet chestnut (*Castanea sativa*), overstood by widely-spaced oak standards (*Quercus robur*) with a canopy cover of about 20%. Regular coppicing on this site took place until about 45 years ago; after that silvicultural management was neglected until nature reserve management took over. Currently one hectare per annum of underwood is coppiced on an 11-year cycle.

West Dean Woods Nature Reserve is leased by the Sussex Wildlife Trust and is run almost completely by volunteers, both in terms of management and ecological survey. The woods were, and are, known in particular for their epiphytic lichen flora, higher plant flora and butterflies. Archaeological, historical and ecological surveys of the whole parish, including these woods, have been carried out as part of the long-term Chilgrove Valley Landscape Project.

The primary aim of the Sussex Trust was to save the wood from grubbing-up, and to return it to a coppicing system for conservation. (There was only 5% by area of woodland in the parish in medieval times, of which the present Nature Reserve site was a large part; the rest was chalk downland, chalk heath, rabbit warren and open-field arable (Tittensor and Tittensor, 1985); the area of woodland in the parish has since increased to 33%).

Coppicing at West Dean, probably as elsewhere, was carried out because it was generally agreed to be a 'good thing'. This hearsay reflected the great upsurge of interest in historical ecology in the 1960s and 1970s. Moreover those of us involved in the early years of West Dean Woods' management made a fundamental error which was not to carry out a baseline ecological survey before re-starting coppicing. Nevertheless, a number of studies were subsequently made including: analysis of the height regrowth of the hazel coppice; the floristic changes in selected cants from one year prior to their coppicing to three years after coppicing; monitoring the population structure of the wild daffodil colony (*Narcissus pseudonarcissus*) for 11 years by a local natural history society; the long-term recording of butterfly populations by the Nature Conservancy Council; and the record of a Common Birds Census for 13 years. Lichens, bats, dormice, snails, bryophytes and soils have also been subjects of study.

During the intervening years, professional ecologists throughout the UK have been analysing the effects of coppicing regimes on different taxa in various types of coppice woodland. However, before research ecologists can set up their experiments and monitor their favourite taxa in a coppice woodland there are three important matters to be sorted out. They are as follows (not necessarily in order of priority):

- To determine the priorities of the coppice woodland.
- To start the coppicing regime and to keep it going.
- To formulate an appropriate historical and cultural context.

15.2 PRIORITIES

In any conservation matter – indeed in any business, planning, scientific or educational matter – it is vital to set priorities. Management is then geared to a particular target. However, at West Dean, our priorities seemed self-evident – to restore the original coppice management system. A number of developments have subsequently affected our views. The results of detailed botanical surveys in compartments of age C−1 to C+3 (C being the year of coppicing) showed that there was not merely a response in amount of flowering, but in the types of species that appeared and in their chronological pattern of appearance and disappearance. Secondly, worries about the deleterious effects of opening up the canopy on the spread and dominance of bramble (*Rubus fruticosus*), roe deer (*Capreolus capreolus*) browsing and on lichens, etc., proved needless. The coppice grew, the bramble was regularly shaded out nine to ten years after coppicing, the lichens were still there: indeed still more have been recorded. Some butterflies increased while others decreased or were counted in different areas. Surprises were constantly in store. After the Great Storm of October 1987, which resulted in the thinning of the standard oaks, it became obvious that we had a coppice of small-stooled monoculture hazel and even-aged standards: more a product of 19th century reorganization of ancient silvicultural practice than a medieval coppice.

Environmental archaeology studies in several parts of the UK demonstrated the immensely long history of coppicing and the varied ways in which it has been carried out in the past. What, then, were we trying to emulate? Were we coppicing in a Neolithic, Celtic, Medieval, Tudor or 19th century way? Or if in a modern way, what were we trying to conserve in this coppice woodland? Arguments raged in our Reserve Committee over the relative importance of different conservation criteria, such as the maintenance of an historic management system, the importance of standard trees and the role of amenity. Beneath these arguments lie three fundamental questions:

1. What is a coppice woodland for? ('Wildlife', game, timber, or amenity?)
2. Who is it for? (Ecologists, the landowner, the local population, future generations?)
3. Who decides? (Ecologists, the public, the landowner, the Reserve Committee, the Government, the conservation volunteers?)

Unless some consensus is made about what and who woodland management is for there is no reason to go ahead with coppicing.

15.3 STARTING THE COPPICING REGIME AND KEEPING IT GOING

Let us assume that a decision has been made already to initiate or continue to coppice a particular woodland. Coppicing in the past must have been extremely hard and wet work; it was done because of the essential resources it provided the community. Even today, coppicing is labour-intensive and involves a wide variety of physical activities. The majority of non-commercial, traditional coppices – whether they are Trust Reserves, local authority woodlands, community woodlands, or farm and estate woodlands – depend to some extent on a volunteer labour force. Volunteers do not turn up week after week, month after month, year after year, merely to wear themselves out and spend cold, wet days in sodden trousers and boots.

Maintaining a volunteer work-force in order to carry out coppice conservation means attending to volunteers' aspirations and needs. This in turn means that an agreed management plan, which sets out the priorities and work schedules, must not begin with expectations that are too high. There is nothing more demoralizing for a work-force than to be always behind schedule. Furthermore, maintaining a work-force means careful consideration of the social and personal needs of the conservation volunteers and their physical fitness. It means teaching them various skills, attending to disposal or removal of produce, and trying to fit the coppicing operation into the local and cultural context. It is important to ensure that the whole purpose of management has a meaning in the volunteers' minds, a framework for their personal activities, and a focus for the future which they are producing.

Consideration for the work-force also necessitates compromises between different conservation viewpoints and interests. For instance, ornithologists may be concerned about disturbance to nesting birds if work continues in the woods beyond the end of March; and in spite of trampling and compaction damage to the flora during summer work of any kind, the coppice volunteers like to see the results of their work during the fine weather. It is therefore a good idea to continue a work programme of some kind during the summer months which will hold the labour force together until the following autumn. There are always other items of work to be done: mending fences, filling in cracks in paths, showing round visiting groups of all kinds, selling firewood. The natural history surveyors also like to meet up with the conservation volunteers during the summer.

The sort of skills needed by a good reserve manager are those used by business people in the commercial world: liaison with customers, in demonstrating and explaining things to people, organizing a variety of disparate activities, marketing and selling produce, to name but a few. Foresters and land agents running commercial coppice operations will already understand the labour relations aspects and selling techniques needed to keep coppicing going. They are relevant to a nature conservation coppice woodland too, and it is important that actively involved scientists and ecologists also understand this aspect of their research sites.

West Dean Woods Nature Reserve illustrates some of the complexities faced by the practical side of the operation. The (volunteer) reserve manager has to ensure that the operations stipulated in the management plan are fully finished for the current year; in this case the coppicing of two half-hectare plots per winter. If it is necessary to go on beyond 31st March in any year the manager has to apply to the Nature Conservancy Council for a permit, because of restrictive conditions attached to the reserve as a Site of Special Scientific Interest. A permit is likely to be needed most years because of all the odd jobs mentioned above and the desire of the work-force to be in the woods during spring and summer.

The reserve manager also ensures that the mechanical equipment and hand-tools are in working order. The chain-saw operators must have the correct paperwork and safety gear. They must not overstretch their abilities and skills as a result of their enthusiasm and dedication. The reserve manager prepares a yearly report for the reserve committee, setting out the past 12 months' achievements together with survey results. The report also includes details of visits from students and specialist ecologists that the reserve manager has hosted each year. The manager also endeavours to persuade the reserve committee to apply for grants to analyse the biological data constantly being generated and asks for guidance concerning future recording work. Finally, as a public relations exercise, the County Trust, local authority or landowner may decide it would be a good idea to show off the spring flowers to the general public. Car parking and toilet facilities are just the first problems when 1000 or more people turn up on two days in spring – yet another aspect of conservation management!

Many reserve managers may cope with two full-time jobs, one paid and one voluntary. Scientists and ecologists need to be aware of this, and of the unusual, (even if dedicated) nature of the work-force, when preparing management plans. They should be prepared to adapt, modify or find additional resources, if the programme is impossible to achieve with the initial work-force. Part of keeping the coppicing going is to formulate a proper context for the operations in the absence of any economic justification.

15.4 FORMULATING AN HISTORICAL AND CULTURAL CONTEXT

Coppicing has an enormous historical and cultural heritage value to offer society, together with its nature conservation value. Is West Dean Woods as historically and culturally valuable as nearby Chichester Cathedral? Millions of pounds are spent on cathedrals and hundreds, possibly thousands, on historical coppices. Cathedrals served the spiritual needs of local populations while coppices served their fundamental physical needs, particularly fuel.

Throughout its history the context of coppicing has been as an essential resource. When it was not needed, it was not carried out, as is now patently the case for hazel underwood in central southern England. In a sense, conservation has stepped in as a stop-gap use for traditional coppice to preserve living museums of enormous cultural value. Can they be maintained? Recently, changes in Government Youth Training Schemes have meant the severe curtailment of some countryside management schemes. County Wildlife Trusts are now increasingly finding the financial and labour costs of managing up to 30 nature reserves too great (when those reserves bring in no income), and are reconsidering their existing holdings. Table 15.1 shows how much labour is needed to coppice 1 ha of hazel coppice at West Dean each winter, with the modern labour costs, national insurances and current rents to a landlord, if all were applicable.

Does it matter if for lack of funds, labour or motivation we have to stop coppicing a site? The answers to that question are discussed elsewhere in this volume: they depend on which taxa the coppicing is for, their rarity, ability to survive under other silvicultural regimes, their ecological requirements and a host of other factors. But, from the cultural point of view, unlike a Neolithic flint, Celtic hill fort or Tudor warship, a living habitat

Table 15.1 Costs per hectare of managing hazel coppice in southern England, assuming an 11-year rotation

1.	Volunteer man-days to coppice and extract to roadside 1 ha hazel	275 days
2.	Assume a skilled woodsman would complete ten times faster	27.5 days
3.	1990 daily cost of coppice worker (including paid holidays, national insurance, sick pay, transport costs, etc.), minimum pay	£40.00
4.	Probable labour costs per hectare of coppice = 27.5 × 40	£1100.00
5.	Costs of chain-saw, fuel, small tools	£100.00
6.	Ground rent at £12/ha/year	£132.00
	Minimum break-even costs (administration costs not included)	£1332.00
7.	Maximum sales at West Dean	£1000.00
8.	Minimum deficit per annum	£332.00

depends upon continued human effort if it is to be maintained as a site of cultural importance. How many coppices can the tax payer or Wildlife Trust subscriber afford? Traditional coppices cost money – that is the whole point of this chapter – and they depend upon volunteer effort. For both to be forthcoming, coppice must be presented in a context that is real and urgent.

The rather vague notion of 'conservation' in itself satisfies neither the public, unless they can see masses of spring flowers, hear bird song and are allowed free access for that purpose, nor ecologists, who now know that coppicing affects every species of flora and fauna differently. We feel that a context relevant to modern society needs to be found if coppices on any scale are to continue in the modern countryside. The other alternative is to find markets for the produce or to make markets so that coppice becomes a necessity once more. Attempts at finding significant markets appear to have failed. Despite a combination of suitable circumstances which make West Dean Woods apparently ideal as a marketing venture, our successes have been erratic and small scale. Nobody can or will pay a price that reflects the labour input as well as the overheads.

There appears to be no way in which a coppice like West Dean Woods can ever be an economic proposition. Either it will continue to be dependent upon volunteers for both coppicing and survey, or it will need to be changed into something which is needed and produces an income. Has it a future along either avenue? Which is more feasible and which might have the greater conservation significance? Could we persuade the public that the cultural significance of the reserve compares with, say, a cathedral and charge admission accordingly? This is unlikely, as people are happy to pay to visit beautiful properties but expect their landscapes to be free. On the other hand, if changes of silvicultural emphasis are made in a attempt to develop economically viable woodlands out of derelict coppices nature conservation would then become a vicarious rather than a primary activity. The resulting woodland would acquire cultural status relevant to the new century and beyond. As traditional coppices they therefore come to exist and become identified with the needs of modern society.

REFERENCE

Tittensor, A.M. and Tittensor, R.M. (1985) The rabbit warren at West Dean near Chichester. *Sussex Archaeological Collections*, **123**, 151–85.

16

Coppicing – a conservation panacea?

F.B. GOLDSMITH

16.1 INTRODUCTION

A few years ago I gave a lecture on coppicing which started as follows:

> *Coppicing is widely regarded as a conservation panacea, resulting in high structural diversity and species diversity, but it can also be viewed as producing a dwarf, uniform, even-aged, biologically impoverished, nutrient-depleted woodland. Which view is correct?*

Nineteen references followed but none quite answered the question. This was somewhat strange because ecologists in Britain have been studying coppicing since at least 1911. Some of the papers in this volume suggest that we are still preoccupied with questions such as: which length of coppice cycle is preferred? or what cheaper types of management might simulate coppicing? rather than a fundamental reappraisal of the desirability of coppicing.

16.2 CONSERVATION PRIORITIES

Depending on the values that one attaches to different species, communities of species and to the concept of naturalness, coppice management will be regarded differently by different people. The subject of conservation evaluation is a complex one which I have tried to explore elsewhere (Goldsmith, 1983, 1988) and which most ecologists understandably prefer to avoid. A detailed ecological discussion of this subject is necessary before any reappraisal of the desirability of coppicing can truly be made.

Peterken has tried to set down conservation priorities for British and European woodlands (Peterken, 1977, 1981). Briefly, these revolve around the protection of rare or characteristic species, the inclusion of a range of woodland or stand types, and safeguarding particular types of stand structure and/or management. Size of wood (usually meaning the larger the better), habitat continuity over time, and a range of other conservation criteria are also considered to be important (Ratcliffe, 1977). However, any management carried out on any ecosystem favours some components and disfavours others. There is no such thing as a single optimal management prescription. But if several different prescriptions are put into effect on one site, then excessive fragmentation is likely to occur. In the worst instances we may regard such management as 'gardening', which is already apparent on some nature reserves. Most British ancient woods are already less than 5 ha in size, which is very small by primeval standards (Mitchell and Kirby, 1989). Many overseas ecologists and conservationists would not be able to believe the small scale of some current management prescriptions on many of our national and local nature reserves. In addition, coppicing is not analogous with natural woodland cycles (Chapter 5) because the cycles are short (7–25 years instead of 300–1000) and very regular, removing any unpredictability or stochasticity experienced by natural forest stands. All in all, coppice may be seen as a poor substitute for natural cyclical phenomena such as windthrow and vertebrate grazing.

One possibile consequence of coppicing is that short cycles carried out consistently over centuries may have caused losses of nutrients on some sites, leading to yield reductions and hence longer cutting cycles (Rackham, 1976). However, there is no substantial evidence for nutrient depletion due to coppicing. Because other elements are replaced by aerosols and precipitation, it is probable that phosphorus is the only element likely to be reduced in the ecosystem. Low levels of phosphorus may lead to an increase in the number of flowering plant species and so this could be regarded as a slim argument for coppicing. Alternatively, high phosphorus levels may lead to an increase of undesirable and invasive species such as stinging nettle (*Urtica dioica*) and elder (*Sambucus nigra*).

16.3 SPECIES COMPONENTS OF COPPICE MANAGEMENT

Benefits from coppicing may occur for some species components and disbenefits for others, hence another value judgement has to be made which involves deciding which taxa to favour – butterflies or beetles, birds or lichens, spiders or flowering plants? Further investigation reveals that discrepancies occur within taxa, e.g. which bird groups should be favoured – hole-nesting birds, spring migrants, resident songbirds or raptors?

16.3.1 Plants

The leaf area index of coppice increases rapidly from zero after cutting to five or six in only ten years, and at this level shading is incredibly dense (Chapter 3). As such, in the later stages, it suppresses the growth of most other plants. Coppicing does not add to the total number of woodland plant species in an area but enhances the abundance of some species in the early part of the cycle, especially those with attractive flowers such as primrose (*Primula vulgaris*), bluebell (*Hyacinthoides non-scripta*) and wood anemone (*Anemone nemorosa*) (Salisbury, 1924; Ash and Barkham, 1976; Ford and Newbould, 1977). It may be that coppicing results in more species in the first two years, but the additions are mostly non-woodland ruderal species, often present as buried seed which are usually considered to have a lower conservation value than, say, ancient woodland indicator species.

Seed of many species remains viable for up to 30 years (Brown, 1981; Brown and Oosterhuis, 1981), and even after 70 years of neglect half the species may survive in the seed bank (Chapter 8). Whatever management is carried out a different proportional composition of flowering plants will result. Most of the high value* ancient woodland indicator species will survive just as well, or better, under high forest management as under coppicing (Chapter 7). Most species of lichens and bryophytes favour old trees, i.e. standards or high forest and not coppice. These have been little studied but most specialists would agree that repeated and frequent increases in light intensity cause desiccation, and may be damaging to many species of these groups.

16.3.2 Insects

Some butterflies are high canopy species such as the purple emperor (*Apatura iris*), purple hairstreak (*Quercusia quercus*) and white letter hairstreak (*Strymonidia w-album*) which are not favoured by coppicing. Some have food plants such as violets (*Viola* spp.) or nettles which may be found in coppice but are far from exclusive to that form of management. Again, honeysuckle (*Lonicera periclymenum*) is known to be important for white admiral but is abundant in a variety of woodland and other habitats. Many species of butterfly favour woodland edges, rides, open scrub and the early phases of coppicing, e.g. gatekeeper (*Pyronia tithonus*), white admiral (*Ladoga odoratum*), comma (*Polygonia c-album*) and speckled wood (*Pararge aegeria*) (Warren, 1985), but not the later stages of closed coppice. Long coppice cycles as often occur today are not ideal, except perhaps for the wood white (*Leptidea sinapis*), and the optimum management will vary from species to species. Heath fritillary (*Mellicta athalia*) is a rare species which is often cited as depending on coppicing, but a mosaic of scrub and grass would also favour its food plant common cow-wheat (*Melampyrum pratense*), and hence the butterfly itself (Warren *et al.*, 1984; Warren, 1986; Chapter 13). This species is unlikely to occur in most woods, however intensively we coppice. A great deal of effort is spent managing one national

*Highly valued by conservationists

nature reserve for it, but elsewhere in the country it occupies a different habitat (Cressy, 1987).

As with individual butterfly species, different groups of invertebrates favour different types of management. Coppicing will favour some species but disfavour many others. Many of the rarer beetles favour dead wood which is a feature of high forest and pasture woodland rather than coppice (Chapter 6). This is especially true of invertebrates favouring dead wood in the larger diameter size classes. Again, there is some evidence that some species favour coppice while others favour high forest (see Sterling and Hambler, 1988). Recommendations are often made for leaving piles of dead wood as suitable habitat for spiders and other invertebrates in coppice, but dead wood is not a feature of actively coppiced woods (Chapter 6). Mature birch has twice as many leaf miners as young birch (Chapter 14) so, for this group at least, older trees are beneficial.

16.3.3 Birds

Coppice may favour a few species but the case for coppicing may have been overstated. For example, Nightingales (*Luscinia megarhynchos*) need a mosaic of habitats and are absent from many coppice woods (Chapter 9). Hole-nesting birds, e.g. woodpeckers, favour large old trees which are only found in high forest, standards or hedgerows (Simms, 1971; Flegg and Bennett, 1974). Treecreeper (*Certhia familiaris*), nuthatch (*Sitta eurapaea*), owls, sparrowhawk (*Accipiter nisus*) and other 'high-value' conservation species do not favour coppice. Some birds such as blackcap (*Sylvia atricapilla*) and tree pipit (*Anthus trivialis*) need standards as sun posts and would not use pure stands of coppice. Wren (*Troglodytes troglodytes*), robin (*Erithacus rubecula*), blue tit (*Parus caeruleus*) and blackbird (*Turdus merula*) have high frequencies in coppice (Fuller *et al.*, 1989) but these are mostly common species which are also abundant in high forest. Willow warbler (*Phylloscopus trochilus*) and other spring migrants are more abundant in the early stages of coppice regrowth (3–7 years). As these two groups are songbirds that we enjoy hearing, is this one of the reasons why so many people favour coppicing?

16.3.4 Mammals

The dormouse (*Muscardinus avellanarius*) is an arboreal specialist and favours 25–30 year coppice especially with oak trees (Chapter 11) and may be regarded as a 'high value' species. Most other mammal species (mice, shrews, voles, hedgehog (*Erinaceus europaeus*), squirrel (*Sciurus* spp.), badger (*Meles meles*), fox (*Vulpes vulpes*), deer and wild board (*Sus scrofa*) are found in varying proportions, whatever form of woodland management is practised, including neglect. Peak numbers of small mammals occur three years after coppicing but these are relatively common in most woodlands and therefore may be interpreted as 'lower value' species. Only old broadleaved trees are likely to have large hollows within them which are suitable for bats (Mitchell and Kirby, 1989). These are less likely to occur in pure coppice but may be a feature of coppice with over-mature standards.

16.4 CONCLUSIONS

To continue a traditional management is 'safe', and nothing is likely to be lost. But we still have to decide which tree species to favour (many conservationists would agree that the exotic chestnut (*Castanea sativa*) is 'bad', while hazel (*Corylus avellana*) and ash (*Fraxinus excelsior*) are 'good' but would find it difficult to quantify such assertions. Moreover, are species which cast deep shade, such as hornbeam (*Carpinus betulus*) also 'good'? Other questions to be considered are: what frequency of cutting is preferred, should cut wood be removed, and, with current large deer populations, should we reduce the extent of deer browsing?

Coppicing is very hard work and destructive, therefore we must decide: is it really worth the effort? Decision makers and reserve managers need fundamentally to reappraise their views and avoid 'gardening'. Such intensive management can hardly be deemed 'natural'. Where carried out, coppicing should at least be associated with standards in order to diversify stand structure.

So what is our overall conclusion? This volume addresses the question 'what effects does coppice have?' and to a lesser extent 'how do we do it?' but (with the exception of Chapter 15) hardly touches the question 'why do we do it?' We should be conscious that the appeal of coppicing is partly aesthetic: it enhances the attractiveness of the vernal flora after our grey winters, together with the concurrent delights of the song of spring migrants. Intrinsic appeal is a legitimate criterion of conservation value officially listed by the Nature Conservancy Council (Ratcliffe, 1977; Goldsmith, 1991) but is not a 'scientific' or 'ecological' one.

On the other hand, coppicing enthusiasts will find enough evidence in these pages to be able to justify their actions and continue. Others may question the need for coppicing and its use as a universal panacea. If reserve wardens slip behind with their coppicing schedules they should not despair: it is unlikely that species losses will occur as a result, and the reserve will acquire a little extra 'naturalness' – a valuable attribute in its own right. A little more 'nature' will have been added to the nature conservation value of their site.

REFERENCES

Ash, J.E. and Barkham, J.P. (1976) Changes and variability in the field layer of a coppiced woodland in Norfolk, England. *Journal of Ecology*, **64**, 697–712.

Brown, A.H.F. (1981) The recovery of ground vegetation in coppice wood: the significance of buried seed, in *Forest and Woodland Ecology*, ITE Symposium 8 (eds F.T. Last and A.S. Gardiner), Institute of Terrestrial Ecology, Cambridge, pp. 41–4.

Brown, A.H.F. and Oosterhuis, L. (1981) The role of buried seed in coppiced woods. *Biological Conservation*, **21**, 19–38.

Cressy, J. (1987) *Ecology and conservation of cow wheat*, MSc thesis in Conservation, University College London.

Flegg, J.J.M. and Bennett, T.J. (1974) The birds of oak woodlands, in *The British Oak: its history and natural history* (eds M.G. Morris and F.H. Perring) Botanical Society of the British Isles, E.W. Classey, Faringdon, pp. 324–40.

Ford, E.D. and Newbould, P.J. (1977) The biomass and production of ground vegetation and its relation to tree cover through a deciduous woodland cycle. *Journal of Ecology*, **65**, 201–12.

Fuller, R.J., Stuttard, P. and Ray, C.M. (1989) The distribution of breeding songbirds within mixed coppiced woodland in Kent, England, in relation to vegetation age and structure. *Annales Zoologicii Fennici*, **26**, 265–75.

Goldsmith, F.B. (1983) Evaluating nature, in *Conservation in Perspective* (eds A. Warren and F.B. Goldsmith), Wiley, Chichester, pp. 233–46.

Goldsmith, F.B. (1988) Threats to woodland in an urban landscape: a case study in Greater London. *Landscape and Urban Planning*, **16**, 221–8.

Goldsmith, F.B. (1991) The selection of protected areas, in *The Scientific Management of Temperate Communities for Conservation* (eds I. Spellerberg, F.B. Goldsmith and M.G. Morris), Blackwell Scientific, Oxford, pp. 273–91.

Mitchell, P.L. and Kirby, K.J. (1989) *Ecological Effects of Forestry Practices in Long-established Woodland and their Implications for Nature Conservation, Oxford Forestry Institute Occasional Paper*, 39, Department of Plant Sciences, University of Oxford.

Peterken, G.F. (1977) Habitat conservation priorities in British and European woodlands. *Biological Conservation*, **11**, 223–36.

Peterken, G.F. (1981) *Woodland Conservation and Management*, Chapman and Hall, London.

Rackham, O. (1976) *Trees and Woodland in the British Landscape*, Dent, London.

Ratcliffe, D.A. (1977) *A Nature Conservation Review*, Volume 1, Cambridge University Press, Cambridge.

Salisbury, E.J. (1924) The effects of coppicing as illustrated by the woods of Hertfordshire. *Transactions of the Hertfordshire Natural History Society*, **18**, 1–21.

Simms, E. (1971) *Woodland Birds*, Collins, London.

Sterling, P.H. and Hambler, C. (1988) Coppicing for conservation: do hazel communities benefit? in *Woodland Conservation and Research in the Clay Vale of Oxfordshire and Buckinghamshire* (eds K.J. Kirby and F.J. Wright) Nature Conservancy Council, Peterborough, pp. 69–80.

Warren, M.S. (1985) The influence of shade on butterfly numbers in woodland rides, with reference to the wood white, *Leptidea sinapsis*, *Biological Conservation*, **33**, 147–64.

Warren, M.S., Thomas, C.D. and Thomas, J.A. (1984) The status of the heath fritillary butterfly, *Mellicta athalia* Rott., in Britain. *Biological Conservation*, **29**, 287–305.

Glossary

ancient woodland a site continuously wooded for at least the past 400 years

basal area the sum of the cross-sectional areas of all tree stems, measured at 1.3m above ground, per unit area

blackheart a dark staining of the wood of ash, causing significant loss in timber value

canopy closure the growth stage of the woodland after cutting at which a continuous leaf canopy forms over the forest floor

canopy turnover rate the rate at which gaps form in forest canopies and are subsequently closed by regeneration: often used to describe natural disturbance regimes in unmanaged forests

clear felling the felling of moderate or large areas of woodland, often involving replanting rather than natural regeneration or coppice regrowth

coppice the cutting of the stems of young trees or shrubs close to the ground, causing them to resprout and to re-establish the canopy: or an area so treated

coppice with standards a system of woodland management in which timber sized trees are grown over a coppice or **underwood** crop

coupe a felled or coppiced area of woodland

disturbance a natural or man-induced event causing a change in environmental conditions in the canopy or below it, or both

diversity the number and variety of plant or animal species or taxa present within a given area, dependent upon prevailing natural environmental conditions or management

early successional species a species dependent on, or favouring conditions present in the young forest growth stages following felling

field capacity the water content of a soil following an initial thorough wetting of the profile and two or three days of free drainage

gap formation rate the rate at which disturbances create gaps in the forest canopy (see **canopy turnover rate**)

glade an open area within a woodland, usually grassy and maintained permanently by grazing

group felling the felling of small groups of trees (e.g. 0.1–0.5ha in extent) at varying time intervals, creating an unevenaged forest structure

high forest a forest management system which allows the trees to grow to at least two-thirds of their full height, as opposed to earlier cutting or coppicing, producing **underwood**

late successional species a species dependent upon, or favouring areas of mature forest growth or old forest stands

maiden tree a tree which has never been previously felled or coppiced

neglected coppice coppice uncut for several years beyond its normal **rotation** age, and tending to revert to high forest (see also **stored coppice**)

normal forest a forest in which all growth stages of the trees are present, allowing an even supply of wood and timber of all sizes to be produced from it each year

overmature a term (usually coined by foresters) to describe older forest stands where the productivity or yield has fallen below an optimum level

panel an area or compartment of underwood which is cut and then allowed to regrow. Locally also called a 'cant' or 'hag'

PAR photosynthetically active radiation, i.e. that part of the solar radiation spectrum used by plants for photosynthesis

pits and mounds the pattern of forest floor microtopography created by the uprooting of mature trees by periodic storms

pole stage the point in a forest rotation after **canopy closure**, and usually after first thinning, when the supression of lower branches produces a **stand** of well-defined single terms or poles

pollard a tree whose shoots are repeatedly cut 2–3 metres above ground in order to avoid browsing damage, as in **wood pasture**

regeneration the re-establishment of tree cover, either from seed shed from an adjacent canopy, or from the formation of new coppice shoots

ride a broad trackway or extraction route separating two adjacent management units of woodland

rotation the period for which trees are grown before they are cut for produce

seed bank the quantity of germinable buried seed present in the upper soil profile of woods and other habitats

shake a defect in freshly felled timber resulting from the splitting of wood along an annual ring (ring shake) or radiating from the pith (star shake)

snag dead standing trees, stumps or large attached branches

soil matric potential the sum total of forces retaining water within the soil, such as capilliary attraction and the adsorptive forces largely associated with the clay and organic fractions

soil-water deficit a soil moisture deficit caused by the removal of water through surface evaporation or transpiration, measured in millimetres from **field capacity**

soil-water energy status the sum of gravitational potential, matric potential and osmotic forces acting upon the movement of water within the soil

stand an (often uniform) area or tract of woodland, or other type of vegetation

standard large trees, generally intended for construction timber, overstanding coppiced **underwood** (see **coppice with standards**)

stocking the density per unit area of trees used or selected for planting, natural regeneration or thinning operations

stool the cut tree base from which coppice shoots arise

stored coppice A **stand** of coppice origin which has been allowed to grow on beyond its normal **rotation** age

thicket a dense, young-growth stage of coppice or woodland in which the individual tree crowns coalesce to form a closed canopy, while still maintaining most of their lower branches

underwood general name for a wood consisting of coppice shoots, root suckers and pollard poles, grown for wood rather than timber

wood pasture a system in which the woodland is permanently used for pasture, with the trees **pollarded** to prevent browsing damage of young shoots

Bibliography

Nomenclature of plant and animal species follows the authorities listed below, where appropriate:

Vascular plants

Clapham, A.R., Tutin, T.G. and Moore, D.M. (1987) *Flora of the British Isles*, Cambridge University Press, Cambridge.

Mosses

Smith, A.J.E. (1978) *The Moss Flora of Britain and Ireland*, Cambridge University Press, Cambridge.

Birds

Voous, K.H. (1973) List of recent Holarctic bird species. Non-passerines. *Ibis*, **115**, 612–38.

Voous, K.H. (1977) List of recent Holarctic bird species. Passerines. *Ibis*, **119**, 223–50 and 376–406.

Mammals

Corbet, G.B. and Harris, S. (1990) *The Handbook of British Mammals*, 3rd edn, Blackwell Scientific, Oxford.

Invertebrates

Bradley, J.D. and Fletcher, D.S. (1979) *A Recorder's Log-book or Label List of British Butterflies and Moths*. Curwen, London.

Kloet, G.S. and Hincks, W.D. (1964) A check list of British insects, 2nd edn, Part 1: Small order and Hemiptera. *Handbooks for the Identification of British Insects*, **11**, (1), Royal Entomological Society, London.

Kloet, G.S. and Hincks, W.D. (1977) A check list of British insects, 2nd edn. Part 3: Coleoptera and Strepsiptera. Revised by R.D. Pope and edited by A. Watson, *Handbooks for the Identification of British Insects*, **11**, (3), Royal Entomological Society, London.

Kloet, G.S. and Hincks, W.D. (1978) A check list of British insects, 2nd edn. Part 4: Hymenoptera. Revised by M.G. Fitton, M.W.R. de V. Graham, Z.R.J. Boucek, N.D.M. Fergusson, T. Huddleston, J. Quinlan and O.W. Richards, *Handbooks for the Identification of British Insects*, **11**, (4), Royal Entomological Society, London.

Merrett, P., Locket, G.H. and Millidge, A.F. (1985) A check list of British spiders. *Bulletin of the British Arachnological Society*, **6**, 381–403.

Skinner, B. (1984) *Colour Identification Guide to the Moths of the British Isles*, Viking, Hamondsworth.

Scientific names index

Subject index